YEAST - INDUSTRIAL APPLICATIONS

Edited by **Antonio Morata** and **Iris Loira**

Yeast - Industrial Applications

http://dx.doi.org/10.5772/intechopen.69360
Edited by Antonio Morata and Iris Loira

Contributors

Shakira Ghazanfar, Muhammad Imran, Iftikhar Ahmed, Nauman Khalid, Lina María Agudelo-Escobar, Solange I. Mussatto, José A. Teixeira, Mariana Peñuela, Guillermin Aguero-Chapin, Deborah Galpert Cañizares, Sara Del Río García, Francisco Herrera, Evys Ancede Gallardo, Agostinho Antunes, Antonio Morata, Iris Loira, María Antonia Bañuelos, Carlos Escott, Jose Antonio Suárez Lepe, Alejandro Ruiz Marín, Yunuen Canedo López, Asteria Narváez García, José Del Carmen Zavala Loría, Juan Carlos Robles Heredia, Ramón Martínez-Peniche, Jesús A. Aldrete-Tapia, Sofía M. Arvizu-Medrano, Montserrat H. Iturriaga, Dalia E. Miranda Castilleja, Lourdes Soto Muñoz, Ewa Żymańczyk-Duda, Małgorzata Brzezińska-Rodak, Magdalena Klimek-Ochab, Maciej Duda, Agata Zerka, Dimitris Tsaltas, Minas Mina, Dimitris Anagnostopoulos, Despina Bozoudi, Zeynep Yilmazer Hitit, Suna Ertunc, Baran Ozyurt, Edwil Gattas, Natália Manuella Strohmayer Lourencetti, Flávia Danieli Ibelli, Maria Priscila Franco Lacerda, Maria José Soares Mendes Giannini, Cleslei Fernando Zanelli, Ana Marisa Fusco Almeida, Naïma El Ghachtouli, Wifak Bahafid, Nezha Tahri Joutey, Hanane Sayel, Meryem Asri, Nabil Tirry

Published by InTech
Janeza Trdine 9, 51000 Rijeka, Croatia

Publishing Process Manager Romina Rovan
Technical Editor SPi Global
Cover InTech Design team

First published November, 2017
Printed in Croatia
Legal deposit, Croatia: National and University Library in Zagreb

Additional hard copies can be obtained from orders@intechopen.com

Yeast - Industrial Applications, Edited by Antonio Morata and Iris Loira
p. cm.
Print ISBN 978-953-51-3599-9
Online ISBN 978-953-51-3600-2

For EU product safety concerns:
IN TECH d.o.o., Prolaz Marije Krucifikse Kozulić 3, 51000 Rijeka, Croatia, info@intechopen.com or visit our website at intechopen.com.

Meet the editor

Antonio Morata is a professor of Food Science and Technology at the Universidad Politécnica de Madrid (UPM), Spain, specializing in wine technology where he also is the coordinator of the Master in Food Engineering Program. Dr. Morata is a professor of enology and wine technology in the European Master of Viticulture and Enology, Euromaster Vinifera-Erasmus. He is the Spanish delegate at the group of experts in wine microbiology of the International Organisation of Vine and Wine (OIV). Dr. Morata authored more than 60 research articles, 3 books, and 6 book chapters.

Iris Loira is an assistant professor of Food Science and Technology at the Universidad Politécnica de Madrid (UPM), Spain. She is the author of 19 research articles, 1 edited book, and 2 book chapters.

Contents

Preface

Yeast and yeast by-products as enzymes, metabolites, and cell wall polysaccharides are used extensively in industry for many years. Fermentative applications are as ancient as human culture; the use of yeast in the production of wine, beer, and bread is an old and essential application for humanity, but it can be also used to ferment milk products like kefir or in the production of fermented vegetables. Moreover, yeast has been used for a long time as a cell model to understand eukaryotic cells from the beginning of cellular biology and even now is a model in synthetic biology for the artificial synthesis of yeast chromosomes and the final production of synthetic yeasts. Another interesting application is the use of yeast enzymes to transform or to produce many compounds that can be used in chemical, pharmaceutical, food, or agricultural industries. The production of alternative combustibles from natural sources as vegetal by-products or residues is also a trending activity in which yeasts are strongly involved. Production of bioethanol from vegetal residues using yeasts is a natural way to get a cleaner energy source, renewable and with less impact on global warming and global CO_2 production. The use of yeasts in bioremediation or in the recovering of degraded environments is another key application. The ability of yeast either to transform toxic environmental molecules in less harmful derivatives or to adsorb toxic compounds and heavy metals makes them really useful in environmental chemistry.

The traditional use of the well-known *Saccharomyces cerevisiae* yeast species is currently being complemented with the use of many yeast genera and species because of their improved performance in some applications that opens new possibilities in the industry. The production of new enzymes or their higher expression in fermentative processes, the release of different metabolites with useful industrial applications, the production of specific cell wall biopolymers, and many other applications open new attractive possibilities in industry. Some non-*Saccharomyces* species as *Torulaspora delbrueckii, Lachancea thermotolerans, Dekkera bruxellensis, Saccharomycodes ludwigii, Schizosaccharomyces pombe, Metschnikowia pulcherrima, Kloeckera apiculata,* and many others are now extensively used in industry for food, chemical, energetic, or environmental applications.

The conventional use of yeast selection to improve fermentative properties or the production of metabolites has been improving yeast performance for many years, and currently it is also a powerful tool to get the best strains of *Saccharomyces* and non-*Saccharomyces* yeasts. Synthetic biology and genetic engineering open also new possibilities to improve the performance of some specific yeast or to remove either undesirable properties or activities.

The biological use of microorganisms and their enzymes or metabolites for their antimicrobial capacity is also a trend in biotechnology. Many times, yeast and yeast bioactive compounds can be a way to reduce harmful microorganisms in foods, or they can even have medical applications.

This book is a useful reading to understand better the industrial application of yeasts; it contains 12 chapters written by experts from different universities and research centers with long experience and expertise.

Antonio Morata and Iris Loira
Universidad Politecnica de Madrid,
Madrid, Spain

Section 1

Introduction

Chapter 1

Yeast as a Versatile Tool in Biotechnology

Ewa Żymańczyk-Duda,
Małgorzata Brzezińska-Rodak,
Magdalena Klimek-Ochab, Maciej Duda and
Agata Zerka

Additional information is available at the end of the chapter

http://dx.doi.org/10.5772/intechopen.70130

Abstract

Yeasts represent a very diverse group of microorganisms, and even strains that are classified as the same species often show a high level of genetic divergence. Yeasts biodiversity is closely related to their applicability. Biotechnological importance of yeast is almost immeasurable. For centuries, people have exploited its enzymatic potential to produce fermented food as bread or alcoholic beverages. Admittedly, yeasts application was initially instinctual, but with science and technology development, these microorganisms got the object of thorough scientific investigations. It must be recognized that yeast represents an excellent scientific model because of its eukaryotic origin and knowledge of genetics of yeast cells as well as metabolism examined in detail. In 1996, the genome of baker yeast *Saccharomyces cerevisiae* has been elucidated, what opened the opportunity for the global study of the expression and functioning of the eukaryotic genome. Also, currently, an international team is working on the synthesis of the 16 yeast chromosomes by synthetic biology tools, and the results are expected till the end of the year. Nowadays, yeast is regarded as a versatile tool for biotechnological purposes.

Keywords: fermentation, SCP, biocatalysis, molecular biology applications—fundamentals

1. Introduction

1.1. Yeasts: commercial applications

Yeasts have a wide range of applications mainly in food industry (wine making, brewing, distilled spirits production, and baking) and in biomass production (single-cell protein [SCP]). More recently, yeast has also been used in the biofuel industry and for the production of heterologous compounds. Obviously, their main application arises from the metabolic capacity to carry out the transformation of sugars into ethyl alcohol and carbon dioxide under anaerobic conditions. Moreover, a large number of secondary flavor compounds are created what implies on organoleptic attributes of particular food products. However, it would be misguided to trivialize their metabolic capacities only to fermentative activity. The main factors influencing yeast metabolism are the oxygen availability and the type of carbon source. Many yeast strains can function under both anaerobic as well as aerobic conditions of environment, switching their metabolism types easily [1]. Obviously, the courses of main metabolic pathways are conserved, but some regulative mechanisms attract the attention, denoting unusual metabolism flexibility [2]. In food industry, *Saccharomyces cerevisiae* is the genus of yeast most frequently used, whereas *Candida, Endomycopsis,* and *Kluyveromyces* are crucial for SCP production. Yeast strains of industrial importance are carefully selected from the immense natural biodiversity and their properties improved according to process outcome. Both classical approaches as well as modern strategies of gene manipulations are applied to generate variants relevant to work under industrial specific conditions [3, 4]. Strains alterations concern not only the route of fermentations step and its direct yield but also facilitation of product recovery procedures and finally the best quality of particular end product. These nonpathogenic strains with both genotypic and phenotypic stability should have short-generation time and low nutritional requirements. They should perform the fermentation process quickly to minimize the contamination risk. Additionally, they should be tolerant of a wide range of physiological stresses, such as low pH, high ethanol concentration, and high osmotic stress.

The enzymatic power of yeast is central to beer manufacturing (**Figure 1**). Industrial species are carefully selected, nonsporulated polyploides that do not perform sexual reproduction process [5].

Brewer's yeasts are divided into two separate categories: top-fermenting yeast (ale) and bottom-fermenting one (lager). Both yeast types have similar cell morphology, but they differ in some physiological and metabolic features [6], what closely corresponds to the process conditions and type of end product. Fermentation of ale yeast is carried out at room temperature and results in beers with a characteristic fruity aroma. In the case of lager yeast, the fermentation temperature is lower, and therefore, this step takes longer than fermentation with ale strains [7]. The share of aerobic respiration in yeast metabolism is higher in case of ale strains than in lager yeast. Both top-fermenting and bottom-fermenting yeast belong to the genus *Saccharomyces*. Ale-brewing yeasts are genetically more diverse and classified as *Saccharomyces cerevisiae*, whereas the taxonomy of the lager strains has undergone several changes. Initially, bottom-fermenting yeast was classified as *Saccharomyces carlsbergensis*, but due to the application of modern taxonomic approaches, the species *S. carlsbergensis* were included as a part of *Saccharomyces pastorianus* taxon. On the basis of genetic studies, *S. pastorianus* strains are now considered as allopolyploid interspecies hybrids of *S. cerevisiae* and *S. bayanus* [8]. From technological point of view, beside the high fermentative activity of yeast cells and tolerance for several environmental stresses, the aptitude for asexual aggregation known as flocculation seemed to be especially important, because this ability causes the yeast to sediment to the bottom of the fermenter at the end of fermentation step, what simplifies downstream processing [9, 10].

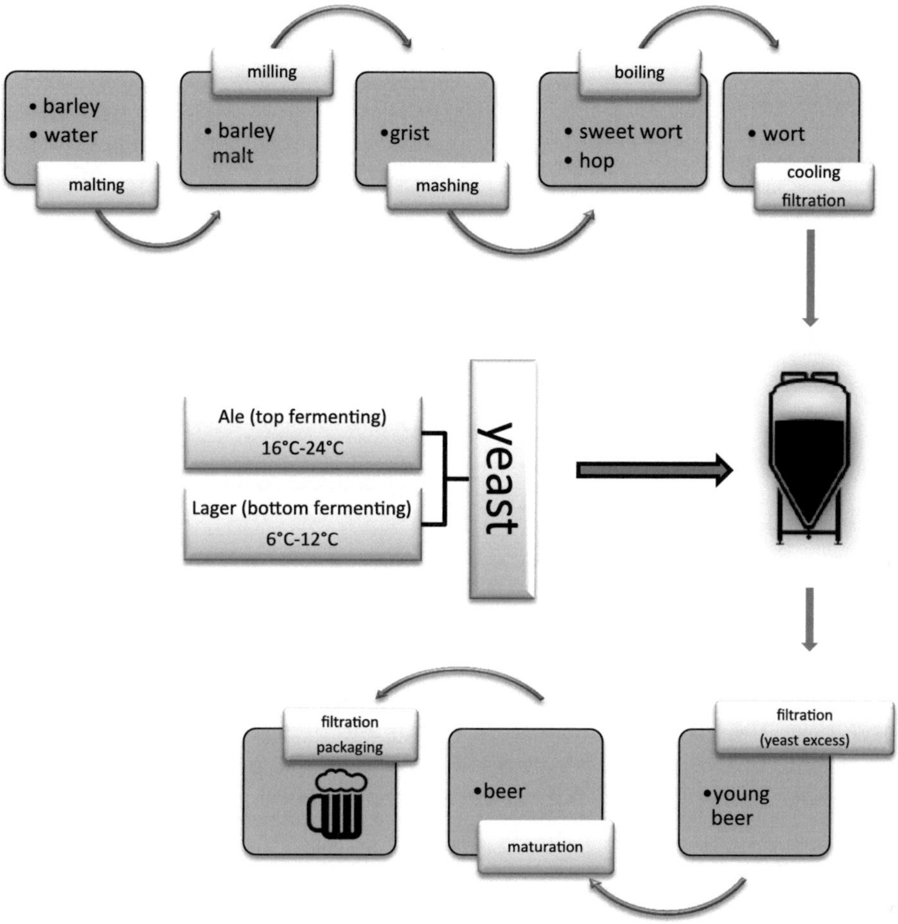

Figure 1. Beer production process.

Various yeast species constitute the predominant microbial group of natural microbiota of fruits ecosystems, what is the reason of fast and spontaneous fermentation of juices or musts resulted in wine production. Yeast colonizing various fruits belongs to the genera *Saccharomyces, Brettanomyces* (its sexual form (teleomorph) *Dekkera*), *Candida, Cryptococcus, Debaromyces, Hanseniaspora* (anamorph *Kloeckera*), *Hansenula, Kluyveromyces, Pichia, Rhodotorula, Torulaspora, Schizosaccharomyces,* and *Zygosaccharomyces* [11]. It goes without saying that most fermented products are generated by a mixture of microbes. These microbial consortia perform various biological activities responsible for the nutritional, hygienic, and aromatic qualities of the product [12]. Doubtlessly, yeast plays a principal role in wine making both in domestic environment as well as in commercial scale (**Figure 2**). Nonconventional yeast (*non-Saccharomyces species*) dominates during early and middle stages of fermentation process, whereas the latter phase of natural process is mediated by *Saccharomyces cerevisiae*. The raw material of particular importance in global wine making is grapes that are harvested at specific stages of ripeness depending on the style of wine to be made.

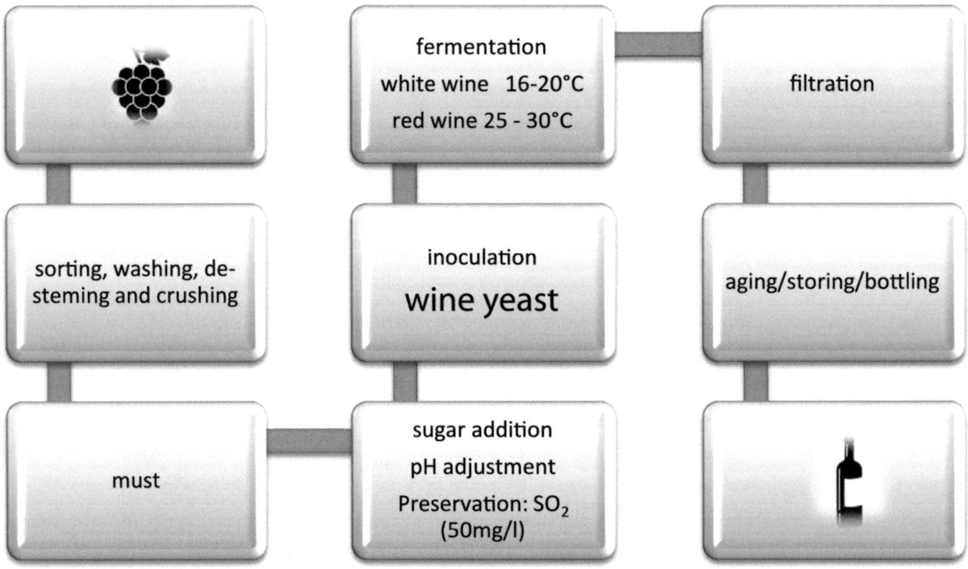

Figure 2. Wine production process.

Once again yeast enzymatic power is crucial for the product chemical profile—except for ethanol biosynthesis, the creation of flavor compounds and fragrances from substrates abundant in fruits implies the organoleptic features of particular wine product. During the degradation of grape sugars, amino acids, fatty acids, terpenes, and thiols, some by-products like glycerol, carboxylic acids, aldehydes, higher alcohols, esters, and sulfides are formed, and their synthesis is largely dependent on the peculiarities of the strains used [13, 14]. In 1965, the first two commercially active dried wine yeasts called Montrachet and Pasteur Champagne were produced for a large Californian winery [15]. Nowadays, modern winegrowers routinely use selected yeast starters in practice. These microorganisms dominate native yeast species and give desired direction of chemical transformations occurring in musts and allow to obtain product of predictable quality [16]. Presently exploited commercially available starters have been created as a result of naturally occurring phenomenon called "genome renewal" as well as planned processes of genetic improvement [17] followed by a careful selection for their good fermentation performance. Genome renewal hypothesis in the standard version assumes that infrequent sexual cycles, characterized by a high degree of selfing, can help to purge deleterious alleles and fix beneficial alleles, thus helping to facilitate adaptation in yeast [18]. This hypothesis had to be re-evaluated [19] due to the fact that in the case of many environmental isolates, very high levels of genomic heterozygosity had been observed [20, 21]. Presently, the majority of commercial wine yeast comprises strains of *Saccharomyces cerevisiae*, including those described by enologists as *S. bayanus*, which has been re-identified in most cases as *S. cerevisiae* [22, 23]. The growing demand for more diversified wines or for specific characteristics has led to the exploration of new species for wine making [24–26]. This non-conventional yeast may contribute to the wine's organoleptic characteristic by producing a broad range of unique secondary metabolites and secreting particular enzymes or exhibiting others substantial features (release of mannoproteins, contributions to color stability etc.) [27]. The wine industry currently proposes starters of a few nonconventional yeast (*Torulaspora delbrueckii, Metschnikowia pulcherrima, Pichia kluyveri, Lachancea thermotolerans*, etc.) [28]. Yeast variants used as starters must qualify for commercial application. The list of desired properties is very long and includes both fermentation characteristics as well as flavor characteristics (e.g., low sulphide/dimethyl sulphide/thiol formation, liberation of glycosylated flavor precursors, low higher alcohol production, etc.), metabolic properties with health implications (low formation of sulphite, biogenic amines, or urea) and technological properties (e.g., high sulphite tolerance, low foam formation, flocculation properties, etc.) [29].

Another example of *Saccharomyces cerevisiae* strains of commercial interest encompasses distiller's yeasts, applied in industry where alcohol fermentation is followed by distillation. In this industry type, spontaneous fermentations are not practiced, and specific yeast is inoculated into fermenter. Starters used for distilled beverages (aquavit, gin, vodka, whisky, rum, tequila, cognac, brandy, and kirsch) commercial production should exhibit exquisitely intensive anaerobic metabolism, what should result not only in high ethanol productivity but also in the high level of cellular tolerance to this product. Additionally, thermostable variants tolerant to acids and increased osmotic pressure are in greater demand for distillers [30]. Since many years, there has been a rising interest in new variants that are able to degrade starch. Many attempts have been made to produce ethanol from starch using recombinant haploid strains that express amylolytic enzymes, due to the simplicity of genetic manipulation of these strains. However, it is difficult to use laboratory haploid strains in practice because their fermentation characteristics are not as good as those of industrial polyploidy strains [31, 32].

In addition to alcoholic beverages production, enzymatic power of yeasts is also essential for baking industry, where concentrated yeast biomass is used as a starter in dough fermentation in order to produce bread and other bakery products. Commercially available baker's yeast forms include fresh compressed biomass, dehydrated cells (dry yeast), and lyophilized cells (instant). Fermentation of dough substrates leads to ethanol production as well as number of volatile and nonvolatile compounds that have an important contribution to the flavor of bread [33]. As a result of carbohydrates (maltose mainly), fermentation carbon dioxide is generated what increases the dough volume and is responsible for crumb texture. Baker's yeast is simply brewery yeast produced *via* submerged fermentation process carried out in the presence of oxygen (**Figure 3**). Aerobic conditions favor yeast cells production, which is not of interest to ethanol producers, but is important when large amount of cells mass must be produced.

The main ingredient of industrial production medium used in yeast production factories are beet or cane molasses, mainly because of the low cost of this waste products and high sucrose content. In most cases, the industrial production is a multistage process carried out under batch or fed-batch conditions with sequential stages differing in fermenter size, performed under controlled intense aeration [34]. Aeration is generally considered as the most important single factor to increase yeast yield and numerous studies have been carried out to investigate the optimization of particular technological solutions [35]. It should be underlined, the particular uniqueness of yeast metabolism—baker's yeast must exhibit efficient respiratory metabolism during yeast manufacturing, which determines biomass yield, but at the same time, cells must

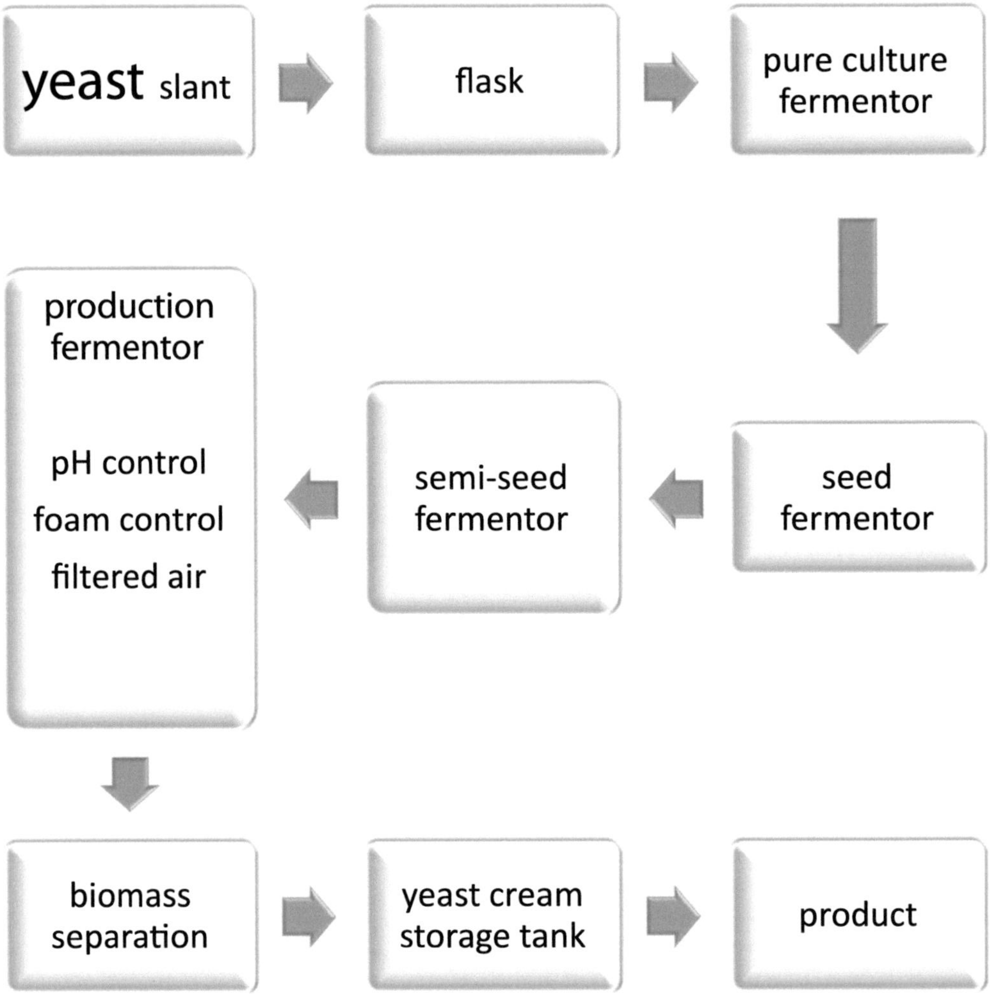

Figure 3. Baker's yeast production process.

possess strong fermentative potential in order to produce excellent bakery products. During the fermentation of dough yeast cells is exposed to numerous environmental stresses (*baking associated stresses*) such as freeze thaw, high sugar concentrations, air drying, and oxidative stresses [36]. Nor should it be surprising that industrial starters should be characterized by appropriate stress tolerance. Yeasts certainly have evolved some mechanisms of adaptation, but if the level of stress will increase too much, their enzymatic potential will be restricted. It has been demonstrated that two molecules: trehalose and proline are extremely important for yeast stress tolerance, so the engineering of their metabolism is a promising approach to the development of stress-tolerant yeast strains relevant to industrial use [37].

Next long-standing industrial processes involving yeast are the production of single-cell protein (SCP)—alternative source of high nutritional value proteins used as a food or feed supplements. Idea of such protein concentrate production was born in response to growing human population in the world and worldwide protein deficiency [38]. SCP manufacture includes simply the cell mass obtaining by way of the application of cheap, waste raw materials to cultivate various nonpathogenic microorganisms (bacteria, fungi, algae) under conditions of submerged (rarely solid-state) fermentation. Besides its high protein content (about 60–82% of dry matter), SCP also contains fats, carbohydrates, nucleic acids, vitamins, and minerals [39]. In practice, several technologies were evaluated, products commercialized and currently obtained SCP found the application primarily in animal feeding. However, baker's yeast mentioned above can be considered also as an example of particular SCP preparation. Many fungal species are used as protein-rich food. Most popular among them are the yeast species *Candida, Hansenula, Pichia, Torulopsis,* and *Saccharomyces* [40], but also, marine yeast is considered now as a valuable source of protein [41, 42]. Yeast-derived SCP has desired nutritional value and contains essential amino acids—above all sulfur containing amino acids. It is also significant that people accept yeast as a food source, because of its historical impact. The main disadvantage of this preparation limiting the utilization as food is the nucleic acid content what is associated with additional purification step in downstream processing [43] as well as with external mannoprotein layer of yeast cell wall what impedes the digestion.

Yeast can also be considered as an alternative source of lipids. Some species are capable of synthesis and accumulation of over 20% of biomass in form of neutral lipids and for that reason are called "oleaginous." Under optimal growth conditions and/or as a result of genetic improvement, the level of lipid accumulation can reach even 70%. Oleaginous yeast includes species of *Candida, Cryptococcus, Pichia (Hansenula), Lipomyces, Pseudozyma, Rhodosporidium, Rhodotorula, Trichosporon, Trigonopsis, Yarrowia,* and *Saccharomycopsis* [44]. The ability of these yeast species to accumulate high quantities of lipids offers the commercial potential for production of single-cell oil (SCO). Microbial oils can serve both as alternative edible oils for food industry as well as substrates used in synthesis of the oleochemicals such as fuels, soaps, plastics, paints, detergents, textiles, rubber, surfactants, lubricants, additives for the food and cosmetic industry, and many other chemicals [45]. Microorganisms regarded as an alternative oils source cannot presently compete directly with plants, but their use has many advantages like shorter process cycle, independence of season and climate, facility for genetic improvement and the possibility to manufacture lipid of unique structure, and nonsynthesized by plants. It should be mentioned that the composition of yeast oils is similar to vegetable products, and predominant fraction of them is made of triacylglycerols (TGA) rich in polyunsaturated fatty acids [46]. Lipids are stored intracellularly mainly as granular forms called "lipid body," and their content and profile of fatty acids differs between species [47]. The main fatty acids formed by oleaginous yeast are the myristic (C14:0), palmitic (C16:0), stearic (C18:0), oleic (C18:1), palmitoleic (C16:1), and linoleic (C18:2) acids [48]. Two different pathways are involved in lipid accumulation by oleaginous yeast: *de novo* lipid synthesis and *ex novo* lipid accumulation. Fundamental differences at the biochemical level exist between *de novo* lipid accumulation from hydrophilic substrates and *ex novo* lipid accumulation from hydrophobic substrates. In contrast, *ex novo* lipid production is a growth-associated process that occurs simultaneously with cell growth and is entirely independent of nitrogen exhaustion of the culture medium. The synthesis of *ex novo* lipids is the modification of fats and oils by oleaginous microorganisms [49]. The production of microbial oils became more important in the light of biodiesel production. The oleaginous yeast is considered as potential candidate for the production of "2nd generation" biodiesel deriving from lipid produced by oleaginous microorganisms growing on wastes or agro-industrial residues like sewage sludge, hemicelluloses, hydrolysates, waste glycerol, cheese whey, etc. [50, 51]. SCO may be produced *via* submerged fermentation performed under aerobic conditions through the mode of batch, fed batch, or continuous operation. To achieve both sufficient biomass formation and proper lipid accumulation level, yeast must be cultivated under carefully evaluated conditions. Many factors have been described as influencing lipid accumulation process and proper medium composition seemed to be one of the most important ones [52]. After biomass had been formed, to promote lipid accumulation process, stress conditions must be induced. Lipid storage is triggered by a nutrient limitation combined with an excess of carbon. Mostly nitrogen limitation is used to trigger lipid accumulation, but also other nutrients as phosphorus and sulphur have been shown to induce lipid storage. In spite of the applicative potential, the commercialization of microbial products is limited to oils containing polyunsaturated fatty acids like docosahexanoic acid (DHA), arachidonic acid (ARA), and eicosapentaenoic acid (EPA) [53, 54]. Producing other microbial oils either for human consumption, industrial use, or for biofuel production is still cost inhibitory—currently, the production of microbial oils is more expensive than that of vegetable oils [55]. To extend the industrialization of yeast as alternative source of oils, all efforts should be focused on genetic improvement of cell factories combined with optimization of nutrient supply and cultivation conditions and modifications of downstream technology [56].

2. Yeasts as whole-cell biocatalysts

Different genera of yeasts are convenient biocatalysts applied in many fields of chemistry, especially for the synthesis of chiral building blocks and fine chemicals. They are interesting catalytic tools, not only for their varied enzymatic activities but also for their microbiological features such as simplicity of cultivation, low nutritional requirements, and adaptive capacities. These capacities result from their flexible metabolism, which responds to the environmental impacts, so the direction (also stereoselectivity) and the effectiveness of the biotransformations can be driven by the physical chemical parameters of the process. What is also important, they are susceptible to the engineering of the reaction media (e.g., water, organic) and to the biocatalysts form (e.g., permeabilization, immobilization). This significantly broadens the field of their application by overcoming the limitations such as low solubility of the bioconversion substrates. It can be said that yeasts are used for decades and are one of the first whole-cell biocatalysts applied in industrial processes.

Literature data proved that the core applications of yeasts are connected with their extraordinary reductive abilities. Since it was proven that whole-cell biocatalysis is (as enzymatic one) chemoselective, regioselective, and stereoselective and able to regenerate dehydrogenases cofactors under biocatalytic conditions, yeasts were extensively examined as reductive catalysts for chiral building blocks synthesis—especially chiral alcohols of defined absolute configuration. The activity of a number of yeasts genera has been tested toward structurally different ketones, and in the most cases desired, alcohols have been obtained as pure enantiomers [57]. Alcohols drawn below (**Figure 4**) represent both, simple, and more complicated—unusual structures obtained thanks to the whole-cell biocatalysis driven by yeasts.

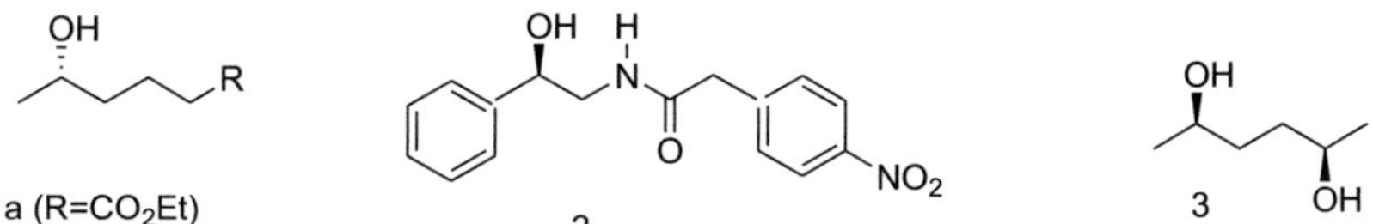

1a (R=CO_2Et)
1b (R=CN)
2
3

Figure 4. Alcohols synthesized using whole yeasts cells.

Pichia methanolica is one of the most important biocatalysts employed in industry because of its reductive ability and low substrate specificity. Thus, *Pichia methanolica* mediated reduction of ethyl-5-oxo-hexanoate and 5-oxohexanenitrile led to S-alcohols: 1a and 1b (**Figure 4**) obtained with conversion degree up to 90 and 97% of *ee* (enantiomeric excess). What is also important, bioreduction of compound 1a (**Figure 4**) was performed successfully on the gram scale with the similar yield. *Pichia methanolica* is employed for the reduction purposes by Bristol-Mayers Squibb Company (USA). Alcohol 2 (**Figure 4**) is important for pharmaceutical industry, this molecule is considered as supporting drug in the diabetes type 2 therapy by elevating the rate of metabolisms. Application of *Candida sorbophila* MY 1833 allowed receiving pure product on the large scale level with up to 60% of conversion degree and 98% of *ee*. Optically pure diols are also important, mainly as chiral building blocks for pharmaceuticals and fine chemicals synthesis. As an example—compound 3 (**Figure 4**)—2*S*, 5*S*–hexanediol can be received by diketone reduction performed with *Saccharomyces cerevisiae* on preparative scale, with the complete substrate reduction and with the *de* (distereomeric excess) of 96%. Except to mentioned diol, such procedure can be applied also for 5-hydroxyhexane-2-on formation, which is a substrate for chiral tetrahydrofuranes synthesis—chemicals of significant meaning for obtaining of the biodegradable polymers, drugs, and perfumes. Among different chiral compounds with hydroxyl functionalities, there are some of special interest—they are building platforms for the synthesis of series compounds. Such importance can be attributed to the (*S*)-4-chloro-3-hydroxybutanoate ester (*S*)-CHBE (**Figure 5**) [58]. A number of fungal catalysts, e.g., *Geotrichum* sp., *Candida* sp., *Aureobasidium* sp. are active toward appropriate ketone, but the reaction proceeds with an average stereoselectivity. Among others, *Pichia stipites* CBS 6054 was found as one allowed obtaining chiral product with *ee* of 80%. The other products of yeast-mediated enantioselective bioreduction are chiral bicyclic alcohols, e.g., synthons of (*S*)-pramipexole (anti-Parkinson drug) (**Figure 6**) and also its (*R*) isomer—considered as an anti-amyotrophic lateral sclerosis (ALS) agent [59]. Chirality is introduced into these molecules by the prochiral bicyclic ketone (**Figure 6**) reduction mediated by *Sachcaromyces cerevisiae,* and this is the crucial step of the sequences of the reaction leading to the final enantiomers of pramipexole. Discussed objective is also an example of the yeast biotransformation of the compounds with heteroatoms in their structures—here, sulphur atom is an element of the bicyclic.

Reductive activity of *Saccharomyces cerevisiae* is also applied for some nontypical reactions such as geraniol into citronellol hydrogenation achieved with resting yeast cells on preparative level (**Figure 7**) [60]. Process productivity reaches 2.38 g/L for the reaction carried out in the continuous-closed-gas-loop bioreactor.

Entirely a different activity of yeasts as biocatalysts found some practical application, lately. Successful experiments with *Saccharomyces cerevisiae* were performed according to **Figure 8**

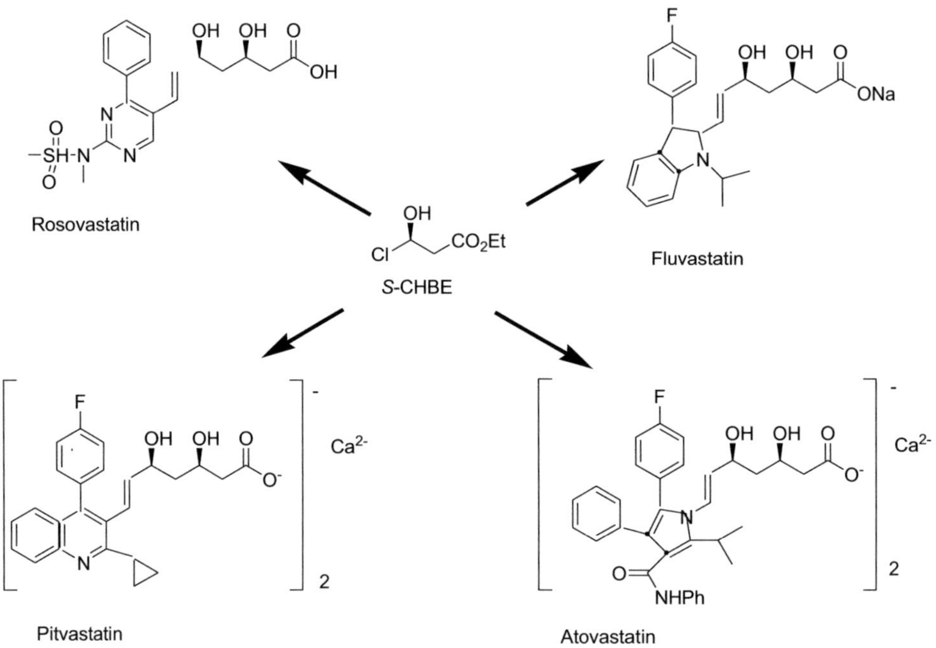

Figure 5. (*S*)-CHBE as chiral building platform.

prochiral substrate

(*R*)-pramipexole. R^1=H, R^2=$NHCH_2CH_2CH_3$
(*S*)-pramipexole. R^1=$NHCH_2CH_2CH_3$, R^2=H

Figure 6. Pramipexole (2-amin-6-propylamino-4,5,6,7-tetrahydrobenzothioazole).

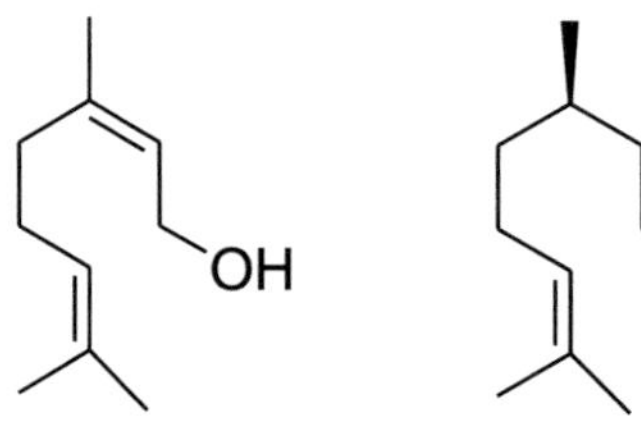

Geraniol

R-(+)-β–citronellol

Figure 7. Structures of geraniol and citronellol.

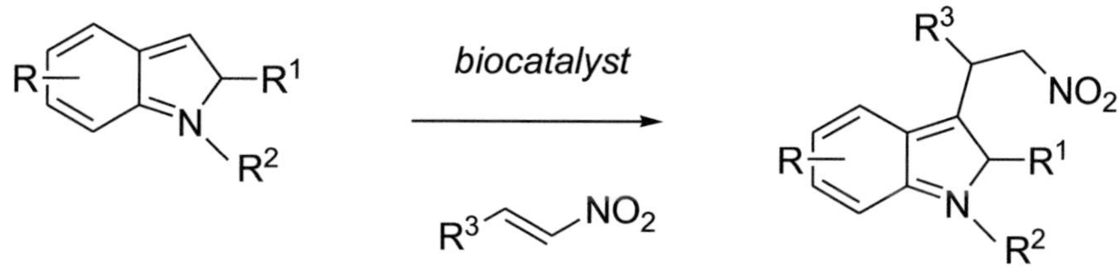

Figure 8. Biocatalyzed addition of indoles to nitroolefins.

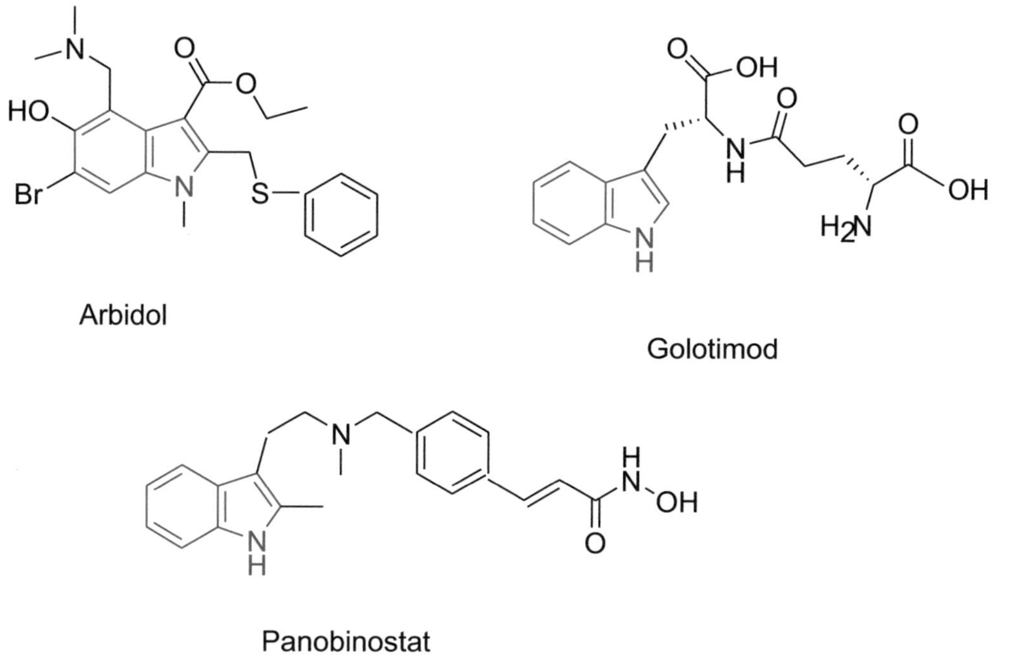

Figure 9. Compounds with the indol-motif.

[61], which illustrates the addition of diverse 3-substituted indol derivatives to nitroolefines. Chemical synthesis of such substituted indols requires some hazardous organic solvents, while biological synthesis significantly reduces this problem and offers easy operated and effective system (yield of the reaction range between 72 and 90%).

Such indol motifs are crucial part of the biologically active compounds such as arbidol (influenza A and B virus treatment and prophylactic); golotimod (immunostimulating, antimicrobial, and antineoplastic agent); and panobinostat (acute myeloid leukemia treatment; **Figure 9**).

3. Yeast's enzymes applications

Biocatalysis includes both biotransformations (e.g., the conversion of xenobiotics using whole cells or resting cell systems) and enzyme catalysis (e.g., the conversion of xenobiotics using cell-free extracts or purified enzymes) [62]. Although both whole cells and isolated enzymes can be used as biocatalysts, whole cells are very often preferable because they are more stable and cheap sources than purified enzymes, without the need for purification and coenzyme addition. However, in the case of single-step biotransformation, isolated enzymes can be considered as a better choice and can be used as a free or immobilized biocatalyst either in aqueous or organic media [63]. Yeasts, especially *Saccharomyces* species, are primarily known from whole cell reductive activity [64, 65] and are used in the food industry for the production of alcoholic beverages as well as for bread fermentation [3]. However, yeasts are a source of enzymes such as: lipases, dehydrogenases, or invertase.

3.1. Yeast's lipases

Lipases are widely distributed in nature and are produced by plants, animals, and microorganisms. Microbial enzymes are more useful than the other ones because of the diversity of catalytic activities, simplicity of manipulations, and low cost production (extracellularly during rapid growth on inexpensive media) [66]. Additionally, microbial enzymes are free from problems associated with contamination with hormones, viruses, and can be used in food processing and pharmaceutics productions for vegetarian or kosher diets [67]. Microbial lipases (EC 3.1.1.3) are suitable enzymes for organic synthesis because they are active toward broad range of nonphysiological substrates and are stable in biphasic systems or pure organic media. Lipases can be applied for either of lipid modifications and synthesis of special compounds: pharmaceuticals, polymers, biodiesels, and biosurfactants [68]. Under physiological conditions, lipases catalyze hydrolysis of ester bond in triacylglycerol to glycerol and free fatty acids. Under nonaqueous conditions, they catalyze the reverse process—esterification. The term transesterification refers to exchange the group between an ester and acid, ester and alcohol, or at least between two esters (**Figure 10**) [69].

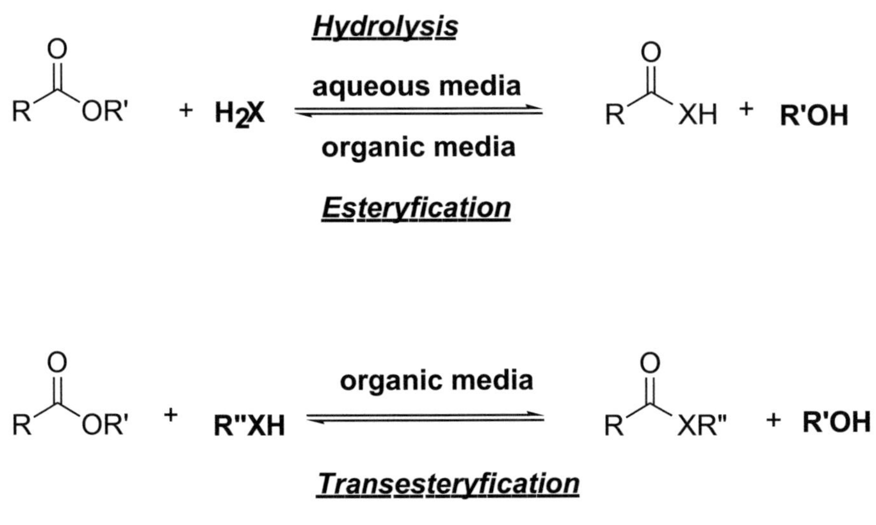

Figure 10. Types of reactions carried out by lipases.

Mentioned features make them significant biocatalyst for various applications. There are a certain number of yeast species able to produce lipases, most of them belong to *Candida* genus: *Candida utilis, C. rugosa* (*cylindracea*), *C. antarctica, C. viswanathii,* and additionally, *Yarrowia lipolytica* [70].

3.1.1. *Hydrolysis*

For pharmaceutical industry, lipases are used to resolve racemic mixtures of alcohols or carboxylic acids through asymmetric hydrolysis of acyl derivatives. *Candida antarctica* lipase, isoform B (CAL-B) is one of the most employed psychrophilic lipases for many different applications (kinetic resolutions, desymmetrization, aminolysis, etc.) [71]. Commercial CAL-B is available either in free, lyophilized, and immobilized forms (onto Lewatit VP OC 1600 (poly(methyl methacrylate-*co*-divinylbenzene)—Novozyme 435, Chirazyme L2-C2). Novozym 435 is a suitable catalyst for both small organic molecules [72, 73] and for polymerization reactions [74, 75]. Also, the immobilization of CAL-B onto different supports may result in different activity and enantioselectivity (**Table 1**) and may be a tool of control of selectivity of the hydrolysis. This feature was used for the resolution of racemic mixture of 2-*O*-butyryl-2-phenylacetic acid—precursor of both enantiomers of mandelic acid (**Figure 11**) and for the enantioselective hydrolysis of 3-phenylglutaric dimethyl diester—precursor in the drug synthesis (e.g., HIV inhibitor) [76]. As it is shown in the **Table 1**, it is possible to change the enantioselectivity of biocatalyst just by simple replacement of one support material to another one (**Figure 11**).

Another possible way to change the enantioselectivity of hydrolysis is the addition of the organic co-solvent to reaction medium. In organic media, the conformation of enzyme appears to be more rigid which may influence the enantioselectivity of the reaction. For the *Candida rugosa,* lipase-catalyzed hydrolysis of various substituted phenoxypropionates, the addition of 30–70% dimethyl sulfoxide (DMSO) or sodium dodecyl sulfate (SDS) improved the enantioselectivity, (E = 4 to >100) [77, 78]. For the same lipase (*C. rugose*), complete

Biocatalyst	Activity $\mu mol g_{cat}^{-1} h^{-1}$	*ee* p (%)[1]	Stereochemical preference	E[2]
Novozym 435	0.75	>99	S	>100
NOVO-CAL-B	0.56	>99	S	>100
Lewatit-CAL-B	0.55	>99	S	>100
Octyl-agarose-CAL-B	0.19	95	R	49
Octadecyl-sepabeads-CAL-B	0.35	90	R	23
Butyl-agarose-CALB-B	0.16	72	R	7.5

Optical purity of product.
Enantioselectivity of the enzyme towards substrate.

Table 1. The influence of different methods of immobilization on the activity and enantioselectivity of CAL-B applied for hydrolytic resolution of 2-*O*-butyryl-2-phenylacetic acid.

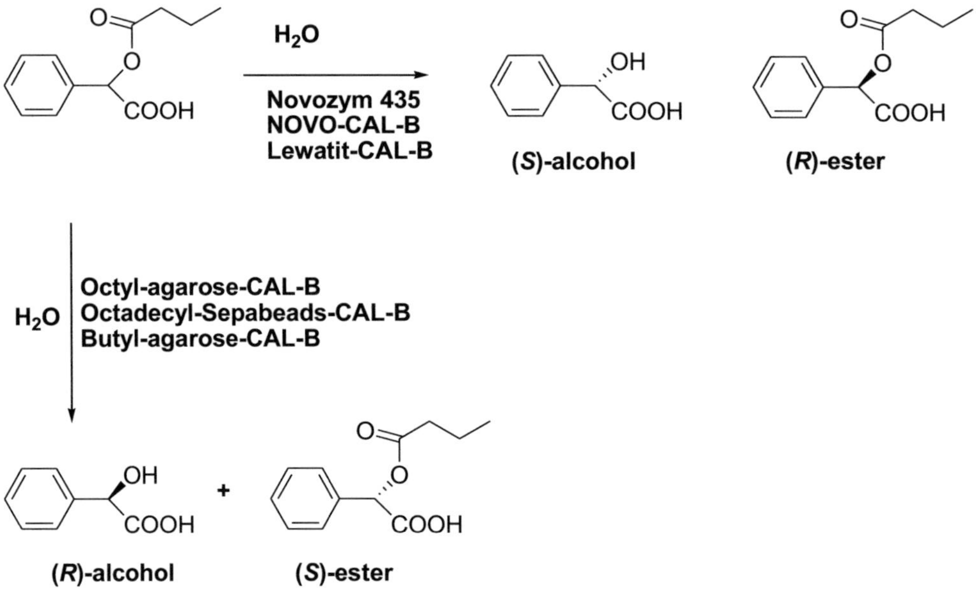

Figure 11. Kinetic resolution of 2-O-butyryl-2-phenylacetic acid by immobilized CAL-B on different supports.

reversal of enantioselectivity of hydrolysis of 1,4-dihydropyridines was observed in different organic solvents saturated with water [79], this allowed to obtain both enantiomers of 1,4-dihydropiridine (**Figure 12**).

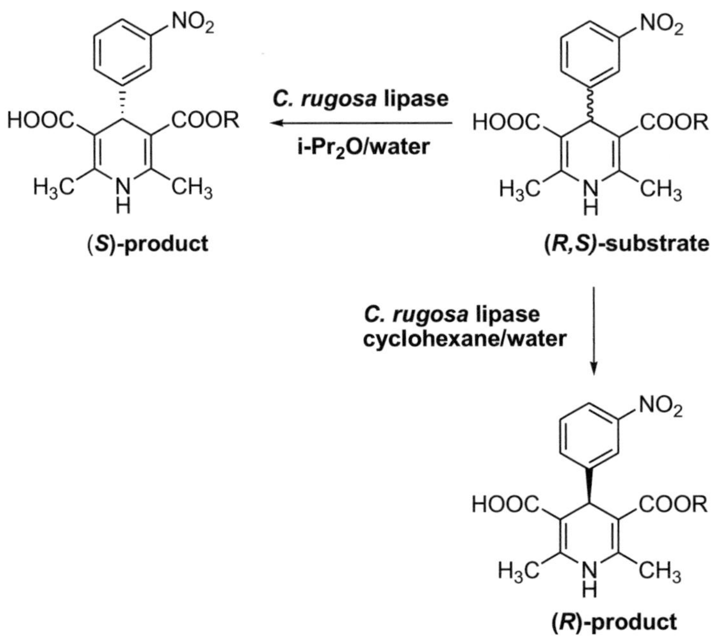

Figure 12. Reversal of enantioselectivity in hydrolysis catalyzed by *Candida rugosa* lipase in organic solvent saturated with water [79].

Candida antarctica produces two isoforms of lipases (A and B). However, more attention has been directed to the application of CAL-B, but in the last few years, also CAL-A has found remarkable applications. The most surprising biochemical property of CAL-A is its high termostability of over 90°C [71, 80]. It is quite strange that rather psychrophilic microorganism produces thermostable enzyme. This feature allows using the CAL-A under unique reaction conditions [81] and towards unusual substrates, especially sterically hindered tertiary alcohols and their derivatives or bulky cyclic compounds. Bioconversion of tertiary alcohols can be useful for the removal of *tert*-butyl protecting group even under high temperature (**Table 2**) [82]. Several examples of such applications are summarized in **Table 2**.

Another interesting application of CAL-A is the regioselective hydrolysis of cyclic diacetates, useful building blocks for the synthesis of vitamin D_3 derivatives [83], or hydrolysis of different sterically hindered carboxylic acids [84].

3.1.2. Esterification and transesterification

The most important application of lipases in organic synthesis is esterification important for the resolution of racemic mixtures of secondary alcohols and carboxylic acids. Chiral secondary alcohols serve as intermediates for pharmaceutical synthesis [85–87]. Lipase-catalyzed

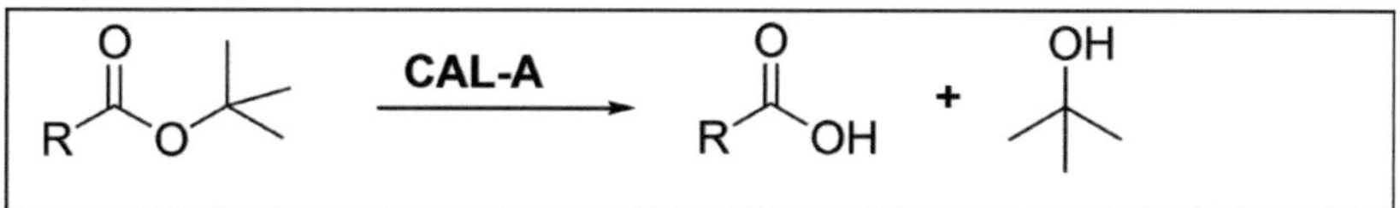

R	Yield (%)
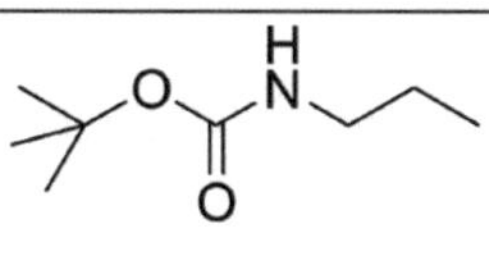	50
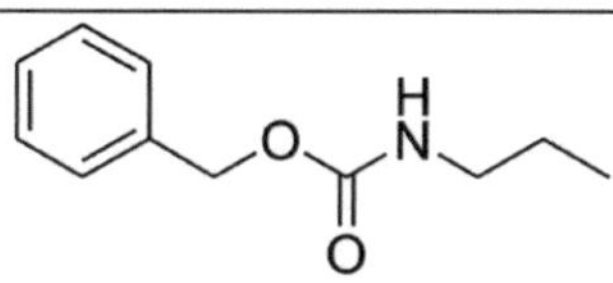	80
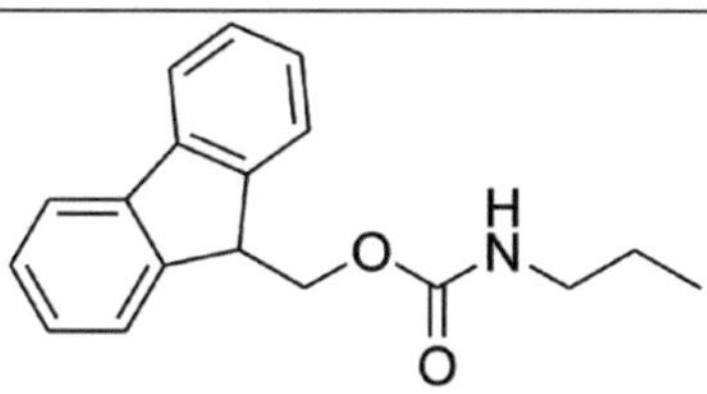	40
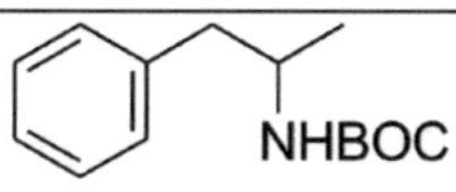	68

Table 2. Examples of substrate structures accepted by CAL-A–protecting group removal.

methods available for the preparation of enantiopure compounds are kinetic resolution (KR), dynamic kinetic resolution (DKR), and desymmetrization. Enzymatic kinetic resolution is based on the difference between the reaction rates of the enantiomers of a racemate at the presence of chiral catalyst—enzyme. Dynamic kinetic resolution combines kinetic resolution with the *in-situ* racemization of the unreacted enantiomer. Racemization can be performed chemically or enzymatically. The kinetic resolution of secondary alcohols and esters is carried out in organic solvents with lipase catalyzed acylation and alcoholysis. It leads to the formation of one enantiomer obtained as an alcohol and the other one as an ester. The maximum theoretical yield for each enantiomer is 50%. *C. antarctica* enzymes are often used for the resolution of secondary alcohols, mostly with bulky group [88, 89] but also for resolution of aliphatic compounds [90]. As an example—CAL-B was applied for the resolution of racemic mixture of 2-pentanol in heptane—media (yield (49,6%) and *ee* >99%) (**Figure 13**). *S*-(+)-2-pentanol is a key chiral intermediate for synthesis of anti-Alzheimer drugs, which inhibit *β*-amyloid peptide release and/or its synthesis [91, 92].

C. rugosa is the producer of lipase employed for the resolution of profens (2-aryl propionic acids) in enantioselective transesterification process. Profens are an important group of nonsteroidal anti-inflammatory drugs, and their biological activity depends on the optical purity of the compounds, mainly (*S*)-enantiomer [93]. For instance, (*S*)-ibuprofen ((*S*)-2(4-isobutylphenyl) propionic acid) is 160 times more effective than (*R*)- isomer in the inhibition of prostaglandins synthesis. Optically, pure profens can be synthesized by asymmetric chemical synthesis, catalytic kinetic resolution, and chiral chromatography [69], but enzymatic enantioselective esterification seems to be the best method (**Figure 14**). Discussed reaction was carried out in saturated cyclohexane with 1-propanol or 2-propanol as acyl agents and completed with good conversion degree and excellent enantiomeric excess of (*S*)-ibuprofen [94].

Also, enantiomerically pure amines constitute a class of compounds with possible biological properties and industrial applications [95]. *Candida antarctica* lipase B is one of the most effective catalyst in the preparation of enantiomerically pure nitrogenated compounds (e.g., amines, amides, amino acids, amino alcohols, etc.). This is achieved by enantioselective acetylation [96]. For example, resolution of aminoalkylpyridines was most effective (conversion 50%, time 4h, *ee* of product, and substrate >99%) with the use CAL-B and ethyl acetate as an acyl donor in the *tert*-butyl methyl ether (TBME-medium) (**Figure 15**) [97].

CAL-A isoform of *C. antarctica* lipase is able to selectively acylate cyclic, sterically hindered structures via kinetic resolution (**Figure 16**) alicyclic *β*-aminocarboxylic acids esters—

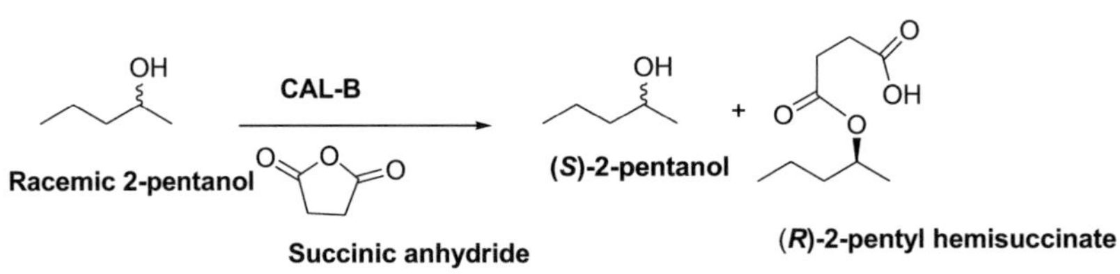

Figure 13. Kinetic resolution of 2-pentanol by esterification.

http://dx.doi.org/10.5772/intechopen.70130

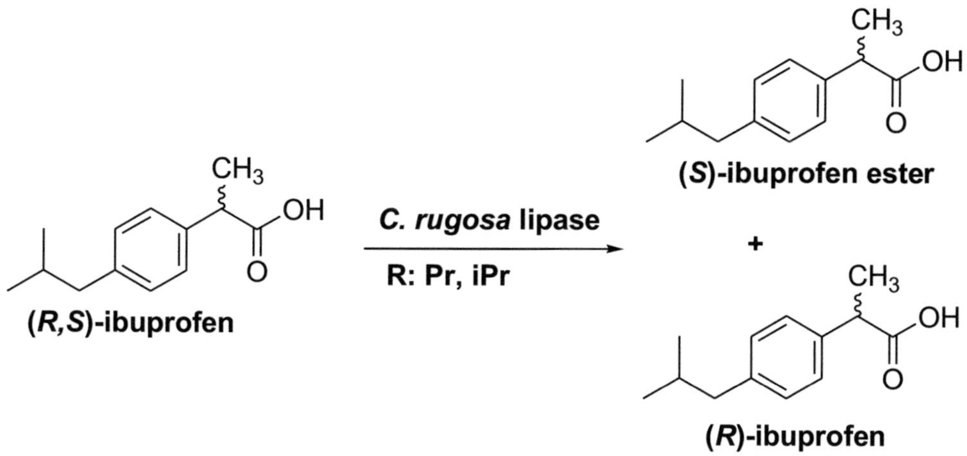

Figure 14. Enantioselective esterification of (*R,S*)-ibuprofen.

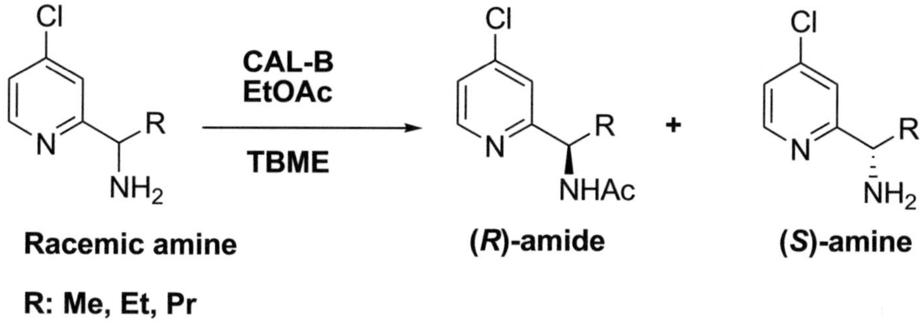

Figure 15. Kinetic resolution chloro-aminopyridines.

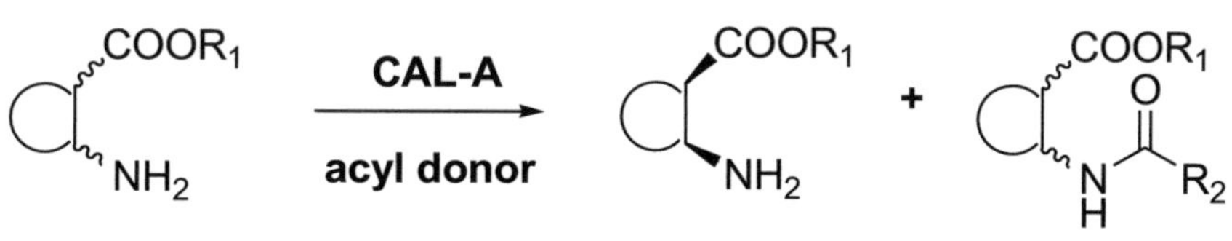

Figure 16. Kinetic resolution of alicyclic amino acids.

building blocks for the synthesis of various pharmaceutical important heterocycles [81] are synthesized this way. The best activity and enantioselectivity were observed in diethyl ether or diisopropyl ether with 2,2,2-trifluoroethyl hexanoate as an acyl donor.

CAL-A is also active towards sterically hindered tertiary alcohols. This feature is quite unique among hydrolases. The first example of enantioselective kinetic resolution of racemic mixture of tertiary alcohol was acylation of 2-phenylbut-3-yn-2-ol. The reaction was quite enantioselective, but the yield was rather moderate (25%) because of the steric hindrance. Another interesting application of CAL-A is selective acylation of sterols [98], furyl substituted allyl alcohol [99], or cyanohydrins [100, 101].

3.2. Yeast's invertase

Invertase (β-fructofuranosidase- EC 3.2.1.26) catalyses hydrolysis of the glycoside bond from the terminal nonreducing beta-fructofuranoside side in disaccharide [102]. It is also widely distributed in the environment, mainly in plants and microorganisms. The most important application of invertase is production of invert syrup—equimolar mixture of fructose and glucose, released from sucrose (**Figure 17**), which is used in food and beverage industries. Monosaccharides mixture is sweeter than sucrose and hygroscopic, it mainly is used for production of soft-centered candies and fondants. Invertase is also applied for the manufacture of artificial honey, plasticizing agents for cosmetics, pharmaceutical and paper industries, and enzyme electrodes for the detection of sucrose [103, 104]. Additionally, it can be applied for the synthesis of probiotic oligosaccharides like non-digestible oligosaccharides (NDO), e.g., lactosucrose [105]. Commercially invertase is produced mainly by *Saccharomyces cerevisiae* (Baker's yeast) or *Saccharomyces carlsbergensis*. In yeast cells, invertase is produced either in intracellular or extracellular form [106].

3.3. Yeast's oxidoreductases

Enantiometrically pure alcohols including α- and β-hydroxyesters are important and valuable intermediates in the synthesis of pharmaceuticals and other fine chemicals [107]. Enantioselective ketone reductions are one of the most common methods applied for optically pure alcohols productions. Because reactions catalyzed by dehydrogenases/reductases require cofactors (NADH or NADPH), the use of whole cells rather than isolated enzymes is preferred, to decrease the cost of enzyme purification and cofactor regeneration. However, isolated dehydrogenases employment decreased product purification problems (**Figure 18**).

Generally, α-ketoesters are reduced with lower enantioselectivities by whole yeast cells, so pair of purified reductases are selected to produce both enantiomers of (*S*)-ethyl-3-hydroxybutyrate

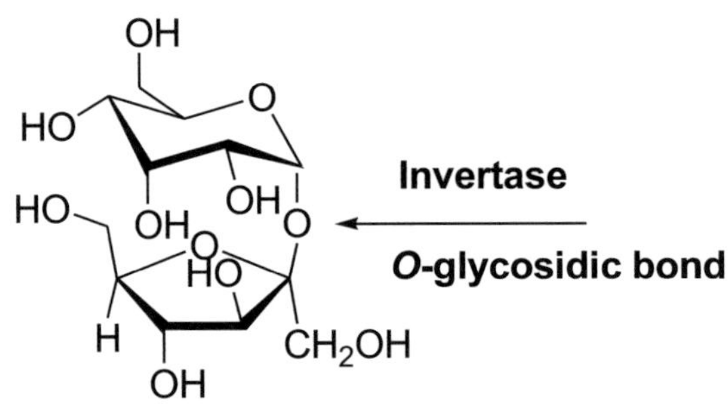

Figure 17. Hydrolysis of sucrose by invertase.

http://dx.doi.org/10.5772/intechopen.70130

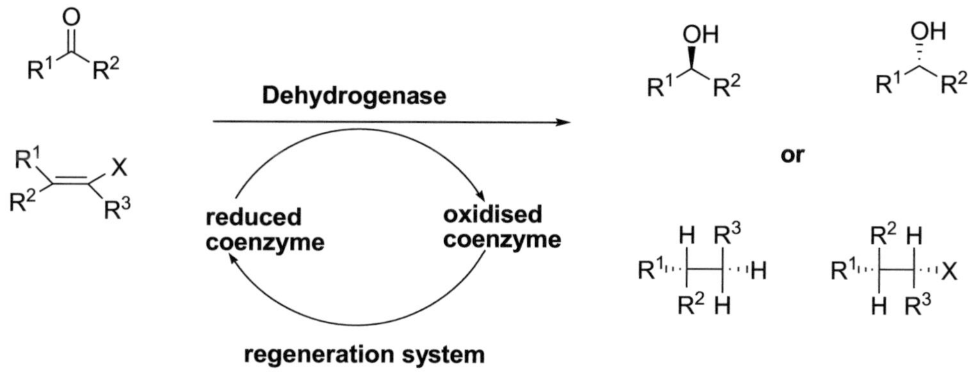

Figure 18. Reduction reactions catalyzed by dehydrogenases. Cofactor regeneration circle.

(pharmaceutical building block) in optically pure forms on preparative scale [108]. Reduction of β-ketoesters depends on both structure of substrate and specificity of the enzyme and usually yields in desired enantiomer of high optical purity [109].

Saccharomyces are also known as a producers of old yellow enzyme (OYE), the first discovered and characterized flavoprotein [110], which can be used for double bond reduction (**Figure 19**) or for dismutation reactions toward cyclic substrates (**Figure 20**) [111] and also for the reduction of nitrate esters [112] with the addition of coenzyme—NADPH.

Other example of reductase is selective carbonyl reductase from *Candida magnolia,* active toward the structurally different ketones, reduced to the corresponding optically pure (*R*)- aryl and aliphatic alcohols. This enzyme also catalysis reduction of ketones with anti-Prelog enantioselectivity which is an unusual feature of bioreductions (**Figure 21**) [107]. Configuration of obtained aryl alcohols is mostly *R*—enantiomers but also strongly dependent on R group structure (**Figure 21**).

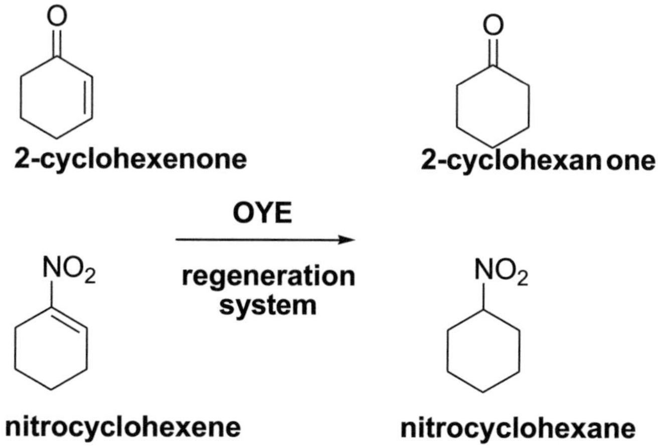

Figure 19. Example of reduction of double bonds catalysed by OYE.

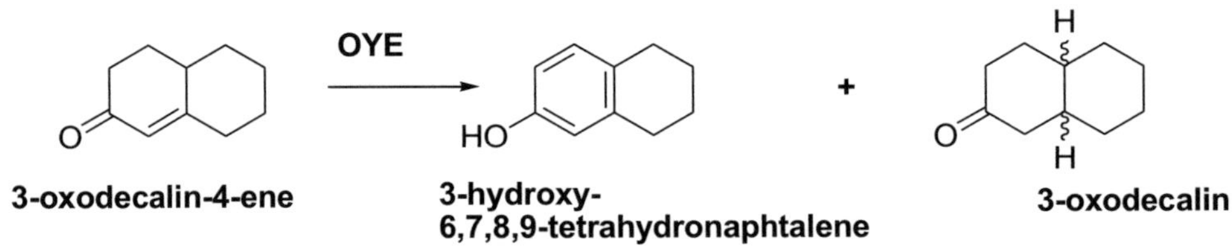

Figure 20. Dismutation reaction calalysed by OYE.

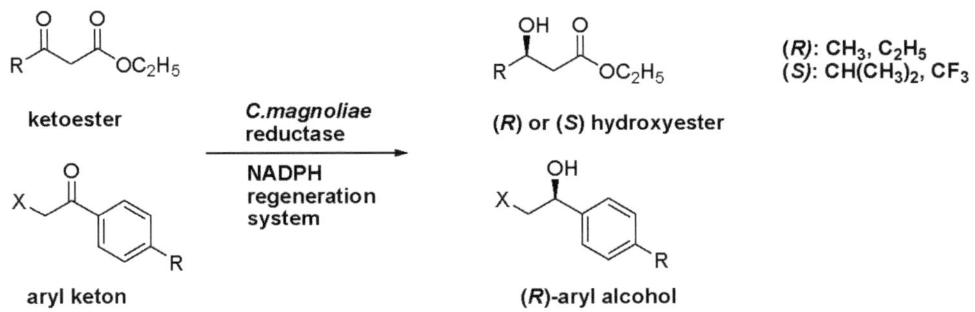

Figure 21. Anti-Prelog reduction of ketones by *C. magnoliae* reductase.

4. Yeast's applications in molecular biology

Yeasts of the *Saccharomyces* genus, in particular *S. cerevisiae*, are one of the fundamental models for eukaryotic organisms, commonly used in genetic and molecular biology studies. *S. cerevisiae* is a unicellular organism that can be grown on defined media, which gives the complete control over its chemical and physical environment. Culturing yeast is simple, economical, and rapid and can be conducted under aerobic and anaerobic conditions. As a nonpathogenic and nontoxic organism, they are safe for laboratory work, without any special precautions. Big accessibility as well as easy culturing on both liquid and solid medium makes yeast cheap and handy organism with significant biotechnological capabilities.

Although yeast and humans have been evolving along separate paths for 1 billion years, still a substantial amount of yeast genes exhibit high homology to mammalian ones. Since the basic cellular mechanics of replication, recombination, cell division, and metabolism are generally conserved between yeast and larger eukaryotes, they constitute a good model for studying different processes such as aging, regulation of gene expression, signal transduction, cell cycle, metabolism, apoptosis, neurodegenerative disorders, and many more [113]. Furthermore, its protein expression systems have more in common with higher organisms than with prokaryotic ones, mainly due to the posttranscriptional and posttranslational processing, which makes it a great candidate for acquiring a number of industrially or medically significant biomolecules, such as recombinant proteins for pharmaceutical purposes [114].

Life cycle of *Saccharomyces cerevisiae* strains include haploid and diploid phase, both of which typically grow asexually by budding. The cell cycle consists of four distinct phases (G1, S, G2, and M) and is regulated in a similar way to that of the cell cycle in larger eukaryotes [115]. Haploid yeast cells can be either mating type a or α and under normal condition can mate together to generate a/α diploids. The diploid cells cannot mate but can reproduce asexually by budding like haploids. However, under specific circumstances, like unfavorable environment conditions (lack of nutrients), diploid cell can undergo meiosis to produce haploid spores. Subsequently, the newly produced haploid nuclei are packaged into four spores that contain modified cell walls, resulting in structures that are very resistant to environmental stress [116]. Each single haploidic spore from tetrad arising after meiosis can be isolated and analyzed by various micromanipulation methods. It provides a unique opportunity to study the coupling between genes among many others. Haploid states cell can be also used for recessive mutation studies, while diploid strains can be exploited for complementation tests.

S. cerevisiae have also been first eukaryotes whose genome has been fully sequenced and published in 1996. Its nucleus genome constitutes of 12,068 kb organized into the haploid set of 16 chromosomes ranging in size from 200 to 1600 kb. The characteristic feature is that yeast genome is much more compact in comparison to other eukaryotic relatives, with genes representing 70% of total sequence. It possesses around 5885 potential protein-encoding genes, approximately 140 genes specifying ribosomal RNA, 40 genes for small nuclear RNA molecules, and 275 transfer RNA genes. Currently, an international, multidisciplinary team is involved in the production of 16 chromosomes of *S. cerevisiae* by synthetic biology tools, and the results are expected at the end of the year [117]. Another highly unique and unusual, as for eukaryotes, feature of *S. cerevisiae* genome is the presence of DNA plasmids that enables a variety of genetic manipulations and are of great importance for modern molecular biology [118].

Techniques used for yeast transformation and specific selection have been well described. For this purpose, shuttle vectors are commonly used due to the fact that they can transform both yeast and bacteria, such as *Escherichia coli*. Various yeast strains carry different auxotrophic markers that can be generated by genetic engineering methods, for instance, by gene deletion in amino acid biosynthesis pathways. Scientists have developed a number of bifunctional vectors that are easy to isolate and can autonomously replicate in each yeast and bacteria cells.

4.1. Yeast's plasmid vectors

Saccharomyces cerevisiae is a very important microorganism in modern biotechnology, not only for its contribution to brewing and bread-making industry, but also for showing great potential in the field of molecular biology and biomedicine due to its unique form of genetic material and protein expression systems. *S. cerevisiae* is one of very few eukaryotic organisms that contain circular DNA in the form of plasmids. Almost every strain of this yeast has the 2-μm plasmid, which can constitute the outstanding basis for cloning vectors (**Figure 22**), as it is 6 kb in size and is equipped with four following elements: origin of replication; genes *REP1* and *REP2* that code for proteins involved in replication process; *FLP* gene that is utilized by the plasmid to switch between isoforms, and gene *D* which role is not established yet [119]. In order for

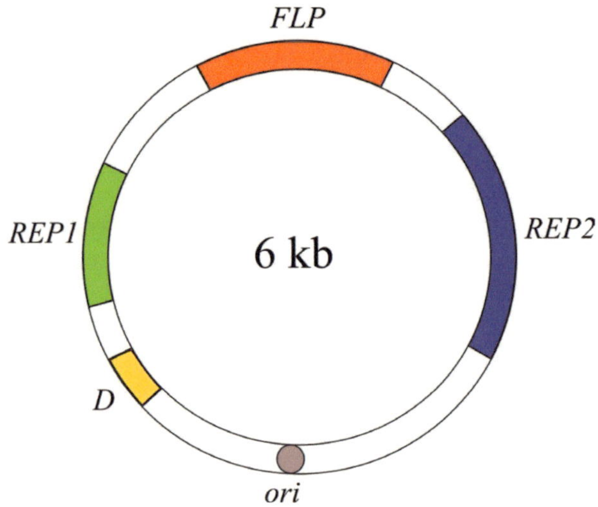

Figure 22. Genetic structure of 2 μm yeast plasmid.

2-μm plasmid to work as a fully functional cloning vector, there has to be an incorporated element, called a selective marker, which allows for the transformed cells to be identified after cloning experiment.

Most of the bacterial vectors are provided with genes-encoding resistance to various kinds of antibiotics, such as ampicillin (*ampR*) or tetracycline (*tetR*). Therefore, upon culturing transformed cells in the medium with the addition of such antibiotic, the colonies that were unable to grow did not carry the resistance gene (hence, the uptake of the vector did not occur); thus, the only cells that survive are the transformed ones carrying the properly inserted vector.

Cloning techniques with yeast differ mostly in the strategic approach of the selective markers. In this case, usually a special kind of organism is required as the host, namely an auxotrophic mutant that is unable to obtain or synthesize a pivotal compound of one of its metabolic pathways. A good example is *leu2*$^-$ yeast that has an inactive form of *LEU2* gene; hence, it cannot synthesize leucine and can only grow in a medium that is supplied with this amino acid. To properly use that organism as the host, a vector with *LEU2* active gene has to be prepared. The cells are then transformed with the plasmid and cultured in minimal medium that lacks leucine. This way the only colonies that will grow will be the transformed cells [120].

There are few kinds of yeast cloning vectors, but all of them are so-called shuttle vectors, which means that they can replicate and be selected in both bacteria and yeast. Shuttle vectors were developed mostly because plasmid preparation from yeast only is highly ineffective; hence, the large-scale DNA propagation and convenient genetic manipulation are performed in bacterial organism, such as *Escherichia coli*.

Yeast cloning vectors based on 2-μm plasmid are called yeast episomal plasmids (YEps) (**Figure 23(a)**). Depending on the kind of YEp, they can either contain most of the 2-μm plasmid or just the origin of replication, their backbone is usually constructed from *E. coli* vectors, such

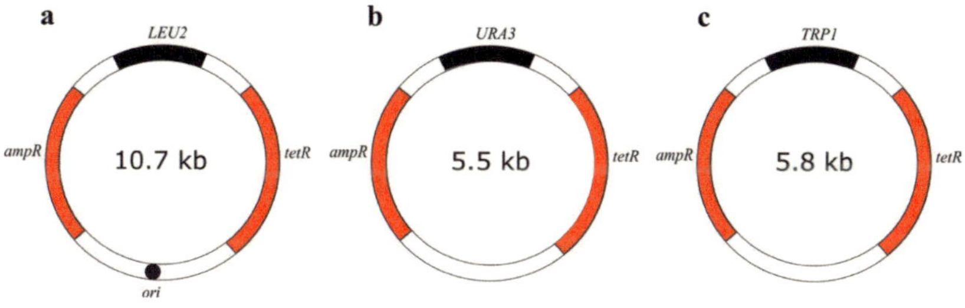

Figure 23. Genetic structure of exemplary yeast vectors. *AmpR* and *tetR* are the antibiotic resinstance genes from *E. coli* pBR322 plasmid, whereas *LEU2, URA3, TRP1* fragments represent yeast chromosomal DNA: (a) yeast episomal plasmid YEp13, (b) yeast integrative plasmid YIp5, (c) yeast replicative plasmid YRp7.

as pBR322 that contains genes encoding resistance for antibiotics, like ampicillin or tetracycline. As the name suggests, YEps can replicate independently, or they can be integrated into one of the yeast chromosomes. The most common reason for YEps integration is that they carry selective marker gene, which is very similar to the mutant chromosomal DNA of the host organism; for example, the already mentioned *LEU2* gene or others, such as *URA3, HIS3, TRP1,* or *LYS2* all involved in biosynthesis pathways of pyrimidine nucleotides, histidine, tryptophan, or lysine, respectively. YEps are considered as high copy number vectors, yielding up to 200 copies per cell with the transformation frequency between 10,000 and 100,000 transformed cells per μg. Unfortunately, the major drawback is that their recombinants are highly unstable, which makes it very difficult and time consuming to achieve conclusive and reliable results when working with YEps.

The other important type of yeast vectors are yeast integrative plasmids (YIps) (**Figure 23(b)**). They mainly consist of *E. coli* plasmid, such as pBR322 and a selective marker (usually one of the mentioned above). What is important is they do not contain yeast origin of replication; therefore, they cannot replicate in any other way than through the process of integration with chromosomal DNA. In terms of transformation frequency, YIps come at the very last place with the number significantly lower than 1000 transformed cells per μg and usually only one copy per cell. On the other hand, their recombinants are very stable, usually making them the top pick for the experimental purposes.

Yeast replicative plasmids (YRps) (**Figure 23(c)**) are another type of yeast cloning vectors. They contain a backbone from *E. coli* vector, the yeast origin of replication in close proximity to the selective marker. Such structure suggests that YRps can replicate independently with relatively high transformation frequency ranging from 1000 to 10,000 transformed cells per μg and a number of copies between 5 and 100 per cell. YRps recombinants are as unstable as the YEps ones, making them one of the last choice for laboratory work.

Along with the time, there was a growing demand for much larger pieces of DNA to be manipulated through the techniques of genetic engineering. It was at this point that the last type of yeast cloning vectors was developed, namely the yeast artificial chromosomes (YACs). The general idea behind those constructs was that yeast chromosomes usually carry several hundred kilobases of genetic material, so why not imitate the native DNA? YACs were thought to contain the three key elements of a chromosome:

- centromeres required for the proper chromosome positioning during the cell division;
- origins of replication, which are the places on chromosome where the replication of genetic material starts;
- telomers as the defenders of the chromosomes against exonucleases.

Several types of YACs have been developed over the years, they usually consist mostly of a *E. coli* pBR322 plasmid with some yeast genes, such as selective markers (usually at least two located oppositely), origin of replication, gene *CEN4* coding for centromere region, the two telomeres fragments *TEL*, and an additional selective marker with a restriction enzyme site, such as *SUP4* gene that compensates for a mutation-causing accumulation of a red pigment (**Figure 24**). YACs are equipped with a restriction enzyme site between two *TEL* sites, so that upon cleavage, an "artificial" linear chromosome is created, subsequently *SUP4* is cut, and the chromosome is divided into two arms, between which a DNA fragment to be cloned is ligated, thus recreating the single-line chromosome structure. The next step is to transform double auxotrophic mutant organisms that will not be able to survive in the minimum medium without the properly received YAC. Additional experimental control can be then tested by simple optical inspection—colonies with disrupted *SUP4* gene will appear as white, meaning they are transformed, any other color means that the colony has not been properly transformed [121].

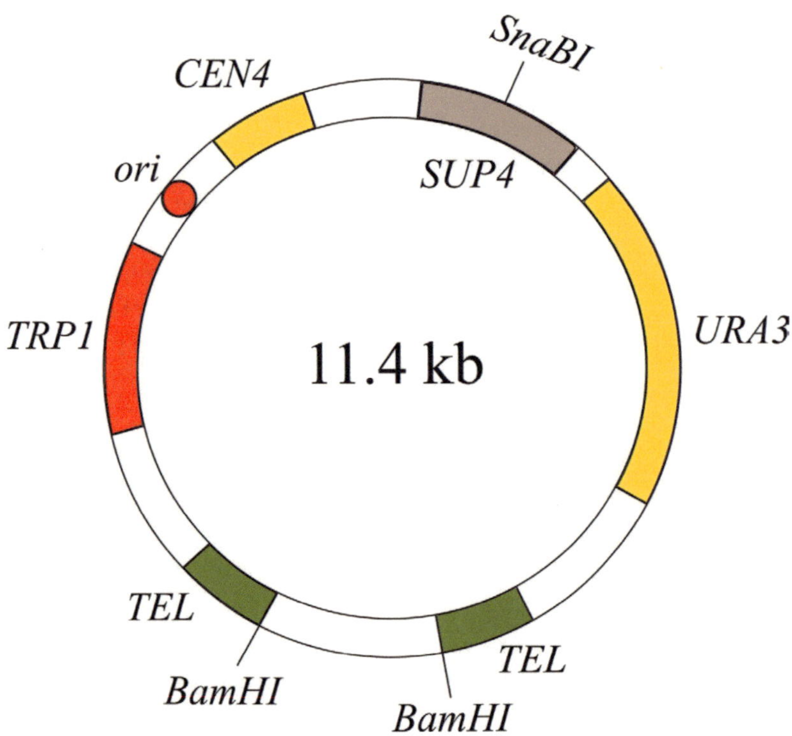

Figure 24. Genetic structure of yeast artificial chromosome pYAC3.

4.2. Yeast expression system

Recombinant proteins are the biomolecules of great importance, because among other things, they are able to mimic the functions of native proteins; hence, they are extensively studied in biotechnology and biopharmaceutical research. The critical point of target protein production is the choice of efficient expression system which enables obtaining functional product with high yield.

Yeast expression system constitutes a good alternative for widely used bacterial and higher eukaryote expression systems. They are genetically well defined and are known to perform many posttranslational modifications, including proper protein folding, disulfide bond formation, and glycosylation [122]. The culturing of yeast is also easy, rapid, and cheap, which is their big advantage over the insect or mammalian cells. They easily undergo genetical manipulation and adapt to fermentation processes; therefore, using yeasts as a cell factory is convenient and enables to obtain a fair amount of target protein. In contrast to bacteria, recombinant proteins obtained in yeast expression systems are free of endotoxins that make this system safer, especially in terms of medical and food application [114]. In fact, about 20% of all biopharmaceuticals are produced by *S. cerevisiae*. Among them, the most dominant are insulin, human serum albumin, hepatitis vaccines, and virus-like particles used for vaccination against human papillomavirus [123].

However, yeast cells are limited in the production of human-like glycoproteins by their inability to produce complex *N*-linked glycans. In addition, *S. cerevisiae* produce the hypermannosylated *N*-linked glycans with the mannose residues being attached to the chitobiose core (a dimmer of β-1,4-linked glucosamine units). Hypermannosylation also results in a short half life *in vivo* and thereby compromises the efficacy of most therapeutic glycoproteins [124]. To overcome this issue, great deal of effort has been put into altering the glycosylation pathways in *P. pastorsis* to produce strains possessing human-like *N*-glycosilation patterns [125, 126]. This achievement has contributed to the increased usefulness of yeast in industry for the production of stable and recombinant glycoproteins.

There is no ultimate procedure for yeast expression system that could work equally well for the production of all kinds of proteins. Optimization of whole process is the critical step to obtain sufficient amounts of pure, properly folded and secreted protein of interest. While small and simple in structure proteins are easy to obtain, the big and multi-domain protein could require certain chaperones to facilitate the folding process [127]. The advantage of yeast expression system is that it allows extracellular secretion of produced protein when proper signaling sequence has been attached to the structural gene [114]. It significantly facilitates the recombinant protein purification process from the culturing medium and allows to optimize the culturing conditions. In order to increase protein secretion level, a few strategies have been developed. One of them is protein engineering of a desired product, for instance by modifying protein coding sequences and signaling sequences [128–130]. Since this methodology is highly specific against each protein, the conditions optimized for one protein do not always work for another. Different approach is to engineer the host strains and tune-up folding and secretory machinery by overexpression or deletion genes that are critical for the protein secretion [131, 132]. Additionally, it has been shown that expression in low temperatures enhances the level of secretion [133].

There are numerous varieties of expression vectors available for producing heterologous proteins in yeast, and these are the derivatives of YIp, YEp, and YRp plasmids described previously. The DNA coding for the protein of interest is inserted into the vector. The type of selective marker and promoter strength are key factors that determine the plasmid copy number and the mRNA level of the recombinant protein. Varieties of inducible and constitutive promotors have been applied for gene expression in yeasts in the past. The first of these allow the controllable gene expression. Most of inducible promoters are responsive to catabolite repression or react to other environmental conditions, like in-cell iron concentration, stress, or lack of essential amino acids. *GAL* promoter, which is induced by adding galactose, provides a straightforward system for expression regulation of the cloned foreign gene [134]. Another good example can be *CUP1* promoter, which is induced by copper [135] or heat shock factor promoter, induced by heat stress at 39°C [136]. There are also some other groups of promoters that initiate strong and constitutive expression. *TEF1* promoter, as an example of *S. cerevisiae,* is a widely used representative of this group, as it can drive high gene expression in both high and low glucose conditions [137]. Selection of a suitable promoter depends on specific process requirements and the properties of the target protein to be produced.

Additionally, yeasts are recognized as a generally recognized as safe (GRAS) organism, which only strengthens its position as the most frequently used microbial eukaryote for recombinant protein synthesis.

4.3. Yeast's two hybrid system

Ever since the Field and Song discovery described in 1989, a new approach toward the examination of protein-protein interactions emerged, it was named as the yeast two-hybrid system. It allows to detect the interaction of two proteins in the yeast cell, and it can be used to select an interacting partner of a known protein. This technique takes the advantage of the fact that majority of eukaryotic transcriptional factors, such as Gal4p, consist of two independent, functional DNA domains: binding domain (BD) and transcription activation domain (AD). While the two domains are normally on the same polypeptide chain, the transcription factor also functions when these two domains are brought together by noncovalent protein-protein interactions. In yeast hybrid system, each of these domains is connected to the one from the studied protein. As a result, two fusion proteins are created: one combined to DNA-binding domain (BD) and the other joined to activation domain (AD) [138]. The genes coding for both fusion proteins are carried by different plasmids, but each plasmid undergoes expression in the same yeast cell. If the interaction between studied proteins occurs, BD and AD domains are close enough to activate transcription of a reporter gene that is regulated by the transcription factors.

General idea of the yeast two-hybrid system can be represented by an example of transcriptional factors and a gene coding for β-galactosidase, wherein the interaction between studied proteins may potentially lead to the expression of the reporter gene coding for β-galactosidase in *E. coli lacZ* reporter gene. The presence of this enzyme can later be verified by simple reaction with X-gal, which yields a blueish insoluble product, thereby confirming or denying the association of studied proteins. Since 1989, yeast two-hybrid system has been studied extensively and further developed to find countless new applications, some of which are summarized and generally described in Ref. [139].

5. Conclusions

Discussed unique yeasts features, which are fundamental for their versatile applications are still examined and after finishing the "Synthetic Yeast Genome Project (Sc2.0)" the new perspectives of the applying them will be opened as well as in the molecular biology and in the industrial applications.

Acknowledgements

The project was supported by the Wroclaw Center of Biotechnology program, the Leading National Research Center (KNOW) for the years 2014–2018.

Author details

Ewa Żymańczyk-Duda[1]*, Małgorzata Brzezińska-Rodak[1], Magdalena Klimek-Ochab[1], Maciej Duda[1] and Agata Zerka[2]

*Address all correspondence to: ewa.zymanczyk-duda@pwr.edu.pl

1 Wrocław University of Science and Technology, Wrocław, Poland

2 Institute of Immunology and Experimental Therapy, Wrocław, Poland

References

[1] Otterstedt K, Larsson C, Bill RM, Stahlberg A, Boles E, Hohmann S, Gustafsson L. Switching the mode of metabolism in the yeast *Saccharomyces cerevisiae*. EMBO reports. 2004;**5**:532-537. DOI: 10.1038/sj.embor.7400132

[2] Barnett JA, Entian K-D. A history of research on yeasts 9: Regulation of sugar metabolism. Yeast. 2005;**22**:835-894. DOI: 10.1002/yea.1249

[3] Steensels J, Snoek T, Meersman E, Nicolino MP, Voordeckers K, Verstrepen KJ. Improving industrial yeast strains: Exploiting natural and artificial diversity. FEMS Microbiology Reviews. 2014;**38**:947-995. DOI: 10.1111/1574-6976.12073

[4] Hahn-Hägerdal B, Karhumaa K, Larsson CU, Gorwa-Grauslund M, Görgens J, van Zyl WH. Role of cultivation media in the development of yeast strains for large scale industrial use. Microbial Cell Factories. 2005;**4**(31):1-16. DOI: 10.1186/1475-2859-4-31

[5] Krogerus K, Magalhães F, Vidgren V, Gibson B. Novel brewing yeast hybrids: Creation and application. Applied Microbiology and Biotechnology. 2017;**101**:65-78. DOI: 10.1007/s00253-016-8007-5

[6] Piskur J, Rozpedowska E, Polakova S, Merico A, Compagno C. How did *Saccharomyces* evolve to become a good brewer? Trends in Genetics. 2006;**22**(4):183-186. DOI: 10.1016/j.tig.2006.02.002

[7] Berlowska J, Kregiel D, Rajkowska K. Biodiversity of brewery yeast strains and their fermentative activities. Yeast. 2015;**32**:289-300. DOI: 10.1002/yea.3041

[8] Bond U. Chapter 6: The genomes of lager yeasts. Advances in Applied Microbiology. 2009;**69**:159-182. DOI: 10.1016/S0065-2164(09)69006-7

[9] Soares EV. Flocculation in *Saccharomyces cerevisiae*: A review. Journal of Applied Microbiology. 2010;**110**:1-18. DOI: 10.1111/j.1365-2672.2010.04897.x

[10] Verstrepen KJ, Derdelinckx G, Verachtert H, Delvaux FR. Yeast flocculation: What brewers should know. Applied Microbiology and Biotechnology. 2003;**61**:197-205. DOI: 10.1007/s00253-002-1200-8

[11] Pretorius IS, Van der Westhuizen TJ, Augustyn OPH. Yeast biodiversity in vineyards and wineries and its importance to the South African wine industry. South African Journal of Enology and Viticulture. 1999;**20**:61-74

[12] Liu Y, Rousseaux S, Tourdot-Marechal R, Sadoudi M, Gougeon R, Schmitt-Kopplin P, Alexandre H. Wine microbiome: A dynamic world of microbial interactions. Critical Reviews in Food Science and Nutrition. 2017;**57**(4):56-873. DOI: 10.1080/10408398.2014.983591

[13] Cordente AG, Curtin CD, Varela C, Pretorius IS. Flavor-active wine yeasts. Applied Microbiology an Biotechnology. 2012;**96**:601-618. DOI: 10.1007/s00253-012-4370-z

[14] Styger G, Prior B, Bauer FF. Wine flavor and aroma. Journal of Industrial Microbiology and Biotechnology. 2011;**38**:1145-1159. DOI: 10.1007/s10295-011-1018-4

[15] Degre AR. Selection and commercial cultivation of wine yeast and bacteria. In: Fleet GH, editor. Wine Microbiology and Biotechnology. Reading: Harwood Academic; 1993. pp. 421-447

[16] Ciani M, Capece A, Comitini F, Canonico L, Siesto G, Romano P. Yeast interactions in inoculated wine fermentation. Frontiers in Microbiology. 2016;**7**(555):1-7. DOI: 10.3389/fmicb.2016.00555

[17] Perez-Torrado R, Querol A, Guillamon JM. Genetic improvement of non-GMO wine yeasts: Strategies, advantages and safety. Trends in Food Science & Technology. 2015;**45**: 1-11. DOI: 10.1016/j.tifs.2015.05.002

[18] Mortimer RK, Romano P, Suzzi G, Polsinelli M. Genome renewal: A new phenomenon revealed from a genetic study of 43 strains of *Saccharomyces cerevisiae* derived from natural fermentation of grape musts. Yeast. 1994;**10**(12):1543-1552. DOI: 10.1002/yea.320101203

[19] Magwene PM. Revisiting Mortimer's Genome renewal hypothesis: Heterozygosity, homothallism, and the potential for adaptation in yeast. Advances in Experimental Medicine and Biology. 2014;**781**:37-48. DOI: 10.1007/978-94-007-7347-9_3

[20] Magwene PM, Kaykç Ö, Granek JA, Reininga JM, Scholl Z, Murray D. Outcrossing, mitotic recombination, and life-history trade-offs shape genome evolution in *Saccharomyces cerevisiae*. Proceedings of the National Academy of Sciences of the USA. 2011;**108**(5):1987-1992. DOI: 10.1073/pnas.1012544108

[21] Borneman AR, Desany BA, Riches D, Affourtit JP, Forgan AH, Pretorius IS, Egholm M, Chambers PJ. Whole-genome comparison reveals novel genetic elements that characterize the genome of industrial strains of *Saccharomyces cerevisiae*. PLoS Genetics. 2011;**7**(2):e1001287: 1-10. DOI: 10.1371/journal.pgen.1001287

[22] Moreno-Arribas MV, Polo MC. Winemaking biochemistry and microbiology: Current knowledge and future trends. Critical Reviews in Food Science and Nutrition. 2005;**45**:65-286. DOI: 10.1080/10408690490478118

[23] Eldarov MA, Kishkovskaia SA, Tanaschuk TN, Mardanov AV. Genomics and biochemistry of *Saccharomyces cerevisiae* wine yeast strains. Biochemistry (Moscow). 2016;**81**(13):1650-1668. DOI: 10.1134/S0006297916130046

[24] Hong Y-A, Park H-D. Role of non-*Saccharomyces* yeasts in Korean wines produced from campbell early grapes: Potential use of *Hanseniaspora uvarum* as a starter culture. Food Microbiology. 2013;**34**:207-214. DOI: 10.1016/j.fm.2012.12.011

[25] Wang C, Mas A, Esteve-Zarzoso B. Interaction between *Hanseniaspora uvarum* and *Saccharomyces cerevisiae* during alcoholic fermentation. International Journal of Food Microbiology. 2015;**206**:67-74. DOI: 10.1016/j.ijfoodmicro.2015.04.022

[26] Steensels J, Daenen L, Malcorps P, Derdelinckx G, Verachtert H, Verstrepen KJ, Kevin J. *Brettanomyces* yeasts—From spoilage organisms to valuable contributors to industrial fermentations. International Journal of Food Microbiology. 2015;**206**:24-38. DOI: 10.1016/j.ijfoodmicro.2015.04.005

[27] Padilla B, Gil JV, Manzanares P. Past and future of non-*Saccharomyces* yeasts: From spoilage microorganisms to biotechnological tools for improving wine aroma complexity. Frontiers in Microbiology. 2016;**7**(411):1-20. DOI: 10.3389/fmicb.2016.00411

[28] Masneuf-Pomarede I, Bely M, Marullo P, Albertin W. The genetics of non-conventional wine yeasts: Current knowledge and future challenges. Frontiers in Microbiology. 2016;**6**(1563):1-15. DOI: 10.3389/fmicb.2015.01563

[29] Pretorius IS. Tailoring wine yeast for the new millennium: Novel approaches to the ancient art of winemaking. Yeast. 2000;**16**:675-729. DOI: 10.1002/1097-0061(20000615)16:8<675::AID-YEA585>3.0.CO;2-B

[30] Walker GM, Hill AE. *Saccharomyces cerevisiae* in the production of whisk(e)y. Beverages. 2016;**2**(38):1-15. DOI: 10.3390/beverages2040038

[31] Kim HR, Im Y-M, Ko H-M, Chin J-E, Kim I-C, Lee HB, Bai S. Raw starch fermentation to ethanol by an industrial distiller's yeast strain of *Saccharomyces cerevisiae* expressing glucoamylase and α-amylase genes. Biotechnology Letters. 2011;**33**:1643-1648. DOI: 10.1007/s10529-011-0613-9

[32] Stewart GG, Hill AE, Russell I. 125th anniversary review: Developments in brewing and distilling yeast strains. Journal of the Institute of Brewing. 2013;**119**(4):202-220. DOI: 10.1002/jib.104

[33] Cho IH, Peterson DG. Chemistry of bread aroma: A review. Food Science and Biotechnology. 2010;**19**(3):575-582. DOI: 10.1007/s10068-010-0081-3

[34] Perez-Torrado R, Gamero E, Gomez-Pastor R, Garre E, Aranda A, Matallana E. Yeast biomass, an optimized product with myriad applications in the food industry. Trends in Food Science & Technology. 2015;**46**:167-175. DOI: 10.1016/j.tifs.2015.10.008

[35] Gelinas P. Aeration and foam control in Baker's yeast production: Mapping patents. Comprehensive Reviews in Food Science and Food Safety. 2015;**15**:371-391. DOI: 10.1111/1541-4337.12188

[36] Attfield PV. Stress tolerance: The key to effective strains of industrial baker's yeast. Nature Biotechnology. 1997;**15**:1351-1357. DOI: 10.1038/nbt1297-1351

[37] Shima J, Takagi H. Stress-tolerance of baker's-yeast (*Saccharomyces cerevisiae*) cells: Stress-protective molecules and genes involved in stress tolerance. Biotechnology and Applied Biochemistry. 2009;**53**:155-164. DOI: 10.1042/BA20090029

[38] Suman G, Nupur M, Anuradha S, Pradeep B. Single cell protein production: A Review. International Journal of Current Microbiology and Applied Sciences. 2015;**4**(9):251-262

[39] Patelski P, Berlowska J, Dziugan P, Pielech-Przybylska K, Balcerek M, Dziekonska U, Kalinowska H. Utilisation of sugar beet bagasse for the biosynthesis of yeast SCP. Journal of Food Engineering. 2015;**167**:32-37.DOI: 10.1016/j.jfoodeng.2015.03.031

[40] Nasseri AT, Rasoul-Amini S, Morowvat MH, Ghasemi Y. Single cell protein: Production and process. American Journal of Food Technology. 2011;**6**:103-116. DOI: 10.3923/ajft.2011.103.116

[41] Chi Z, Yan K, Gao L, Li J, Wang X,Wang L. Diversity of marine yeasts with high protein content and evaluation of their nutritive compositions. Journal of the Marine Biological Association of the United Kingdom. 2008;**88**(7):1347-1352. DOI: 10.1017/S0025315408001938

[42] Sarlin PJ, Philip R. Marine yeasts as feed supplement for Indian white prawn *Fenneropenaeus indicus*: Screening and Testing the Efficacy. International Journal of Current Microbiology and Applied Sciences. 2016;**5**(1):55-70. DOI: 10.20546/ijcmas.2016.501.005

[43] Anupamaa, Ravindra P. Value-added food: Single cell protein. Biotechnology Advances. 2000;**18**:459-479. DOI: 10.1016/S0734-9750(00)00045-8

[44] Liang MH, Jiang JG. Advancing oleaginous microorganisms to produce lipid *via* metabolic engineering technology. Progress in Lipid Research. 2013;**52**:395-408. DOI: 10.1016/j.plipres.2013.05.002

[45] Probst KV, Schulte LR, Durrett TP, Rezac ME, Vadlan PV. Oleaginous yeast: A value-added platform for renewable oils. Critical Reviews in Biotechnology, 2016;**36**(5):942-955. DOI: 10.3109/07388551.2015.1064855

[46] Papanikolaou S, Chevalot I, Komaitis M, Aggelis G, Marc I. Kinetic profile of the cellular lipid composition in an oleaginous *Yarrowia lipolytica* capable of producing a cocoa-butter substitute from industrial fats. Antonie van Leeuwenhoek. 2001;**80**:215-224. DOI: 10.1023/A:1013083211405

[47] Beopoulos A, Nicaud JM, Gaillardin C. An overview of lipid metabolism in yeasts and its impact on biotechnological processes. Applied Microbiology and Biotechnology. 2011;**90**:1193-1206. DOI: 10.1007/s00253-011-3212-8

[48] Uemura H. Synthesis and production of unsaturated and polyunsaturated fatty acids in yeast: Current state and perspectives. Applied Microbiology and Biotechnology. 2012;**95**(1):1-12. DOI: 10.1007/s00253-012-4105-1

[49] Papanikolaou S, Aggelis G. Lipids of oleaginous yeasts. Part I: Biochemistry of single cell oil production. European Journal of Lipid Science and Technology. 2011;**113**(8):1031-1051. DOI: 10.1002/ejlt.201100014

[50] Papanikolaou S, Aggelis G. Lipids of oleaginous yeasts. Part II: Technology and potential applications. European Journal of Lipid Science and Technology. 2011;**113**:1052-1073. DOI: 10.1002/ejlt.201100015

[51] Leiva-Candia DE, Pinzi S, Redel-Macías MD, Koutinas A, Webb C, Dorado MP. The potential for agro-industrial waste utilization using oleaginous yeast for the production of biodiesel. Fuel. 2014;**123**:33-42. DOI: 10.1016/j.fuel.2014.01.054

[52] Ageitos JM, Vallejo JA, Veiga-Crespo P, Villa TG. Oily yeasts as oleaginous cell factories. Applied Microbiology and Biotechnology. 2011;**90**:1219-1227. DOI: 10.1007/s00253-011-3200-z

[53] Groenewald M, Boekhout T, Neuveglise C, Gaillardin C, van Dijck PWM, Wyss M. *Yarrowia lipolytica*: Safety assessment of an oleaginous yeast with a great industrial potential. Critical Reviews in Microbiology. 2014;**40**(3):187-206. DOI: 10.3109/1040841X.2013.770386

[54] Ratledge C. Fatty acid biosynthesis in microorganisms being used for single cell oil production. Biochimie. 2004;**86**:807-815. DOI: 10.1016/j.biochi.2004.09.017

[55] Koutinas AA, Chatzifragkou A, Kopsahelis N, Papanikolaou S, Kookos IK. Design and techno-economic evaluation of microbial oil production as a renewable resource for biodiesel and oleochemical production. Fuel. 2014;**116**:566-577. DOI: 10.1016/j.fuel.2013.08.045

[56] Huang C, Chen X, Xiong L, Chen X, Ma L, Chen Y. Single cell oil production from low-cost substrates: The possibility and potential of its industrialization. Biotechnology Advances. 2013;**31**:129-139. DOI: 10.1016/j.biotechadv.2012.08.010

[57] Goldberg K, Schroer K, Lutz S, Liese A. Biocatalytic ketone reduction-a powerful tool for the production of chiral alcohols-part II: Whole cell reductions. Applied Microbiology and Biotechnology. 2007;**76**: 249-255. DOI: 10.1007/s00253-007-1005-x

[58] Ye Q, Ouyang P, Ying H. A review-biosynthesis of optically pure ethyl (*S*)-4-chloro-3-hydroxybutanoatw ester: Recent advances and future perspectives. Applied Microbiology and Biotechnology. 2011;**89**:513-522. DOI: 10.1007/s00253-010-2942-3

[59] Ferraboschi P, Ciceri S, Ciuffreda O, De Mieri M, Romano D, Grisenti P. Baker's yeast catalysed preparation of a new enantiomerically pure synthon of (*S*)-pramipexole and its enantiomer (dexpramipexole). Tetrahedron: Asymmetry. 2014;**25**:1239-1245. DOI: 10.1016/j.tetasy.2014.07.011

[60] Arifin A, Don M, Uzir M. Baker's yeast mediated biotransformation geraniol into citronellol using a continuous-closed-glass-loop bioreactor (CCGLB) system. Bichemical Engineering Journal. 2011;**56**:219-224. DOI: 10.1016/j.bej.2011.07.001

[61] Mane A, Lohar T, Salunkhe R. Baker's yeast as an efficient biocatalyst for regioselective 1,4-conjugate addition of indoles to nitroolefines in aqueous medium. Tetrahedron Letters. 2016;**57**:2341-2346. DOI: 10.1016/j.tetlet.2016.04.057

[62] Straathof AJJ, Panke S, Schmid A. The production of fine chemicals by biotransformations. Current Opinion in Biotechnology. 2002;**13**:548-556. http://dx.doi.org/10.1016/S0958-1669(02)00360-9

[63] Forti L, Di Mauro S, Cramarossa MR, Filippucci S, Turchetti B, Buzzini P. Non-conventional yeasts whole cells as efficient biocatalysts for the production of flavors and fragrances. Molecules. 2015;**20**:10377-10398. DOI: 10.3390/molecules200610377

[64] Schaefer CA, Silva VD, Stambuk BU, Nascimento M. Use of *Saccharomyces cerevi*siae yeasts in the chemoselective bioreduction of (1E,4E)-1,5-Bis(4-Methoxyphenyl)-1,4 pentadien-3-one in biphasic system. Journal of the Brazilian Chemical Society. 2103;**24**:1116-1122. http://dx.doi.org/10.5935/0103-5053.20130140

[65] Katz M, Frejd T, Hahn-Hagerdal B, Gorwa-Grauslund MF. Efficient anaerobic whole cell stereoselective bioreduction with recombinant *Saccharomyces cerevisiae*. Biotechnology and Bioengineering. 2003;**84**(5):573-582. DOI: 10.1002/bit.10824

[66] Hasan F, Shah AA, Hameed A. Methods for detection and characterization of lipases: A comprehensive review. Biotechnology Advanced. 2009;**27**(6):782-798. http://dx.doi.org/10.1016/j.biotechadv.2009.06.001

[67] Vakhlu J, Kour A. Yeast lipases: Enzyme purifications, biochemical properties and gene cloning. Electronic Journal of Biotechnology. 2006;**9**:69-81. DOI: 10.2225/vol9-issue1-fulltext-9

[68] Kazlauskas RJ, Bornscheuer UT. Biotransformations with lipases. In: Rehm J, Reed G, Puchler A, Stadler P, editors. Biotechnology. Weinheim: Wiley-VCH; 1998. pp. 37-191

[69] Sharma S, Kanwar SS. Organic solvent tolerant lipases and applications. The Scientific World Journal. 2014;**2014**:15p. http://dx.doi.org/10.1155/2014/625258

[70] Gonçalves FAG, Colen G, Takahashi JA. *Yarrowia lipolytica* and its multiple applications in the biotechnological industry. The Scientific World Journal. 2014;**2014**:14. p. http://dx.doi.org/10.1155/2014/476207

[71] Anderson EM, Larsson KM, Kirk O. One biocatalyst–many applications: The use of *Candida antarctica* B-lipase in organic synthesis. Biocatalysis and Biotransformations. 1998;**16**:181-204. http://dx.doi.org/10.3109/10242429809003198

[72] Weber N, Weitkamp P, Mukherjee KD. Steryl and stanyl esters of fatty acids by solvent-free esterification and transesterification in vacuo using lipases from *Rhizomucor miehei, Candida antarctica,* and *Carica papaya*. Journal of Agriculture and Food Chemistry. 2001;**49**(11):5210-5216. DOI: 10.1021/jf0107407

[73] Hamam F, Shahidi F. Synthesis of structured lipids containing medium-chain and Omega-3 fatty acids. Journal of Agriculture and Food Chemistry. 2006;**54**:4390-4396. DOI: 10.1021/jf052540r

[74] Gross RA, Kalra B. Biodegradable polymers for the environment. Science. 2002;**297**(5582): 803-807. DOI: 10.1126/science.297.5582.803

[75] Gross RA, Kumar A, Kalra B. Polymer synthesis by in vitro enzyme catalysis. Chemical Reviews. 2001;**10**(7):2097-2124. DOI: 10.1021/cr0002590

[76] Cabrera Z, Fernandez-Lorente G, Fernandez-Lafuente R, Palomo JM, Guisan JM. Novozym 435 displays very different selectivity compared to lipase from *Candida antarctica* B adsorbed on other hydrophobic supports. Journal of Molecular Catalysis B: Enzymatic. 2009;**57**:171-176. http://dx.doi.org/10.1016/j.molcatb.2008.08.012

[77] Ueji SI, Nishimura M, Kudo R, Matsumi R, Watanabe K, Ebara Y. A dramatic improvement of enantioselectivity of lipase in organic solvents by addition of aqueous SDS: A close correlation between enantioselectivity and conformational flexibility of lipase. Chemical Letters. 2001;**30**:912-913. DOI: 10.1002/chin.200206031.

[78] Watanabe K, Ueji SI. Dimethyl sulfoxide as a co-solvent dramatically enhances the enantioselectivity in lipase-catalyzed resolutions of 2-phenoxypropionic acyl derivatives. Journal of the Chemical Society, Perkin Transactions 1. 2001;**I**(12):1386-1390. DOI: 10.1039/B100182P

[79] Hirose Y, Kariya K, Sasaki I, Kurono Y, Ebiike H, Achiwa K. Drastic solvent effect on lipase-catalysed enantioselective hydrolysis of prochiral 1,4-dihydropyridines. Tetrahedron Letters. 1992;**33**:7157-7160. http://dx.doi.org/10.1016/S0040-4039(00)60861-4

[80] Kirk O, Christensen MW. Lipases from *Candida antarctica*: Unique biocatalysts from a unique origin. Organic Process Research & Development. 2002;**6**(4):446-451. DOI: 10.1021/op0200165

[81] Domingues de Maria P, Carboni-Oerlemans C, Tuin B, Bargeman G, van der Meer A, van Gemert R. Biotechnological applications of *Candida antarctica* lipase A: State-of-the-art. Journal of Catalysis B: Enzymatic. 2005;**37**:36-46. http://dx.doi.org/10.1016/j.molcatb.2005.09.001

[82] Schmidt M, Barbayianni E, Fotakopoulou I, Höhne M, Constantinou-Kokotou V, Bornscheuer UT, Kokotos G. Enzymatic removal of carboxyl protecting groups. 1. Cleavage of the *tert*-butyl moiety. Journal of Organic Chemistry. 2005;**70**(9):3737-3740. DOI: 10.1021/jo050114z

[83] Gotor-Fernandez V, Brieva R, Gotor V. Lipases: Useful biocatalysts for the preparation of pharmaceuticals. Journal of Molecular Catalysis. 2006;**40**:111-120. DOI: 10.1016/j.molcatb.2006.02.010

[84] Kingery-Wood J, Johnson JS. Resolution of anti-3-oxotricyclo[2.2.1.0]heptane-7-carboxylic acid by *Candidia antartica* lipase A. Tetrahedron Letters. 1996;**37**(23):3975-3976. http://dx.doi.org/10.1016/0040-4039(96)00759-9

[85] Davies HG, Green RH, Kelly DR, Roberts SM. Recent advances in the generation of chiral intermediates using enzymes. Biotechnology. 1990;**10**:129-152. http://dx.doi.org/10.3109/07388559009068264

[86] Sih CJ, Gu Q-M, Holdgrun X, Harris K. Optically-active compounds *via* biocatalytic methods. Chirality. 1992;**4**:91-97. DOI: 10.1002/chir.530040205

[87] Margolin AL. Enzymes in the synthesis of chiral drugs. Enzyme and Microbial Technologies. 1993;**15**:266-280. http://dx.doi.org/10.1016/0141-0229(93)90149-V

[88] Salzar L, Bermudez JL, Ramirez C, Llama EF, Sinisterra JV. Resolution of 3-α-Naphthoxy-1,2-propanediol Using Candida antarctica Lipase. Tetrahedron. 1999;**10**:3507-3514. http://dx.doi.org/10.1016/S0957-4166(99)00366-3

[89] Orrenius C, Oehrner N, Rotticci D, Mattson A, Hult K, Norin T. Candida antarctica lipase B catalyzed kinetic resolutions: Substrate structure requirements for the preparation of enantiomerically enriched secondary alcohols. Ibid. 1995;**6**:1217-1220. 10.1016/0957-4166(95)00147-h

[90] Patel RN, Banerjee A, Nanduri V, Goswami A, Comezoglu FT. Enzymatic resolution of racemic secondary alcohols by lipase B from *Candida antarctica*. Journal of the American Oil Chemists' Society. 2000;Vol. **77**:Issue (10). 1015-1019. DOI: 10.1007/s11746-000-0161-y

[91] Sauerberg P, Olesen PH. Heterocyclic Compounds and their Preparation and Use. 1995; U.S. Patent 5, **418**:240

[92] Hamilton GS, Li J-H, Steiner J-P. Method of Using Neutrotrophic Sulfonamide Compounds. 1998; U.S. Patent 5, **721**:256

[93] Hutt AJ, Caldwell J. The importance of stereochemistry in the clinical pharmacokinetics of the 2-arylpropionic acid non-steroidal anti-inflammatory drugs. Clinical Pharmacokinetics. 1984;**9**(4):371-373. DOI: 10.2165/00003088-198409040-00007

[94] Siódmiak T, Rumiński JK, Marszał M. Application of lipases from *Candida rugosa* in the enantioselective esterification of (*R,S*)-ibuprofen. Current Organic Chemistry. 2012;**16**(8):972-977. DOI: 10.2174/138527212800194728

[95] Breuer M, Ditrich K, Habicher T, Hauer B, Keßeler M, Stürmer R, Zelinski T. Industrial methods for the production of optically active intermediates. Angewandte Chemie International Edition. 2004;**43**(7):788-824. DOI: 10.1002/anie.200300599

[96] Gotor-Fernandez V, Busto E, Gotor V. Candida antarctica lipase B: An ideal biocatalyst for the preparation of nitrogenated organic compounds. Advanced Synthesis & Catalysis. 2006;**348**:797-812. DOI: 10.1002/adsc.200606057

[97] Torre O, Busto E, Gotor-Fernández V, Gotor V. Enzymatic preparation of novel aminoalkylpyridines using lipases in organic solvents. Advanced Synthesis & Catalysis. 2007;**349**:1481-1488. DOI: 10.1002/adsc.200600634

[98] Borch K, Christensen M. WO 01/53511 A1; 2001

[99] Pihko AJ, Lundell K, Kanerva L, Koskinen AMP. Enantioselective synthesis of a hindered furyl substituted allyl alcohol intermediate: A case study in asymmetric synthesis. Tetrahedron: Asymmetry. 2004;**15**:1637-1643. http://dx.doi.org/10.1016/j.tetasy.2004.04.004

[100] Paizs C, Tähtinen P, Toşa M, Majdik C, Irimie FD, Kanerva LT. Biocatalytic enantioselective preparation of phenothiazine-based cyanohydrin acetates: Kinetic and dynamic kinetic resolution. Tetrahedron. 2004;**60**(46):10533-10540. http://dx.doi.org/10.1016/j.tet.2004.06.136

[101] de Gonzalo G, Lavandera I, Brieva R, Gotor V. Enzymatic acylation reactions on ω-hydroxycyanohydrins. Tetrahedron. 2004;**60**(46):10525-10532. http://dx.doi.org/10.1016/j.tet.2004.06.134

[102] Romero-Gomez S, Augur C, Viniegra-Gonzalez G. Invertase production by *Aspergillus niger* in submerged and solid-state fermentation. Biotechnology Letters. 2000;**22**:1255-1258. DOI: 10.1023/A:1005659217932

[103] Kotwal SM, Shankar V. Immobilized invertase. Biotechnology Advanced. 2009;**27**:311-322. http://dx.doi.org/10.1016/j.biotechadv.2009.01.009

[104] Kulshrestha S, Tyagi P, Sindhi V, Yadavilli KS. Invertase and its applications—A brief review. Journal of Pharmacy Research. 2013;**7**(9):792-797. http://dx.doi.org/10.1016/j.jopr.2013.07.014

[105] Mussatto SI, Mancilha IM. Non-digestible oligosaccharides: A review. Carbohydrate Polymers. 2007;**68**:587-597. DOI: 10.1016/j.carbpol.2006.12.011

[106] Lahiri S, Basu A, Sengupta S, Banerjee S, Dutta T, Soren D, Chattopadhyay K, Ghosh AK. Purification and characterization of a trehalase–invertase enzyme with dual activity from *Candida utilis*. Archives of Biochemistry and Biophysics. 2012;**522**(2):90-99. http://dx.doi.org/10.1016/j.abb.2012.03.026

[107] Zhu D, Yang Y, Hua L. Stereoselective enzymatic synthesis of chiral alcohols with the use of a carbonyl reductase from *Candida magnoliae* with anti-Prelog enantioselectivity. Journal of Organic Chemistry. 2006;**71**:4202-4205. DOI: 10.1021/jo0603328

[108] Seebach D, Sutter MA, Weber RH, Züge MF. Yeast reduction of ethyl acetoacetate: (*S*)-(+)-ethyl 3-hydroxybutanoate. Organic Synthesis. 1985;**63**:1-7. DOI: 10.15227/orgsyn.063.0001

[109] Kawai Y, Tsujimoto M, Kondo S, Takanobe K, Nakamura K, Ohno A. Asymmetric reduction of β-keto esters with an enzyme from Bakers' yeast. Bulletin of the Chemical Society of Japan. 1994;**67**:524-528. http://dx.doi.org/10.1246/bcsj.67.524

[110] Williams RE, Bruce NC. New uses for an old enzyme—the old yellow enzyme family of flavoenzymes. Microbiology. 2002;**148**:1607-1614. http://dx.doi.org/10.1099/00221287-148-6-1607

[111] Świderska MA. Application of Saccharomyces carlsbergensis old yellow enzyme in synthesis of chiral ketones and building blocks for β-amino acids [thesis]. University Of Florida; 2007, Gainesville, Florida 32611, United Stated of America

[112] Meah Y, Massey V. Old yellow enzyme: Stepwise reduction of nitroolefins and catalysis of aci-nitro tautomerization. Proceedings of the National Academy of Science of the United States of America. 2000;**97**:10733-10738. DOI: 10.1073/pnas.190345597

[113] Karathia H, Vilaprinyo E, Sorribas A, Alves R. Saccharomyces cerevisiae as a model organism: A comparative study. PLoS One. 2011;**6**(2):e16015. DOI: 10.1371/journal.pone.0016015

[114] Demain AL, Vaishnav P. Production of recombinant proteins by microbes and higher organisms. Biotechnology Advances. 2009;**27**(3):297-306. DOI: 10.1016/j.biotechadv.2009.01.008

[115] Li F, Long T, Lu Y, Ouyang Q, Tang C. The yeast cell-cycle network is robustly designed. Proceedings of the National Academy of Sciences of the United States of America. 2004;**101**(14):4781-4786. DOI: 10.1073/pnas.0305937101

[116] Duina AA, Miller ME, Keeney JB. Budding yeast for budding geneticists: A primer on the saccharomyces cerevisiae model system. Genetics. 2014;**197**(1):33-48. DOI: 10.1534/genetics.114.163188

[117] http://syntheticyeast,org/

[118] Goffeau A, Barrell BG, Bussey H, Davis RW, Dujon B, Dujon B, Feldmann H, Galibert F, Hoheisel JD, Jacq C, Johnston M, Louis EJ, Mewes HW, Murakami Y, Philippsen P, Tettelin H, Oliver SG. Life with 6000 genes. Science. 1996;**274**:546-567. DOI: 10.1126/science.274.5287.546

[119] Kojo H, Greenberg BD, Sugino A. Yeast 2-micrometer plasmid DNA replication in vitro: Origin and direction. Proceedings of the National Academy of Sciences of the United States of America. 1981;**78**(12):7261-7265. DOI: 10.1073/pnas.78.12.7261

[120] Stearns T, Ma H, Botstein D. Manipulating yeast genome using plasmid vectors. Methods in Enzymology. 1990;**185**:280-297

[121] Brown TA. Gene cloning and DNA analysis: An introduction. 6th ed. Manchester: Wiley; 2010. pp. 105-112

[122] Hou J, Tyo KEJ, Liu Z, Petranovic D, Nielsen J. Metabolic engineering of recombinant protein secretion by *Saccharomyces cerevisiae*. FEMS Yeast Research. 2012;**12**(5):491-510. DOI: 10.1111/j.1567-1364.2012.00810.x

[123] Nielsen J. Production of biopharmaceutical proteins by yeast. Bioengineered. 2013;**4**(4):207-211. DOI: 10.4161/bioe.22856

[124] Burda P, Aebi M. The dolichol pathway of N-linked glycosylation. Biochimica et Biophysica Acta (BBA)—General Subjects. 1999;**1426**(2):239-257. DOI: 10.1016/S0304-4165(98)00127-5

[125] Wildt S, Gerngross TU. The humanization of N-glycosylation pathways in yeast. Nature Reviews Microbiology. 2005;**3**(2):119-128. DOI: 10.1038/nrmicro1087

[126] Hamilton SR, Gerngross TU. Glycosylation engineering in yeast: The advent of fully humanized yeast. Current Opinion in Biotechnology. 2007;**18**(5):387-392. DOI: 10.1016/j.copbio.2007.09.001

[127] Payne T, Finnis C, Evans LR, Mead DJ, Avery SV, Archer DB, et al. Modulation of chaperone gene expression in mutagenized Saccharomyces cerevisiae strains developed for recombinant human albumin production results in increased production of multiple heterologous proteins. Applied and Environmental Microbiology. 2008;**74**(24):7759-7766. DOI: 10.1128/AEM.01178-08

[128] Fukuda K, Jensen MH, Haser R, Aghajari N, Svensson B. Biased mutagenesis in the N-terminal region by degenerate oligonucleotide gene shuffling enhances secretory expression of barley α-amylase 2 in yeast. Protein Engineering, Design and Selection. 2005;**18**(11):515-526. DOI: 10.1093/protein/gzi057

[129] Kjeldsen T, Ludvigsen S, Diers I, Balschmidt P, Sørensen AR, Kaarsholm NC. Engineering-enhanced protein secretory expression in yeast with application to insulin. Journal of Biological Chemistry. 2002;**277**(21):18245-18248. DOI: 10.1074/jbc.C200137200

[130] Rakestraw JA, Sazinsky SL, Piatesi A, Antipov E, Wittrup KD. Directed evolution of a secretory leader for the improved expression of heterologous proteins and full-length antibodies in Saccharomyces cerevisiae. Biotechnology and Bioengineering. 2009;**103**(6):1192-1201. DOI: 10.1002/bit.22338

[131] Harmsen MM, Bruyne MI, Raué HA, Maat J. Overexpression of binding protein and disruption of the PMR1 gene synergistically stimulate secretion of bovine prochymosin but not plant thaumatin in yeast. Applied Microbiology and Biotechnology. 1996;**46**(4):365-370

[132] Smith JD, Tang BC, Robinson AS. Protein disulfide isomerase, but not binding protein, overexpression enhances secretion of a non-disulfide-bonded protein in yeast. Biotechnology and Bioengineering. 2004;**85**(3):340-350. DOI: 10.1002/bit.10853

[133] Homma T, Iwahashi H, Komatsu Y. Yeast gene expression during growth at low temperature. Cryobiology. 2003;**46**(3):230-237

[134] Hovland P, Flick J, Johnston M, Sclafani RA. Galactose as a gratuitous inducer of GAL gene expression in yeasts growing on glucose. Gene. 1989;**83**(1):57-64

[135] Koller A, Valesco J, Subramani S. The CUP1 promoter of *Saccharomyces cerevisiae* is inducible by copper in Pichia pastoris. Yeast (Chichester, England). 2000;**16**(7):651-656. DOI: 10.1002/(SICI)1097-0061(200005)16:7<651::AID-YEA580>3.0.CO;2-F

[136] Li N, Zhang L-M, Zhang K-Q, Deng J-S, Prändl R, Schöffl F. Effects of heat stress on yeast heat shock factor-promoter binding in vivo. Acta Biochimica Et Biophysica Sinica. 2006;**38**(5):356-362

[137] Gatignol A, Dassain M, Tiraby G. Cloning of Saccharomyces cerevisiae promoters using a probe vector based on phleomycin resistance. Gene. 1990;**91**(1):35-41. DOI: 10.1016/0378-1119(90)90159-O

[138] Fields S, Song O. A novel genetic system to detect protein-protein interactions. Nature. 1989;**340**(6230):245-246. DOI: 10.1038/340245a0

[139] Stynen B, Tournu H, Tavernier J, Dijck PV. Diversity in genetic in vivo methods for protein-protein interaction studies: From the yeast two-hybrid system to the mammalian split-luciferase system. Microbiology and Molecular Biology Reviews. 2012;**76**(2):331-382. DOI: 10.1128/MMBR.05021-11

Chapter 2

Big Data Supervised Pairwise Ortholog Detection in Yeasts

Deborah Galpert Cañizares, Sara del Río García,
Francisco Herrera, Evys Ancede Gallardo,
Agostinho Antunes and Guillermin Agüero-Chapin

Additional information is available at the end of the chapter

http://dx.doi.org/10.5772/intechopen.70479

Abstract

Ortholog are genes in different species, evolving from a common ancestor. Ortholog detection is essential to study phylogenies and to predict the function of unknown genes. The scalability of gene (or protein) pairwise comparisons and that of the classification process constitutes a challenge due to the ever-increasing amount of sequenced genomes. Ortholog detection algorithms, just based on sequence similarity, tend to fail in classification, specifically, in *Saccharomycete* yeasts with rampant paralogies and gene losses. In this book chapter, a new classification approach has been proposed based on the combination of pairwise similarity measures in a decision system that consider the extreme imbalance between ortholog and non-ortholog pairs. Some new gene pair similarity measures are defined based on protein physicochemical profiles, gene pair membership to conserved regions in related genomes, and protein lengths. The efficiency and scalability of the calculation of these measures are analyzed to propose its implementation for big data. In conclusion, evaluated supervised algorithms that manage big and imbalanced data showed high effectiveness in *Saccharomycete* yeast genomes.

Keywords: ortholog detection, similarity measures, big data supervised classification, scalability

1. Introduction

Orthologs are genes in different species evolving from a common ancestor while paralogous genes are homologous sequences that evolved in a duplication event derived from a unique sequence [1]. The distinction between orthologs and paralogs is crucial since ortholog genes are considered evolution markers and they help to infer the function of unknown proteins.

The orthology relationship is expressed in terms of pairs with one-to-one relations, or of groups with one-to-many or many-to-many relationships. Thus, the scalability of both gene/protein pairwise comparisons and the classification process constitute a challenge due to the ever-increasing amount of sequenced genomes. The unsupervised ortholog classification problem is presented in [2] starting from the pairwise sequence comparison. Many algorithms built a similarity graph from these comparisons such as the Reciprocal Best Hit (RBH) heuristic [3], Inparanoid [4], OrthoMCL [5], the Comprehensive, Automated Project for the Identification of Orthologs from Complete Genome Data (OMA) [6], the well-known Reciprocal Smallest Distance (RSD) algorithm [7], among others. In this unsupervised learning approach there are also tree-based and hybrid algorithms [8]. On the other hand, [9], a supervised approach form pairwise classification approach is presented where there is no evidence of the imbalance management of the scarce ortholog pairs (minority class) among the majority of non-ortholog pairs. Besides, this approach does not couple with the generalization problem of the built model to external genome pair is not involved in the learning process.

An important issue to tackle with ortholog detection (OD) algorithms is the underlying information of gene/protein features used to classify the sequences. Algorithms just based on sequence similarity tend to fail in classification because OD is negatively affected by genetic and evolutionary events like duplications and gene losses, or changes between genome segments such as duplications, deletions, horizontal gene transfers, fusions, fissions, inversions, and transpositions. Although the high potentiality of the OD methods referenced either in the "omics"[1] tools site or the site of the Quest for Orthologs[2] Consortium, some of their limitations are reported in literature:

- They tend to include paralogs in the orthogroups increasing false positives [8]. In the presence of lots of gene losses caused by whole genome duplications (WGD) in *Saccharomycete* yeasts, a big amount of paralogs can be included in orthogroups, or in orthologous pairs [10].
- They can reduce specificity due to the presence of short sequences or those that is evolved in a convergent way. This kind of affection may arise when genes inherited by horizontal transfers brought about failure in the inference of near phylogenetic relationships between species that are distantly related and have recently exchanged a gene [11].
- They can fail in ortholog detection mainly in the twilight zone (less than the 30% of sequence identity), that is, they can raise false negatives in the presence of distantly related divergent sequences. They can also fail in front of mosaics of proteins [12] since orthogroups not considering hybrid proteins may contain proteins without a common ancestor.

With the aim of reducing these limitations and improving efficacy in ortholog detection, there is a tendency in algorithms to include diverse gene information in addition to the alignment-based features (AB). Some ortholog detection algorithms include information about: (i) synteny [13–15],

[1] http://omictools.com/

[2] http://questfororthologs.org/orthology_databases

(ii) global rearrangements [16, 17], (iii) protein interactions [9], (iv) the architecture of protein mosaics [18], and (v) evolution distances [6, 7]. So far, these attempts have not been enough for increasing efficacy in classification [19]. The combination of AB with other features like the physicochemical profiles, the length of the sequences, and the synteny may be positive for pairwise ortholog detection as well as the learning process from curated classifications in benchmark datasets may allow for a more effective approach.

In order to evaluate OD algorithms, Salichos and Rokas [10] constructed a *Saccharomycete* yeast benchmark dataset having orthogroups deprived of paralogs. Such orthogroups contain "curated or true" orthologs from yeast species that underwent pre- and post-whole genome duplications (WGD) [10]. Actually, when Multiparanoid [20], OrthoMCL, and extended versions of RBH and the RSD were evaluated using this benchmark dataset, they included paralogs in the orthogroups [10]. This fact assures that OD is a bioinformatics field in constant need of new effective algorithms mainly for yeasts.

In terms of efficiency, OD algorithms based on sequence alignments may take from weeks to years of CPU time to compute orthologs of sequenced species, due to the quadratic scaling capacity of these comparisons. That is why in the Quest for Orthologs Consortium meeting [19], efficiency and management of big data were target questions, and some efforts in these senses were presented. They stated that the increased demand of computation for sequence analysis has not been achieved by means of the increasing computation capacity, but it can be achieved with new approaches or algorithm implementations.

In this chapter, a new classification approach has been proposed based on the combination of pairwise protein features in a supervised decision system that consider the extreme imbalance between ortholog and non-ortholog pairs on the benchmark dataset proposed by [10]. Some new gene pair similarity measures considered as pairwise protein features are defined based on protein physicochemical profiles, their membership to conserved regions in related genomes, and their lengths. The efficiency and scalability of the calculation of these measures are analyzed to propose its implementation for big data. In conclusion, evaluated supervised algorithms that manage big and imbalanced data in scalable machine learning libraries as Mahout [21] and MLlib [22] overdid RBH, RSD, and the Comprehensive, Automated Project for the Identification of Orthologs from Complete Genome Data (OMA) [6] in *Saccharomycete* yeast genomes.

2. Methods

2.1. Yeast genomes in experiments

Yeast genomes selected for the experiments are well-studied eukaryote species that shared a common ancestor with human millions of years ago. *Saccharomyces cerevisiae* belongs to *Saccharomycete yeasts* and *Schizosaccharomyces pombe* is a unicellular organism from *Archiascomycete fungus*. These species bring about experimental models for many essential processes of eukaryotes since most of their genes have been predicted as homologs to multicellular eukaryotes ones. Specifically, *S. pombe* genes are more similar to mammals' genes than *S. cerevisiae* ones [23]. However, *S. pombe* is a distant relative of *S. cerevisiae*, accordingly, ortholog detection between them is a difficult task due to the divergence of sequences [13]. On the other hand, other close *Saccharomycete* species as *Kluyveromyces* (*Kluyveromyces lactis* and *Kluyveromyces waltii*) and *Candida glabrata* have been included in one of the ortholog detection benchmarking datasets [10] since some of them underwent a whole genome duplication (WGD) process causing a rampant paralogy and a lot of gene losses. Precisely, the genome pair of *C. glabrata* and *S. cerevisiae* that are post-WGD species constitutes a target study case for OD algorithms. For the studies reported below we have identified each genome pair as follows:

- *S. cerevisiae* – *K. lactis*: ScerKlac
- *K. lactis* – *K. waltii*: KlacKwal
- *C. glabrata* – *K. lactis*: CglaKlac
- *S. cerevisiae* – *C. glabrata*: ScerCgla
- *S. cerevisiae* – *S. pombe*: ScerSpombe.

2.2. Similarity measures as gene pair features in ortholog detection

Five normalized pairwise sequence similarities (global and local alignment scores, protein length, synteny and physicochemical profile) were previously specified in [24]. **Table 1**

Measure	Description of the calculation
S_1 – local and global alignment	The sequence alignment measure averages positives local and global protein alignment scores from the Smith Waterman [32] and the Needleman-Wunsch [33] algorithms calculated with a specified scoring matrix, and "gap open" (GOP) and "gap extended" (GEP) parameters. The scores are normalized by using the maximum score of all (x_i, y_j) protein pairs
S_2 – protein length	Measure S_2 represents the normalized difference for continuous values [34] of the protein sequence lengths
S_3 – membership to locally collinear blocks	This measure is calculated from the distance between pairs of sequences considering their membership to Locally Collinear Blocks (LCBs) which represent truly homologous regions obtained with the Mauve software [35]. The normalized difference is selected for the comparison of the total number of codons in each block
S_4 – physicochemical profile	This measure is based on the spectral representation of sequences from the global protein pairwise alignment by using the linear predictive coding [34]. First, each amino acid that lies in a matching region without "gaps" between two aligned sequences is replaced by its contact energy [36]. The moving average for each spectrum, that is, the average of this physicochemical feature in the predefined window size W is then calculated. Next, the Pearson correlation coefficient with the corresponding significance level between the two spectral representations in a matching region is obtained. Finally, the significant similarities of the regions without "gaps" are aggregated considering the length of each region. From our previous studies presented in [37, 38], three features for the physicochemical profile with W values of 3, 5 and 7 have been considered

Table 1. Protein pair features.

summarizes the calculations for protein pairs (x_i, y_j) from the two proteome representations $P_1=\{x_1, x_2, \ldots, x_n\}$ and $P_2=\{y_1, y_2, \ldots, y_m\}$ with n and m sequences, respectively.

2.3. From parallel to big data calculation of similarity measures

Time complexity analysis of the sequential algorithms for the calculation of pairwise similarity measures in terms of N gene/protein pairs has brought about the need of a parallel version with the scalability target. The protein physicochemical profile turns to be the one with the highest quadratic cost $O(N \times (m + n)^2)$ since it operates over the aligned sequences with m and n as the maximum sequence lengths for two genomes, respectively. The parallel proposal is sketched in **Figure 1** with a fork control flow to distribute the calculation over different processors and a join control flow to terminate the parallel executions.

For the scalability analysis of the parallel physicochemical profile calculation, we estimated the T_S sequential execution time in expression (1) by using constants.

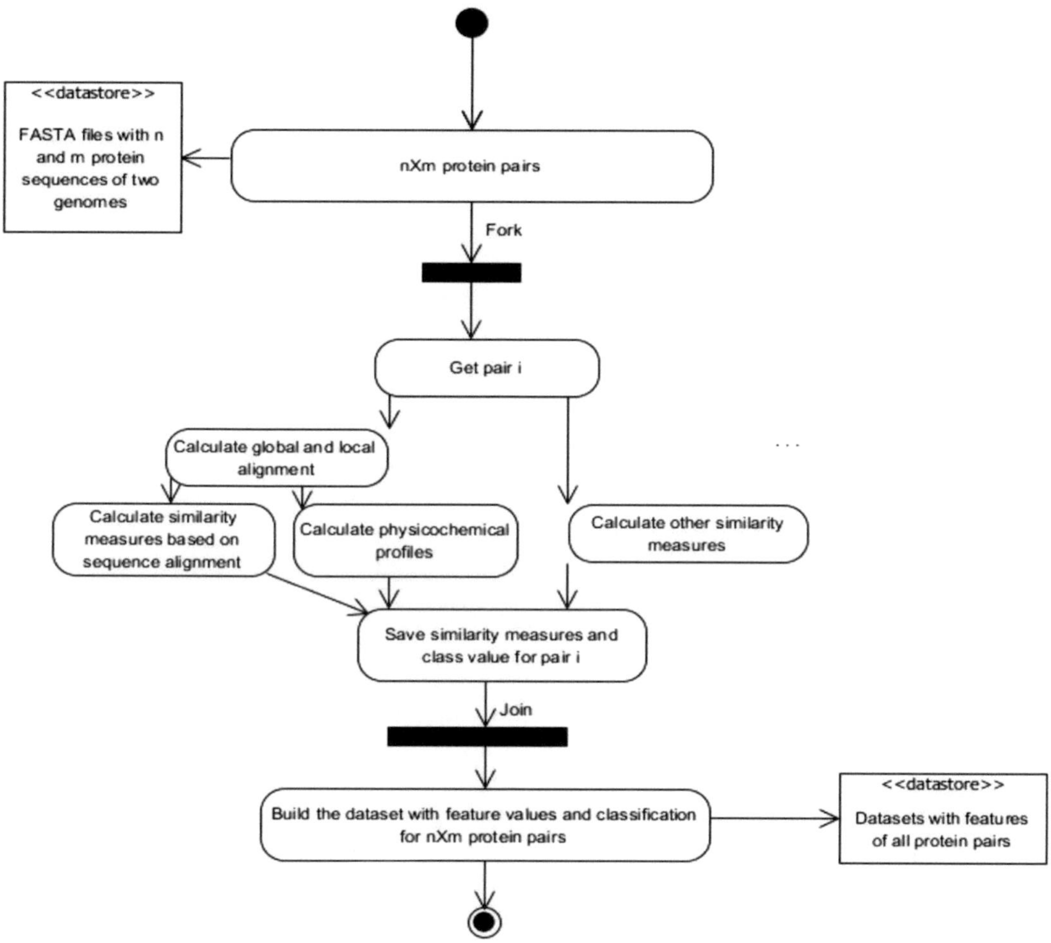

Figure 1. Parallel calculation of protein pair similarity measures.

$$T_S = N \times \left(n \times m + (m+n)^2\right) \times t_c \quad (1)$$

The constant t_c indicates the required time for the arithmetic operations. We assume t_c as optimum value, $t_c = 1$ [25]. Consequently, the parallel execution time T_P is defined in terms of the calculation time and the communication time among processors. In the parallel version, calculation time depends on the distribution of N cycles among P processors. With this distribution: (i) all the processors execute $\lfloor N/P \rfloor$ iterations or (ii) $p < P$ processors execute $\lfloor N/P \rfloor + 1$ iterations. Thus, when we consider the worst case, the calculation time T_{Calc} can be estimated as in expression (2).

$$T_{Calc} = (\lfloor N/P \rfloor + 1) \times \left(n \times m + (m+n)^2\right) \times t_c \quad (2)$$

The communication time among processors is directly related to the information accessed in the sequential section of the algorithm. In each iteration, each processor P needs to receive sequence information. Hence, communication time can be defined as in expression (3).

$$T_{Com} = (\lfloor N/P \rfloor + 1) \times (t_s + t_w \times P \times (m+n)) \quad (3)$$

where t_s represents the time required to establish the communication and to prepare the sending information and t_w is the required time to send a numeric value. These values can be considered as optimum ones, that is, $t_s = 0$ and $t_w = 1$ [25]. In this way, we combine expressions (2) and (3) to obtain an estimation of the parallel execution time for the physicochemical profile as in expression (4).

$$T_P = (\lfloor N/P \rfloor + 1) \times (n \times m + (m+n) \times (m+n+P)) \quad (4)$$

Time complexity of the parallel version of the algorithm can be estimated from the parallel execution time T_P. Expression ($\lfloor N/P \rfloor$+1) can be taken as N/P.

It holds that $m \times n \leq (m+n) \times (m+n+P)$. Therefore, time complexity is $O\left(\frac{N \times (n+m)^2}{P} + N \times (m+n)\right)$, that is lower than $O(N \times (m+n)2)$ as demonstrated in expression (5).

$$\frac{\frac{N \times (n+m)^2}{P} + N \times (m+n)}{N \times (n+m)^2} = \frac{1}{P} + \frac{1}{m+n} \leq 1 \quad (5)$$

By using expressions T_S and T_P, we can analyze the speed-up and the efficiency performance for different samples obtained from N protein pairs and a range of 1–100 processors. **Figure 2A** and **B** shows the speed-up and efficiency values, respectively, in a genome comparison dataset of *S. cerevisiae* and *S. pombe*. The 100% sample represents 5006 sequence pairs; the 75% sample represents 3754 pairs, the 50% represents 2503 pairs, and the 25% represents 1251 pairs. An increased number of processors lead to a saturation in the speed-up. Specifically, when the problem size increase, thus the speed-up and the efficiency values increase for a single number of processors, although both values tend to decrease with an increasing number of processors. The efficiency tends to keep constant when we simultaneously increase both the problem size and the number of processors, consequently, contributing to the scalable conception of the algorithm [26].

http://dx.doi.org/10.5772/intechopen.70479

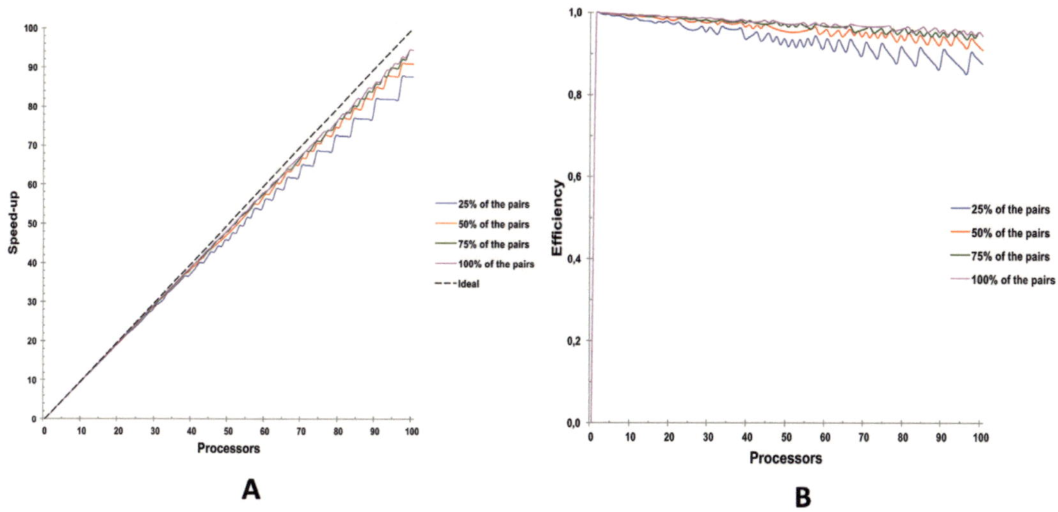

Figure 2. Estimated performance of the speed-up (A) and efficiency (B) in the parallel calculation of the physicochemical profile of proteins for the genome comparison dataset of *S. cerevisiae* and *S. pombe*.

With the sequential and parallel MATLAB (2010) implementations, we calculated real execution time for one, two, three and four processors in a personal computer Intel® Core™ i3 CPU, 2.53 GHz, RAM DDR3 de 4.0 Gb, 64Bits Windows7. With 1921 pairs of *S. cerevisiae* and *S. pombe* representing 100% of the sample, **Figure 3A** and **B** shows the real and estimated values of the speed-up and efficiency, respectively. We can observe a similar performance of estimated and real values for the total and 50% of the sample. The best speed-up was obtained for the execution in four processors, but the efficiency is best with two processors.

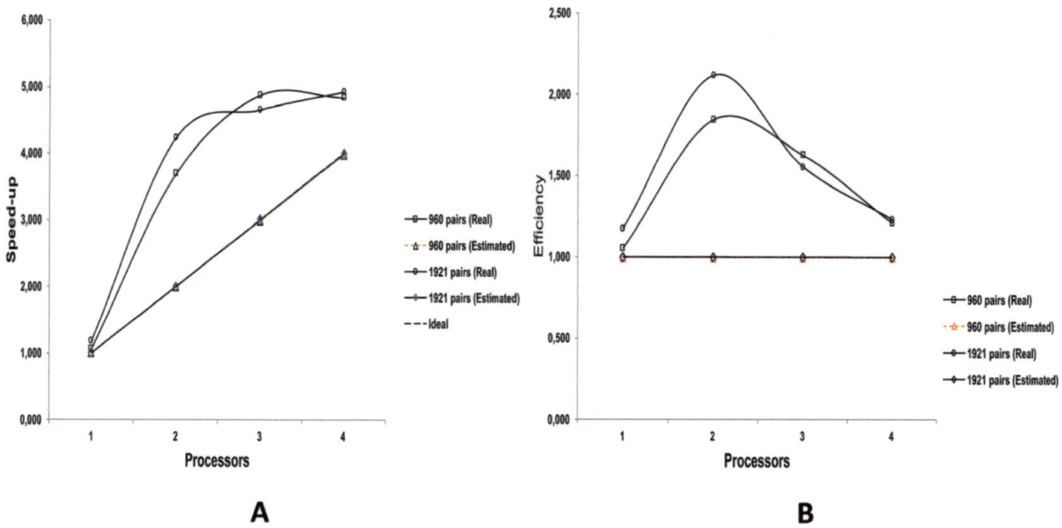

Figure 3. Real and estimated values for speed-up (A) and efficiency (B) in a *S. cerevisiae* and *S. pombe* sample.

Theoretically, the speed-up value is bounded by the number of processors and the efficiency value is in the interval [0, 1]. However, the real speed-up obtained is higher than the number of processors in each execution and the efficiency is higher than one. This event is known as super-linear speed-up. It is a consequence of the fact that each processor consume less time than $\frac{T_S}{P}$. It can be possible since the parallel version may execute less work than the sequential version or may take advantage of resources such as the cache memory [25].

In a further scalability analysis with 133,666,445 pairs of sequences from different yeast genomes, the increased number of processors from 1 to 100 saturates the speed-up. **Table 2** shows the genome data used for this scalability analysis. Datasets are conformed by accumulating gene pairs from different genome pairs with different maximum sequence lengths. **Figure 4A** illustrates the effects of the gene pair accumulation in the speed-up of the parallel physicochemical profile similarity measure calculation while **Figure 4B** depicted the effect in

Datasets	**Total of gene pairs**	**Maximum length for the first genome *m***	**Maximum length for the second genome *n***
ScerSpombe	16,324,500	4910	4924
ScerSpombe + ScerKlac	47,546,047	4910	4924
ScerSpombe + ScerKlac + ScerCgla	78,111,162	4910	4924
ScerSpombe + ScerKlac + ScerCgla + CglaKlac	105,891,467	4910	4924
ScerSpombe + ScerKlac + ScerCgla + CglaKlac + KlacKwal	133,666,445	4915	4924

Table 2. Genome data used in the scalability analysis.

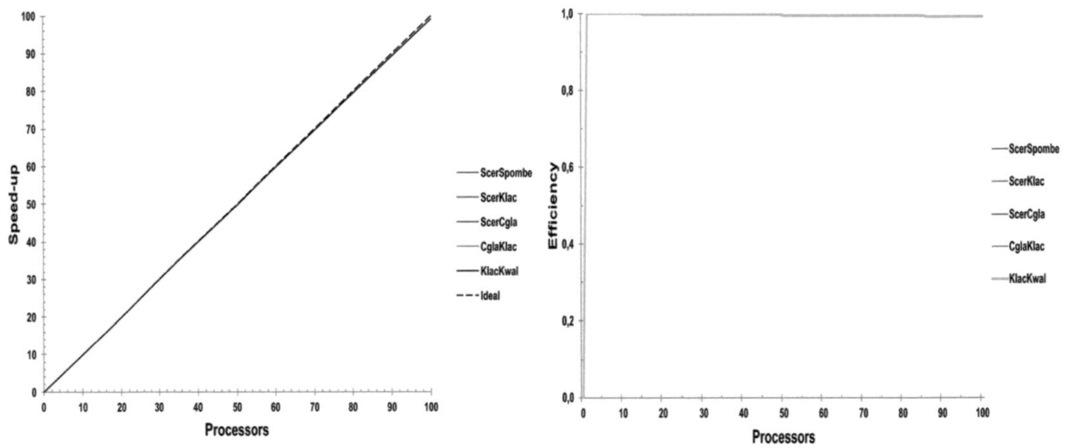

A B

Figure 4. Estimated speed-up (A) efficiency (B) and when the input data and processors increase.

the efficiency of this calculation. When the problem size increases, the speed-up and the efficiency increase for a specific number of processors, although both values tend to decrease when the number of processors increases. Efficiency tends to be constant when both the problem size and the number of processors increase. This is an essential issue to achieve scalability, specifically, the horizontal one.

In sum, the parallel algorithm time complexity considering *P* processors is lower than the cost of the sequential version and the scalability may be achieved with the parallel proposal. Nevertheless, considering implementation hazards of the parallel models, scalability can be improved when these models are executed in a cloud within a big data framework as MapReduce [27, 28]. This framework guarantees reliability, availability, and a good execution performance in a cloud with a distributed file system.

In a MapReduce design, each mapper process should build a subset of the resulting dataset with calculated features of all protein pairs of its partition. **Figure 5** shows three steps such as Initial, Map and Final. The Initial phase divides the protein pair set into independent Hadoop distributed file system (HDFS) blocks; it also duplicates and transfers them to other nodes of the cloud. Then, in the Map step, each mapper calculates the protein features of all the pairs in its partition. Finally, in the Final step, the files created by each mapper would be concatenated to build the final dataset file. In this procedure the Reduce step is omitted since the mappers output are directly combined following the operation reported in [29].

Algorithm 1 shows el pseudo-code of the Map function in the MapReduce process for the protein feature calculations. Step 2 calculates the features for a pair of proteins and Step 5 save the previously calculated data.

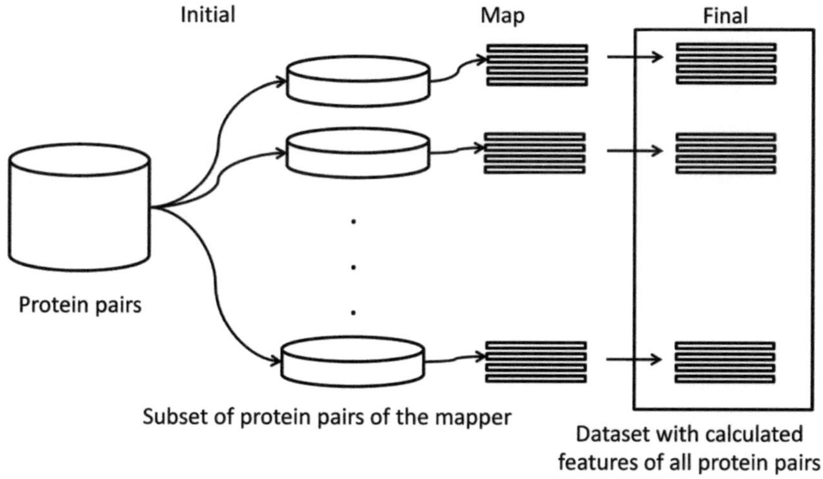

Figure 5. MapReduce design of protein pair feature calculations.

```
MAP (key, value)

Input: <key; value> % key represents the location in bytes, and value, the contents of a protein pair.

Output: <key'; value'> % key' indicates the identification of the pair, and value', the calculated features.

pair ← PAIR_REPRESENTATION(value)
% features variable contains the values of the feature for the pair
features ← CALCULATE_FEATURES(pair)
lkey ← lkey:set(pair:get_id)
lvalue ← lvalue:set(features)
EMIT (lkey, lvalue)
```

Algorithm 1. Map phase for feature calculation.

2.4. Big data supervised classification with imbalance management

Given a set $A=\{S_r(x_i, y_j)\}$ of gene pair features as discrete or continuous values of r gene pair similarity measure functions, previously specified, we represent a pairwise ortholog detection decision system $DS = (U, A \cup \{d\})$, where $U = \{(x_i, y_j)\}, \forall x_i \in G_1, \forall y_j \in G_2$ is the universe of the gene pairs,and $d \notin A$ is the binary decision feature obtained from a curated classification. This decision feature defines the low ratio of orthologs to the total number of possible gene pairs. The big data supervised classification divides *DS* into train and test sets to build a learning/training model and to classify the instances by means of a big data supervised algorithm managing the imbalance between classes. The built model can be generalized to related pairs of genomes/yeast species not used in the learning/training process following the generalization concept in [30]. Thus, the test set is used to compare both supervised and classical unsupervised algorithms. **Algorithm 2** shows the steps required in the training phase as well as **Algorithm 3** shows the classification phase. The general testing/evaluation scheme for supervised and unsupervised algorithms is sketched in **Figure 6** based on the use of a testing external set as reported in [31].

Training phase

Input: $DS=(U, A\cup\{d\})$, $A=\{S_r(x_i, y_j)\}$, $U=\{(x_i, y_j)\}, \forall x_i \in G_1, \forall y_j \in G_2$

Output: Model, Test_set

Train_set, Test_set =
    Selection_of_training_and_testing_sets (*DS*)

Model = Model_building_with_imbalance_management %Learning/Training step

Algorithm 2. Training phase for big data supervised classification with imbalance management.

Classification phase

Input: Model, Test_set

```
Output: Ortholog_pairs_of_the_testing_set
1 Ortholog_pairs_of_the_testing_set = Pairwise_classification (Model,
Test_set)
```

Algorithm 3. Classification phase for big data supervised ortholog detection with imbalance management.

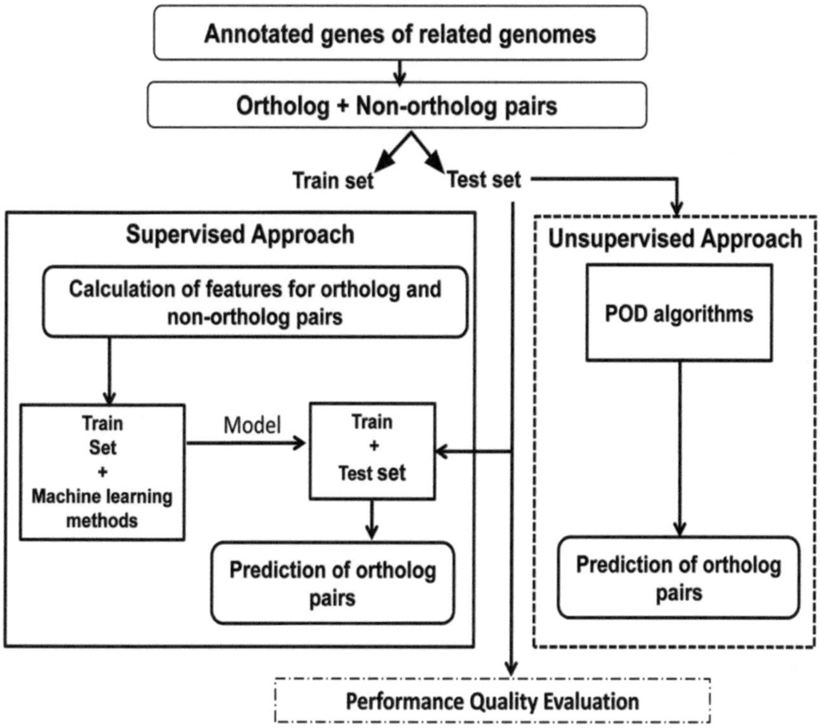

Figure 6. Workflow of the evaluation of supervised versus unsupervised pairwise ortholog detection algorithms.

The general scheme receives annotated genomes (proteomes) from related yeast species since some protein features as the membership to conserved regions require certain evolution closeness. For the evaluation of pairwise ortholog detection algorithms, we can compare the supervised solutions and the unsupervised reference algorithms such as RBH, RSD, and OMA. Firstly, protein pairs are separated into train and test sets and after that pairwise similarity measures for the pairs of both sets are calculated. Test sets are used for the assessment of the unsupervised algorithms while the training set is used to build the supervised models which will be further tested on the previously mentioned test set.

The classification performance will be assessed by evaluation metrics for imbalanced datasets. The Geometric Mean (*G-Mean*) is defined in [32] as:

$$G - Mean = \sqrt{sensitivity * specificity} \tag{6}$$

where sensitivity and specificity are traditionally calculated considering the true positives (*TP*), false negatives (*FN*), false positives (*FP*), and true negatives (*TN*).

The area under the curve (*AUC*) [33] is estimated from plotting of true positive rate (*TPR*) versus false positive rate (*FPR*) values. We approximate the *AUC* by averaging (*TPR*) and (*FPR*) values through the following equation:

$$AUC = \frac{1 + TPR - FPR}{2} \quad (7)$$

where $TPR = \frac{TP}{TP+FN}$ represents the percentage of orthologs (*TP*) correctly classified and $FPR = \frac{FP}{FP+TN}$, the percentage of non-orthologs (*TN*) misclassified. *G-Mean* measure maximizes the accuracy of the two classes (orthologs and non-orthologs) by considering the misclassification costs and *AUC* values in the estimation of the sensitivity and specificity, getting at a balance between them [34].

2.5. Ortholog detection experiments in related yeasts

2.5.1. Settings

For the evaluation of pairwise ortholog detection algorithms in related yeast genomes we carried out (i) three exploratory experiments for an external validation of supervised classification models and (ii) an experiment to evaluate a model built with a pair of yeasts into two external pairs. In Experiment 1, we evaluated the algorithms inside a genome pair by partitioning the complete set of protein pairs at random (75% for training and 25% for testing). Specifically, we divided the *S. cerevisiae – K. lactis* dataset into 16,986,996 pairs for training and 5,662,332 pairs for testing. In Experiment 2, we evaluated the model in the first experiment into 8,095,907 pairs of *S. cerevisiae* and *S. pombe* genomes. In Experiment 3, we tested in *S. cerevisiae* and *S. pombe* as in the second experiment but with models obtained by training with the complete *S. cerevisiae – K. lactis* dataset (22,649,328 pairs). In Experiment 4, we used the model in the third experiment and tested it in two different pairs: 29,887,416 pairs of *S. cerevisiae* and *C. glabrata* and 20,318,472 pairs of *C. glabrata – K. lactis*.

The details for the data used in the experiments are specified in **Table 3**. For each genome pair we built four datasets (BLOSUM50, BLOSUM62_1, BLOSUM 62_2, and PAM250) from combinations of alignment parameter settings shown in **Table 4**. The gene pair features included in the datasets are the average of local and global alignment similarity measures, the length of sequences, the gene membership to conserved regions (synteny), and the physicochemical

Genome pair	Number of features	Instances per class (majority; minority)	Imbalance ratio (*IR*)	Excluded genes
S. cerevisiae – K. lactis	6	(22,646,914; 2414)	9381.489	89 out of 5861 *S. cerevisiae* proteins 37 out of 5215 *C. glabrata* proteins 1403 out of 5327 *K. lactis* proteins
S. cerevisiae – C. glabrata	6	(29,884,575; 2841)	10519.034	
C. glabrata – K. lactis	6	(20,317,232; 1240)	16384.865	
S. cerevisiae – S. pombe	6	(8,090,950; 4957)	1632.227	

Table 3. Genome data used in the experiments.

Dataset	Substitution matrix	Gap open	Gap extended
BLOSUM50	BLOSUM50	15	8
BLOSUM62_1	BLOSUM62	8	7
BLOSUM62_2	BLOSUM62	12	6
PAM250	PAM250	10	8

Table 4. Datasets built with different alignment parameter values.

profiles within 3, 5, and 7 window sizes. We are aggregating global and local alignment similarities to combine structural and functional similarities of proteins.

The *S. cerevisiae* – *S. pombe* dataset represents the union of Inparanoid7.0 and GeneDB ortholog classifications as is described in [35]. On the other hand, the *S. cerevisiae* – *K. lactis* and *S. cerevisiae* – *C. glabrata* datasets contain all ortholog pairs in the gold groups reported in [10]. At the time of building the subset with all possible pairs, we just excluded 89 genes from *S. cerevisiae*, 37 from *C. glabrata*, and 1403 from *K. lactis* because the genome physical location data in the YGOB database [36], required for the LCB feature calculation, was not found.

The amount of pairs in each dataset prevented traditional machine learning methods from training and testing, so big data implementations were selected from scalable MapReduce Mahout [21] and Spark MLlib [37] libraries. The selected algorithms for the evaluation are listed in **Table 5**. Explicitly, we selected the random oversampling (ROS) and the cost-sensitive approaches for imbalance management. Besides, the big data framework details are specified in **Table 6**. On the other hand, the unsupervised algorithms parameter values used for the evaluation are specified in **Table 7**.

The infrastructure used to perform the big data experiments consists of 20 nodes connected via a 40 GB/s Infiniband network. Each node has two Intel Xeon E5-2620 microprocessors (at 2 GHz, 15 MB cache) and 64 GB of main memory working under Linux CentOS 6.5. The head node of the cluster has two Intel Xeon E5-2620 microprocessors (at 2.00 GHz, 15 MB cache) and 32 GB of main memory.

2.5.2. Results

Supervised classifiers that manage big data were compared from two different points of view:

i. The efficacy and the execution time when the training process is performed with one partition of the genome pair dataset or with the complete dataset.

ii. The efficacy and the execution time in the classification process performed on datasets with potential fails for ortholog detection:

- The one of related species (distant relatives), *S. cerevisiae* and *S. pombe*, false negatives can increase in the presence of divergent sequences.
- Other two *Saccharomycete* yeast datasets which complexity is a possible increased number of false positives due to lots of paralog losses produced by the WGD.

Algorithm	Parameter values
RF-BD[1]	Number of trees: 100 Random selected features per node: 3[2] Number of maps: 20
RF-BDCS	Number of trees: 100 Random selected features per node: 3 Number of maps: 20 $C(+\mid -)=IR$ $C(-\mid +)=1$
ROS (100%) + RF-BD	RS[3] = 100%
ROS (130%) + RF-BD	RS = 130%
SVM-BD	Regulation parameter: 1.0, 0.5, and 0.0 Number of iterations: 100 (by default) StepSize: 1.0 (by default) miniBatchFraction: 1.0 (percent of the dataset evaluated in each iteration 100%)
ROS (100%) + SVM-BD	RS = 100%
ROS (130%) + SVM-BD	RS = 130%

[1]BD means big data.
[2]int.($\log_2 N + 1$), where N is the number of features of the dataset.
[3]RS represents resampling size.

Table 5. Supervised algorithms and parameter values selected for the experiments.

Big data framework	Application	Algorithms
Hadoop 2.0.0 (Cloudera CDH4.7.1) with the head node configured as name-node and job-tracker, and the rest, as data-nodes and task-trackers	– MapReduce ROS implementation – A cost-sensitive approach for Random Forest MapReduce algorithm (RF-BD) – MapReduce RF implementation (Mahout Library)	RF-BDCS ROS (100%) + RF-BD ROS (130%) + RF-BD
Apache Spark 1.0.0 with the head node configured as master and name-node, and the rest, as workers and data-nodes	– Apache Spark Support Vector Machines (MLLib)	ROS (100%) + SVM-BD ROS (130%) + SVM-BD

Table 6. Big data framework, applications and algorithms.

Supervised and unsupervised classifiers are compared in terms of the efficacy in complex *Saccharomycete* yeast datasets.

Best mean results of *G-Mean, AUC, TPR,* and *TNR* (true negative rate) in Experiments 1, 2, and 3 are shown in **Tables 8–10**, respectively, with the most outstanding values in bold face. The best performance of *G-Mean, AUC* and the balance between *TPR* and *TNR* mean values were obtained with the cost-sensitive Big Data Random Forest algorithm (RF-BDCS) in Experiment 1 (testing in a partition of *S. cerevisiae – K. lactis*) and with support vector machine-BigData (SVM-BD) (regParam: 0.5) in Experiments 2 and 3 (testing in *S. cerevisiae – S. pombe*). There are no significant changes in classifiers with the best results (those that manage data imbalance)

Algorithm	Parameter values	Implementation
RBH	Soft filter and Smith Waterman alignment E-value = 1e−06	BLASTp program[1] Matlab script
RSD	E-value thresholds: 1e−05, 1e−10 and 1e−20 Divergence thresholds α: 0.8, 0.5 and 0.2.	BLASTp program[1] Python script[2]
OMA	Default parameter values	OMA stand-alone[3]

[1]Available in http://www.ncbi.nlm.nih.gov/BLAST/.
[2]Available in https://pypi.python.org/pypi/reciprocal_smallest_distance/1.1.4/.
[3]Available in http://omabrowser.org/standalone/OMA.0.99z.3.tgz.

Table 7. Unsupervised algorithms and parameter values in the experiments.

Algorithm	*G-Mean*	*AUC*	*TPR*	*TNR*
ROS (RS: 100%) + RF-BD	0.9807	0.9809	0.9621	1.0000
ROS (RS: 130%) + RF-BD	0.9812	0.9813	0.9629	0.9997
RF-BDCS	**0.9889**	**0.9889**	**0.9788**	**0.9997**
ROS (RS: 100%) + SVM-BD (regParam: 1.0)	0.9477	0.9477	0.9369	0.9586
ROS (RS: 100%) + SVM-BD (regParam: 0.5)	0.8791	0.8845	0.93038	0.96259
ROS (RS: 130%) + SVM-BD (regParam: 0.5)	0.8528	0.8629	0.93893	0.95476

Table 8. Best mean *G-Mean, AUC, TPR* and *TNR* results in Experiment 1 in *S. cerevisiae* – *K. lactis*. Highlighted *TPR* and *TNR* with the best stability in these measures.

Algorithm	*G-Mean*	*AUC*	*TPR*	*TNR*
RF-BDCS	0.7103	0.7542	0.5570	0.9997
ROS (RS: 100%) + SVM-BD (regParam: 1.0)	0.8543	0.8641	0.7432	0.9923
ROS (RS: 100%) + SVM-BD (regParam: 0.5)	**0.8795**	**0.8848**	**0.7900**	**0.9796**
ROS (RS: 100%) + SVM-BD (regParam: 0.0)	0.4312	0.5846	0.9781	0.1911

Table 9. Best mean *G-Mean, AUC, TPR* and *TNR* results in Experiment 2 in *S. cerevisiae* – *S. pombe*. Highlighted *TPR* and *TNR* with the best stability in these measures.

Algorithm	*G-Mean*	*AUC*	*TPR*	*TNR*
RF-BDCS	0.6745	0.7294	0.4590	0.9998
ROS (RS: 100%) + SVM-BD (regParam: 1.0)	0.8533	0.8632	0.7332	0.9933
ROS (RS: 100%) + SVM-BD (regParam: 0.5)	**0.8791**	**0.8845**	**0.7895**	**0.9795**
ROS (RS: 130%) + SVM-BD (regParam: 1.0)	0.7956	0.8164	0.6348	0.9979
ROS (RS: 130%) + SVM-BD (regParam: 0.5)	0.8528	0.8629	0.7331	0.9926

Table 10. Best mean *G-Mean, AUC, TPR* and *TNR* results in Experiment 3 in *S. cerevisiae* – *S. pombe*. Highlighted *TPR* and *TNR* with the best stability in these measures.

when we use the four alignment parameter value combinations in **Table 4**. Similarly, good efficacy is obtained when we trained with a partition of *S. cerevisiae – K. lactis* and with the corresponding complete training set.

The SVM algorithm combined with ROS showed the best time results in Experiment 1 (SVM-BD (regParam: 0.5) executed in 14 minutes and 19 seconds) and (SVM-BD (regParam: 0.0) executed in 14 minutes and 21 seconds). These algorithms also showed good classification quality, mainly, in the balance between *TPR* and *TNR*. In Experiments 2 and 3 SVM variants combined with ROS (RS:100) run very rapidly, in particular, SVM-BD (regParam: 0.5) executed in 14 minutes and 29 seconds in Experiment 2 and in 16 minutes and 53 seconds in Experiment 3.

Analyzing measure values of Experiment 4 we noticed that the performance of supervised classifiers is almost the same as in the previous experiments in terms of the stability in the four different alignment parameter value datasets. Best mean results of *G-Mean* and *AUC*, *TPR* and *TNR* for supervised and unsupervised classifiers are shown in **Table 11**. The underlined values highlight the most effective method in the experiment while the bold face values identify the best performing supervised and unsupervised algorithms. Supervised Random Forest classifiers managing data imbalance overdid the unsupervised ones, specifically, when the cost-sensitive version is applied. Among the unsupervised classifiers, RSD reaches the highest *G-Mean* (0.9374) using the recommended parameter values (*E-value* = 1e−05 and $\alpha = 0.8$) in [38]), higher values were also obtained for *AUC* and *TPR*. On the contrary, OMA was the best among the unsupervised algorithms in *S. cerevisiae – S. pombe* datasets.

The results of the Friedman test [39] for the *AUC* measure with the four datasets of *S. cerevisiae – C. glabrata, C. glabrata – K. lactis,* and *S. cerevisiae – S. pombe* and the supervised classifiers that manage imbalance are shown in **Table 12** (first column). The test yielded significant differences among classifiers, being RF-BDCS the one with the high mean rank followed by the classifiers based on Random Forest with ROS preprocessing and the SVM classifier with ROS (RS:100) and 0.5 regulation. In the Friedman test for *G-Mean* (**Table 12** second column) there are also significant

Dataset/algorithm	***S. cerevisiae – C. glabrata***				***C. glabrata – K. lactis***			
	AUC	*G-Mean*	*TPR*	*TNR*	*AUC*	*G-Mean*	*TPR*	*TNR*
RBH	0.8196	0.7995	0.6392	0.9999	0.8052	0.8242	0.6484	0.9999
RSD 0.2 1e−20	0.9238	0.9206	0.8476	0.9999	0.9047	0.9092	0.8185	0.9999
RSD 0.5 1e−10	0.9340	0.9316	0.8680	0.9999	0.9267	0.9294	0.8589	0.9999
RSD 0.8 1e−05	**0.9382**	**0.9362**	**0.8765**	**0.9999**	**0.9354**	**0.9374**	**0.8750**	**0.9999**
OMA	0.9287	0.9259	0.8574	0.9999	0.9125	0.9163	0.8328	0.9999
ROS (RS: 100%) + RF-BD	0.9901	0.9900	0.9805	0.9997	0.9806	0.9811	0.9615	0.9997
ROS (RS: 130%) + RF-BD	0.9901	0.9901	0.9806	0.9997	0.9807	0.9807	0.9617	0.9997
RF-BDCS	**0.9934**	**0.9934**	**0.9876**	**0.9993**	**0.9847**	**0.9862**	**0.9702**	**0.9992**

Table 11. Best mean *G-Mean, AUC, TPR* and *TNR* results in Experiment 4 in *S. cerevisiae – C. glabrata* and *C. glabrata – K. lactis*. Highlighted *TPR* and *TNR* with the best stability in these measures.

Algorithm	*AUC* mean rank	*G-Mean* mean rank	Execution time mean rank
ROS (RS: 100%) + RF-BD	5.96	5.71	8.00
ROS (RS: 130%) + RF-BD	5.96	5.79	9.00
RF-BDCS	7.67	7.67	7.00
ROS (RS: 100%) + SVM-BD (regParam: 1.0)	2.62	6.12	2.33
ROS (RS: 100%) + SVM-BD (regParam: 0.5)	5.50	5.79	2.25
ROS (RS: 100%) + SVM-BD (regParam: 0.0)	5.12	2.00	2.58
ROS (RS: 130%) + SVM-BD (regParam: 1.0)	2.92	4.46	4.33
ROS (RS: 130%) + SVM-BD (regParam: 0.5)	4.79	5.62	4.50
ROS (RS: 130%) + SVM-BD (regParam: 0.0)	4.46	1.83	5.00
Test statistics			
N	12.000	12.000	12.000
Chi-square	31.359	47.850	80.333
df	8.000	8.000	8.000
Asymp. Sig.	0.000	0.000	0.000

Table 12. Results of the Friedman tests comparing supervised classifiers in terms of *AUC*, *G-Mean* and execution time in the four datasets of *S. cerevisiae – C. glabrata, C. glabrata – K. lactis* and *S. cerevisiae – S. pombe.*

differences with higher mean rank values for RF-BDCS and ROS (RS: 100%) + SVM-BD (regParam: 1.0). When we applied the same test to compare execution time (**Table 12** third column) there are significant differences among the algorithms. In particular, SVM classifiers combined with ROS exhibited the best mean ranks.

2.5.3. Discussion

As a general result, experiments showed that supervised classifiers changed only slightly with the selection of different alignment parameters, maybe because of the appropriate combination of scoring matrices and gap penalties in relation to the sequence diversity between the two yeast genomes. The four scoring matrices assessed have been previously recommended to detect homologs in a wide range of amino acid identities. Moreover, gap penalty settings were not low enough to affect the sensitivity of the alignment [40].

The ROS pre-processing method for big data made SVM effective for pairwise ortholog detection and improved the performance of Random Forest for big data even more with a higher value for the resampling size parameter of 130% [41]. Conversely, the experiments showed that the variation in this parameter value from 100 to 130% did not significantly influence on the performance of the SVM big data classifier with different regulation values.

The cost-sensitive classifier RF-BDCS showed the best performance in *S. cerevisiae – C. glabrata, C. glabrata – K. lactis* and *S. cerevisiae – K. lactis* because probably it improved the training from the minority class. The best tree split was selected following the misclassification costs assigned to the instances; such costs were also considered to associate certain class to a leaf [29]. This cost treatment does not imply changes in the sample distribution, and avoids possible overfitting that is commonly found in ROS solutions due to the presenceof duplicated instances. The setting of the cost values (($C(+|-)=IR$) and $C(-|+)=1$)) can also influence on the success of the algorithm.

In the case of SVM for big data classifier, the fixed regularization parameter defines the trade-off between the goal of minimizing the training error and minimizing the model complexity to avoid overfitting. The higher is its value, the simpler the model, however, a better performance in classification may be achieved by setting an intermediate regulation value, or one close to cero [37]. Specifically, the ROS (RS: 100%) + SVM-BD (regParam: 0.5) classifier exhibited the best *AUC* and *G-Mean* values in *S. cerevisiae – S. pombe*, and the best balance between *TPR* and *TNR* in the rest of the datasets.

All methodologies including the proposed supervised big data approach generally declined their performance for ortholog detection in *S. cerevisiae – S. pombe* datasets, probably because of *S. pombe* is a distant relative of *S. cerevisiae* [23]. The supervised classifiers performance was also negatively affected by differences in data distribution between the train and test sets [42]. On the contrary, ROS (RS: 100%) + SVM-BD (regParam: 0.5) remained stable in *S. cerevisiae – C. glabrata*, *C. glabrata – K. lactis* and *S. cerevisiae – S. pombe* datasets when considering the balance between TP_{Rate} and TN_{Rate}. Best results obtained in *S. cerevisiae – C. glabrata* are outstanding where algorithms are vulnerable to produce false positives since both genomes underwent a WGD and a subsequent differential loss of gene duplicates [10].

The initial assumption of RBH, RSD and OMA that the sequences of orthologous genes/proteins are more similar to each other than they are to any other genes from the compared organisms may produce classification errors [12]. The reduced quality shown by RBH, RSD and OMA, mainly in the case of RBH, could be caused by this assumption despite that BLAST parameters can be tuned as has been recommended in [43]. In particular, RBH infer orthology relationships simply based on reciprocal BLAST best hits. In contrast, the RSD procedure is less likely than RBH to be misled by existing close paralogs since it relies on both global sequence alignment and maximum likelihood estimation of evolutionary distances finding many putative orthologs missed by RBH. The OMA algorithm also displays advantages over RBH by using evolutionary distances instead of alignment scores. It allows the inclusion of one-to-many and many-to-many orthologs considering the uncertainty in distance estimations and detects potential differential gene losses.

On the other hand, the success of big data supervised classifiers managing imbalance over RSD and OMA may be explained by feature combinations together with the learning from curated classifications. The assembling of alignment measures with the comparison of sequence lengths, the membership of genes to conserved regions (synteny) and the physicochemical profiles of amino acids improved the detection of homology and certainly the supervised classification results on the test sets, even if both species underwent WGD. Specifically, the aggregation of global and local alignment scores allows us to combine protein structural and functional relationships between sequence pairs, respectively. The periodicity of the physicochemical properties of amino acids can detect similarity among protein pairs with sequences having functional similarities despite their low amino acid sequence identities (<35%). These sequences may affect ortholog detection in *S. cerevisiae – S. pombe* which are moderately related and their orthologs may be diverged. The synteny information lets us to consider that genes belonging to the same conserved segment in genomes of different species will probably be orthologs. Besides, the length of sequences as the relative positions of amino acids within the same protein in different species and in duplicated regions within the same species may also contribute to the enhanced supervised classification results.

3. Conclusions

The combination of alignment measures with other protein pair features such as the sequence lengths, the gene membership to conserved regions, and the physicochemical profiles has complemented the homology detection in the proposed supervised approach for pairwise ortholog detection. Such combined features alongside curated orthologs pairs extracted from a curated dataset have led to an effective and efficient ortholog classification method in a big data scenario with the treatment of the low ratio of orthologs to the total possible gene pairs between two genomes.

The supervised classifiers that manage imbalance outdid the popular unsupervised (RBH, RSD, and OMA) algorithms even when the supervised model was extended to yeast datasets containing "traps" for ortholog detection algorithms. In future research, the introduction of new gene pair features might improve the effectiveness and efficiency of the supervised algorithms.

The scalability analysis of the proposed gene pair feature calculation highlighted the advances that can be achieved in the scalability issue when we parallelize the calculation, and later on, when we implement a big data model for the same calculation.

Acknowledgements

GACh was funded by a Postdoc fellowship (SFRH/BPD/92978/2013) granted by the Portuguese Fundação para a Ciência e a Tecnologia (FCT). AA was partially supported by the Strategic Funding UID/Multi/04423/2013 through national funds provided by FCT and the European Regional Development Fund (ERDF) in the framework of the program PT2020, by the European Structural and Investment Funds (ESIF) through the Competitiveness and Internationalization Operational Program – COMPETE 2020 and by National Funds through the FCT under the project PTDC/AAG-GLO/6887/2014 (POCI-01-0124-FEDER-016845), and by the Structured Programs of R&D&I INNOVMAR (NORTE-01-0145-FEDER-000035 – NOVELMAR) and CORAL NORTE (NORTE-01–0145-FEDER-000036), and funded by the Northern Regional Operational Program (NORTE2020) through the ERDF. The funders had no role in the study's design, data collection and analysis, decision to publish, or preparation of the manuscript.

Author details

Deborah Galpert Cañizares[1], Sara del Río García[2], Francisco Herrera[2], Evys Ancede Gallardo[3], Agostinho Antunes[4,5] and Guillermin Agüero-Chapin[4,5]*

*Address all correspondence to: gchapin@ciimar.up.pt

1 Departamento de Ciencias de la Computación, Universidad Central "Marta Abreu" de Las Villas (UCLV), Santa Clara, Cuba

2 Computer Science and Artificial Intelligence Department, CITIC-UGR (Research Center on Information and Communications Technology), University of Granada, Granada, Spain

3 Centro de Bioactivos Químicos, Universidad Central "Marta Abreu" de Las Villas (UCLV), Santa Clara, Cuba

4 CIMAR/CIIMAR, Centro Interdisciplinar de Investigação Marinha e Ambiental, Universidade do Porto, Porto, Portugal

5 Departamento de Biologia, Faculdade de Ciências, Universidade do Porto, Porto, Portugal

References

[1] Fitch WM. Distinguishing homologous from analogous proteins. Systematic Zoology. 1970;**19**:99-113

[2] Vashist A, Kulikowski C, Muchnik I. Screening for ortholog clusters using multipartite graph clustering by quasi-concave set function optimization. In: Slezak D et al., editors. RSFDGrC 2005. Vol. 3642. LNAI. Berlin Heidelberg: Springer-Verlag; 2005. pp. 409-419

[3] Hirsh AE, Fraser HB. Protein dispensability and rate of evolution. Nature. 2001;**411**:1046-1049

[4] Östlund G, Schmitt T, Forslund K, Köstler T, Messina DN, Roopra S, et al. InParanoid 7: New algorithms and tools for eukaryotic orthology analysis. Nucleic Acids Research. 2010;**38**:D196-D203

[5] Chen F, Mackey AJ, Stoeckert CJ, Roos DS. OrthoMCL-DB: Querying a comprehensive multi-species collection of ortholog groups. Nucleic Acids Research. 2006;**34**(Database issue):D363-D3D8

[6] Dessimoz C, Cannarozzi G, Gil M, Margadant D, Roth A, Schneider A, Gonnet GH. OMA, A Comprehensive, Automated Project for the Identification of Orthologs from Complete Genome Data: Introduction and First Achievements. In: Comparative Genomics: RECOMB 2005 International Workshop, Dublin, Ireland, September 18-20, 2005 Proceedings. Edited by McLysaght A, Huson DH. Berlin, Heidelberg, Germany; 2005:61-72

[7] Wall DP, Fraser HB, Hirsh AE. Detecting putative orthologs. Bioinformatics. 2003;**19**(13): 1710-1711

[8] Kuzniar A, RCHJv H, Pongor S, JAM L. The quest for orthologs: Finding the corresponding gene across genomes. Trends in Genetics. 2008;**30**:1-13

[9] Towfic F, VanderPIas S, OIiver CA, Couture O, TuggIe CK, GreenIee MHW, et al. Detection of gene orthology from gene co-expression and protein interaction networks. BMC Bioinformatics. 2010;**11**(Suppl 3):S7

[10] Salichos L, Rokas A. Evaluating ortholog prediction algorithms in a yeast model clade. PLoS ONE. 2011;**6**(4):1-11

[11] Koonin EV. Orthologs, paralogs, and evolutionary genomics. Annual Review of Genetics. 2005;**39**:309-338

[12] Kristensen DM, Wolf YI, Mushegian AR, Koonin EV. Computational methods for gene orthology inference. Briefings in Bioinformatics. 2011;**12**(5):379-391

[13] Kamvysselis MK. Computational Comparative Genomics: Genes, Regulation, Evolution. Massachusetts: Massachusetts Institute of Technology; 2003

[14] Zheng XH, Lu F, Wang Z-Y, Zhong F, Hoover J, Mural R. Using shared genomic synteny and shared protein functions to enhance the identification of orthologous gene pairs. Bioinformatics. 2005;**21**(6):703-710

[15] Lechner M, Hernandez-Rosales M, Doerr D, Wieseke N, Thévenin A, Stoye J, et al. Orthology detection combining clustering and synteny for very large datasets. PLoS ONE. 2014;**9**(8):e105015

[16] Chen X, Zheng J, Fu Z, Nan P, Zhong Y, Lonardi S, et al. Assignment of orthologous genes via genome rearrangement. IEEE/ACM Transactions on Computational Biology and Bioinformatics. 2005;**2**(4):302-315

[17] Fu Z, Chen X, Vacic V, Nan P, Zhong Y, Jiang T. MSOAR a high-throughput ortholog assignment system based on genome rearrangement. Journal of Computational Biology. 2007;**14**:16

[18] Chen TW, Wu TH, Ng WV. DODO: An efficient orthologous genes assignment tool based on domain architectures. Domain based ortholog detection. BMC Bioinformatics. 2010;**11** (Suppl 7):S6

[19] Sonnhammer ELL, Gabaldón T, Sousa da Silva AW, Martin M, Robinson-Rechavi M, Boeckmann B, Thomas PD, Dessimoz C. Big data and other challenges in the quest for orthologs. Bioinformatics. 2014, **30**(21):1-6

[20] Alexeyenko A, Tamas I, Liu G, Sonnhammer ELL. Automatic clustering of orthologs and inparalogs shared by multiple proteomes. Bioinformatics. 2006;**22**(14):e9-e15

[21] Owen S, Anil R, Dunning T, Friedman E. Mahout in Action, 2011, Manning Publications Co., USA

[22] Krishnan S, Smith V. MLlib. 2013 [enero 2015]. Available from: https://spark.apache.org/docs/

[23] Wood V. *Schizosaccharomyces pombe* comparative genomics; from sequence to systems. In: Comparative genomics 2006 Jan 1 (pp. 233-285). Springer Berlin Heidelberg, Germany

[24] Galpert D, Río Sd, Herrera F, Ancede-Gallardo E, Antunes A, Agüero-Chapin G. An effective big data supervised imbalanced classification approach for ortholog detection in related yeast species. BioMed Research International [serial on the Internet]. 2015;**2015**:748681

[25] Grama A, Gupta A, Karypis G, Kumar V, editors. Introduction to Parallel Computing, Second Edition: Addison Wesley, Boston, USA; 2003

[26] Bonvin N. Linear Scalability of Distributed Applications. Suisse: École polytechnique fédérale de lausanne; 2012

[27] Dean J, Ghemawat S. MapReduce: Simplified data processing on large clusters. OSDI'04: Proceedings of the 6th Symposium on Operating System Design and Implementation. San Francisco, California, USA: USENIX Association; 2004. p. 137-150

[28] Fernández A, Río Sd, López V, Bawakid A, Jesus MJd, Benítez JM, et al. Big Data with cloud computing: An insight on the computing environment, MapReduce, and programming frameworks. WIREs Data Mining and Knowledge Discovery [serial on the Internet]. 2014

[29] Sd R, López V, Benítez JM, Herrera F. On the use of MapReduce for imbalanced Big Data using random forest. Information Sciences. 2014;**285**:112-137

[30] Bishop CM, editor. Pattern Recognition and Machine Learning: Springer Science+Business Media, LLC, NY, USA; 2006.

[31] Kuncheva LI, editor. Combining Pattern Classifiers Methods and Algorithms. Hoboken, New Jersey: John Wiley & Sons, Inc.; 2004

[32] Barandela R, Sánchez JS, García V, Rangel E. Strategies for learning in class imbalance problems. Pattern Recognition. 2003;**36**(3):849-851

[33] Bradley AP. The use of the area under the ROC curve in the evaluation of machine learning algorithms. Pattern Recognition. 1997;**30**(7):1145-1159

[34] He H, Garcia EA. Learning from imbalanced data. IEEE Transactions on Knowledge and Data Engineering. 2009;**21**(9):1263-1284

[35] Koch EN, Costanzo M, Bellay J, Deshpande R, Chatfield-Reed K, Chua G, et al. Conserved rules govern genetic interaction degree across species. Genome Biology. 2012;**13**(7)

[36] Byrne KP, Wolfe KH. The yeast gene order browser: Combining curated homology and syntenic context reveals gene fate in polyploid species. Genome Research. 2005;**15**:1456-1461

[37] Krishnan S, Smith V. Linear Support Vector Machines (SVMs). 2013 [enero 2015]. Available from: https://spark.apache.org/docs/latest/mllib-linear-methods.html#linear-support-vector-machines-svms

[38] DeLuca TF, Wu I-H, Pu J, Monaghan T, Peshkin L, Singh S, et al. Roundup: A multi-genome repository of orthologs and evolutionary distance. Bioinformatics. 2006;**22**(16): 2044-2046

[39] Trawinski B, Smetek M, Telec Z, Lasota T. Nonparametric statistical analysis for multiple comparison of machine learning regression algorithms. International Journal of Applied Mathematics and Computer Science. 2012;**22**(4):867-881

[40] Pearson WR. Selecting the right similarity-scoring matrix. Current Protocols in Bioinformatics. 2013;**43**:3.5.1-3.5.9

[41] Triguero I, del Río S, López V, Bacardit J, Benítez JM, Herrera F. ROSEFW-RF: the winner algorithm for the ECBDL'14 big data competition: an extremely imbalanced big data bioinformatics problem. Knowledge-Based Systems. 2015;**87**:69-79

[42] Moreno-Torres JG, Llorà X, Goldberg DE, Bhargava R. Repairing fractures between data using genetic programming-based feature extraction: A case study in cancer diagnosis. Information Sciences. 2013;**222**:805-823

[43] Hagelsieb GM, Latimer K. Choosing BLAST options for better detection of orthologs as reciprocal best hits. Bioinformatics. 2008;**24**(3):319-324

Section 2

Food Applications

Chapter 3

Growth Kinetics for the Selection of Yeast Strains for Fermented Beverages

Dalia E. Miranda Castilleja, Jesús A. Aldrete Tapia,
Sofía M. Arvizu Medrano,
Montserrat Hernández Iturriaga,
Lourdes Soto Muñoz and
Ramón Á. Martínez Peniche

Additional information is available at the end of the chapter

http://dx.doi.org/10.5772/intechopen.70224

Abstract

Criteria to select autochthonous yeast strains for their use in fermented beverages include their ability to dominate the media and to enhance desired sensorial characteristics and their inability to produce undesired compounds such as biogenic amines or off-odors. One of the key features in yeast selection is its Implantation, surpassing different stresses, and its fermentation performance, which requires setting up the process and monitoring it, involving important amount of resources. Methods to evaluate the tolerance of yeast strains are usually based in the qualitative measure of the growth of the microorganism in a medium containing the limiting compound after a specific time of incubation. However, studying strain growth through optical density measurements permits to estimate quantitative and comparable parameters providing an insight into the fitness of the cell to certain environment, lag phase duration, growth rate, and maximum population, among others. In the last decades, culture-independent methods have been used to evaluate the dynamic of microbial populations during fermentative process. In this chapter, a review of recent advances in the selection of fermentative yeasts as well as the utilization of kinetic evaluation and molecular strategies in conditions associated with fermented beverage for selecting yeast strains is presented.

Keywords: yeast selection, fermentative process, growth evaluation, kinetic parameters, culture-independent methods

1. Introduction

The production of alcoholic beverages is one of the most ancient food traditions. Their elaboration relays on a fundamental stage: the alcoholic fermentation (AF), which is a biochemical conversion of sugars into ethanol through the action of yeasts. The AF can occur in three ways [1, 2]: (i) spontaneously, from the microbiota naturally present in the musts or on surfaces of equipment; (ii) by adding commercial yeasts; and (iii) by inoculating selected native strains. Nowadays, the demand for autochthonous strains has increased worldwide for it is accepted that this option allows preserving the unique and typical character that native microorganisms provide.

Yeasts intervening in AF are considered "fermentative yeasts," and they are divided broadly in two groups: *Saccharomyces* and non-*Saccharomyces* (nS). The *Saccharomyces* possess a high efficiency in the conversion of sugars and tolerate high concentrations of ethanol and SO_2, being the fermentative genus for excellence [3]. Most of the nS are low tolerant to ethanol and include different genera such as *Candida, Kloeckera, Hanseniaspora, Zygosaccharomyces, Schizosaccharomyces, Torulaspora, Brettanomyces, Saccharomycodes, Pichia,* and *Williopsis* [4]. In beverages like wine, the importance of these yeasts lies on their metabolic features, as the wide set of enzymes they offer can improve the typicity and enhance the sensory profile [4–7]. Conversely, in other beverages as Mexican tequila and mescal, as well as Brazilian cachaça, nS are considered the main responsible for the ethanol production [8–10]. Yeast cells are exposed to several stress conditions from the beginning to the end of the fermentation process [11], resulting in the reduction of their growth and survival rate causing a decrease in fermentation efficiency. Yeasts capable of overcoming these conditions with low viability loss are best suited for these purposes [12].

2. Stress tolerance of fermentative yeast

The primary goal of fermentative yeast is to rapidly and efficiently convert simple sugars into ethanol without developing unpleasant flavors [13]. Several factors affect the yeast ability to grow in the fermentative media related with the type and style of beverage produced; therefore, the ability to adapt and to cope with this hostile environment is considered the main feature to select fermentative yeast [12]. Some of the most relevant inhibiting conditions are summarized in this section.

2.1. Limiting conditions associated with musts

2.1.1. Carbohydrates

Carbohydrates are the most important nutrient since they are metabolized to form biomass, ethanol, and different by-products such as volatile compounds, glycerol, and others that will develop the sensorial characteristics [14]. However, they also produce the first stress due to osmotic pressure in the cells after their inoculation. Therefore, tolerance to high sugar concentration is one of the main criteria for yeast selection, especially for those designated for their use in the elaboration of liquorish beverages such as "Sauternes" wine. Yeast cells have developed mechanisms to adjust to high external osmolarity and maintain or reestablish an inside-directed driving force for water. Adaption to this stress usually takes several hours in which yeast cells accumulate glycerol and trehalose [15] and change their cell wall composition [16] to counter loss of water by the osmotic pressure. The stress level will depend on the type and concentration of sugars found. Concentrations range from diluted juices to the high gravity worst containing 16–18% of dissolved solids, rice mash for sake production with 20% of solids, and grape juice with 200 g L^{-1} of sugar content or even more.

Sugar composition of the media strongly impacts on yeast metabolic physiology [17]. Glucose and fructose are widely found in nature as free sugars or as polysaccharides, also in different proportions. Grape juice contains approximately 1:1 of glucose-fructose [18] apple juice 5:8 [19], and agave juice for the elaboration of tequila and mezcal, after thermal processing of inulins from the plant, contains 1:20 ratio [20]; in *pulque* production from raw agave juice, inulin is also predominant [21]. For beer fermentation maltotriose and maltose comprise 50% of sugars; in rum, cachaça, and tafia, produced by sugar cane juice or molasses, mainly sucrose is present [22].

2.1.2. Assimilable nitrogen

Deficiencies in the supply of assimilable nitrogen will lead to sluggish fermentation. The amounts of total nitrogen in musts for the production of fermentative beverages vary from 40 mg L^{-1} in agave juices [23], 50–150 mg L^{-1} in apple juice, 80 mg L^{-1} in sugar cane for cachaça [24], more than 150 mg L^{-1} in beer malt wort [25], and from 100 to 500 mg L^{-1} in grape must [26]. A minimum of 66 mg L^{-1} to sustain the growth of yeasts and to finish fermentation with a total consumption of sugars, and high ethanol yield is considered [26].

In the fermentative media, after yeasts are inoculated, a rapid uptake of nitrogen compounds used for the biosynthesis of macromolecules and storage in vacuoles is carried out if sufficient amount is present in the must. As yeasts have high preference for ammonium ions, it is used as exogenous nitrogen source in fermentations. Glutamate, aspartate, and glutamine are the first amino acids uptaken if they are present in the must. When a favorite nitrogen source is depleted, yeast will use a less preferred, resulting in reduced growth and fermentation rates [12].

A target for the improvement of fermentation is then the selection of yeasts with low nitrogen requirements, which will reduce the necessity of adding supplements to the must that will make the process expensive.

2.1.3. Sulfur dioxide (SO_2)

During the beginning of wine fermentation, native microbiota mainly nS yeasts and bacteria dominate the media with low ethanol production that could negatively affect the quality of the final product. To inhibit them and to favor native or inoculated *Saccharomyces* strains to develop, sulfur dioxide (SO_2) is added at levels ranging from 20 to 50 ppm [27]. *Saccharomyces* also produces SO_2 during metabolism of sulfate ions [28]. If the amount of SO_2 added and produced by *Saccharomyces* strain are high and remain until the end the process, then safety problems arise as this compound causes certain level of toxicity in human population that consume it. The selection of yeast strains resistant to SO_2 and low SO_2 producers is then desirable.

2.2. Fermentative-derived conditions

2.2.1. Interaction with other microorganisms

Different microorganisms interact during the elaboration of fermented beverages. The variety of these interactions and their impacts on efficiency and product quality should be individually determined, as they will depend on the fermentative strains, the native associated microbiota, and the type of beverage. One of the main metabolites exerting a clear effect on yeast growth and performance is ethanol, mainly produced by *Saccharomyces cerevisiae*. Other metabolites as medium-chain fatty acids and high amounts of acetic acid can negatively affect the growth of a co-fermenting yeast species [29]. Cell-to-cell contact as well as oxygen availability appears to be also involved in the interactions between *S. cerevisiae* and other nS species [30].

One special aspect is the "killer phenotype," which refers to those yeasts able to secrete polypeptide toxins which kill sensitive cells and which is believed to be a potential mechanism to prevent a competitor from gaining access to a resource [31]. Killer toxins differ between species and strains, thus varying the modes of action, from changing membrane permeability in sensitive cells to inhibiting DNA replication or stopping cell division at G_1 phase. All killer toxins are usually active and stable at pH 4–5 and 20–25°C; nevertheless each toxin has an optimum pH and temperature at which it manifests its killer character more effectively [32, 33].

2.2.2. Ethanol

Ethanol produced during fermentation is known to inhibit yeast growth, resulting in a primary factor on yeast efficiency; in turn, the viability of yeast cells in the presence of ethanol constitutes a key feature on strain selection for fermentative purposes [34].

Ethanol affects many aspects of yeast survival, as the fluidity of the plasmatic membrane [35], the vacuole morphology [36], the activity of crucial glycolytic enzymes [37], and the mitochondrial DNA [38]. Ethanol also causes the denaturation of hydrophilic and hydrophobic proteins, affecting various transport systems such as the general amino acid permease and glucose uptake processes [18]. Regarding molecular response of *S. cerevisiae* in the presence of ethanol, it increases the expression of genes associated with glycolysis and mitochondrial function and decreases gene expression in energy-demanding growth-related processes; it also induces the production of heat shock-like proteins, lowering the rate of RNA and protein accumulation, enhancing the frequency of small mutations, altering metabolism, denaturing intracellular proteins and glycolytic enzymes, and reducing their activity [39].

Moreover, yeasts have developed diverse strategies to counteract the damages produced by ethanol [40], as the generation of fatty acid unsaturation of membrane lipids in *S. cerevisiae* [35]. Genes involved in intracellular pH homeostasis are also crucial for the resistance to ethanol and other alcohols [41]. This entails in better adapted strains that show a better capacity to activate these mechanisms and endure in the hostile environment formed through alcoholic fermentation.

2.2.3. Organic acids

As previously stated, microorganisms interacting with yeasts during fermentation produce organic acids, and some of them can affect the growth and fermentative efficiency. Lipophilic weak acids, such as acetic, may accumulate inside yeast cells in their undissociated form diffusing into the yeast cells where it dissociates, inducing an acidification of the cytosol [42]. Fatty acids of medium-chain length, as hexanoic, octanoic, and decanoic acids, and also their respective ethyl esters have shown to negatively affect the survival and growth capacity of yeast, being partly responsible for the premature stoppage of fermentations carried by *S. cerevisiae* [43]. These metabolites can be absorbed by the cell membrane; its toxic effect increases in the pH range of 3.0–5.4. This apparent disadvantage of permeability to fatty acids has been exploited, using the denominated "yeast ghosts," which are a commercially available product that can be added when a sluggish or stuck fermentation occurs in order to absorb this kind of inhibitors [44].

2.2.4. pH

During alcoholic fermentation, pH in the media tends to reduce [45], which is known to be a limiting factor on growth of microorganisms, including yeasts. The optimal pH value for yeast growth is around 4.5, and fermented beverages range in pH from 2.5 to 5.5, which by itself does not imply a restrictive condition, but a low pH (<3.5) combined with ethanol, as it usually occurs on fermented beverages, can prematurely inhibit yeast growth and/or fermentation rate [42].

3. Yeast selection workflow

The aspects previously described define the medium in which yeasts will be developed, and they must be kept in mind onward. On the other hand, the selection of fermentative yeasts involves several sequential steps in which the final objective is the identification of strains capable to efficiently ferment the must and obtain a product with optimum sensory qualities [3]. The framework for selection usually starts determining the diversity of yeasts present during the spontaneous fermentation of the beverage. The isolation of yeasts usually takes place along with that stage; afterward, the aspects of interest ranging from tolerance conditions to sensory impact are evaluated in each strain in order to identify the best suited to be selected.

3.1. Determination of yeast diversity

As recently reviewed, in every fermented beverage produced around the world, a particular microbiota will develop as a cause of the raw material from which it starts (fruits, grains, dairy products, parts of plants, etc.), its characteristics (nutrients, pH, type of sugar, etc.), the geographic region, and the modifiable options in each elaboration process, including pre-fermentative manipulations, temperatures, and added substances, among others [45]. Furthermore, yeast diversity that develops will largely determine the sensory profile of the final product; therefore it is first necessary to become acquainted with the species present, their dynamics, and the possible role each one plays during the fermentation. For these studies, advances in molecular techniques have been exploited, being nowadays the area with most scientific contributions on fermented beverages worldwide. Culture-independent methods are the preferred, since they are neither affected by the viability of the microorganism nor by low populations of less abundant species [46]. Some of the preferred techniques are high-throughput or next generation sequencing (HTS or NGS, respectively), denaturing gradient gel electrophoresis (DGGE), and quantitative PCR (qPCR), as it is summarized in **Table 1** [47–49].

In spite of the multiple advantages of the culture-independent techniques, it is necessary to become aware of possible gaps inherent to these approaches, as undetectability of minor populations, preferential amplifications, limited databases, and different effectiveness of lysis protocols on certain species are some of the principal. By applying both, culture-dependent and culture-independent techniques, these drawbacks can be surpassed, along with other benefits, as recovering the isolates needed to characterize individually and select possible starter cultures.

Beverage	**Country**	**Technique**	**Year**	**Main yeasts found**	**Reference**
Caxiri (cassava, corn, and sweet potatoes)	Brazil	DGGE	2015	*S. cerevisiae, P. kluivery, C. tropicalis, D. fabryi*	[50]
Wine (cv. "Grenache")	Spain	NGS	2015	*Hanseniaspora, Saccharomyces, Candida, Issatchenkia*	[51]
Tarubá	Brazil	DGGE	2015	*T. delbrueckii, P. exigua, P. manshurica, P. kudriavzevii, C. tropicalis, C. ethanolica*	[52]
Grape must	Spain	DGGE, qPCR, NGS	2015	*H. uvarum, S. bacillaris, S. cerevisiae*	[53]
Fuzhou Hong Qui (Rice wine)	China	DGGE	2015	*Saccharomycopsis fibuligera, P. guilliermondii, S. cerevisiae, Wickerhamomyces anomalus, C. glabrata*	[54]
Cold soak (wine)	Spain	qPCR, DGGE	2016	*H. uvarum, S. bacillaris, S. cerevisiae*	[55]
Xaj-pitha (rice wine)	India	Whole genome shotgun sequencing	2016	*Meyerozyma guilliermondii, Wickerhamomyces ciferrii, S. cerevisiae, C. glabrata, D. hansenii, Ogataea Parapolymorpha, D. bruxellensis*	[56]
Taberna (palm wine)	Mexico	DGGE	2016	*H. guillermondii, S. cerevisiae, P. kudriavzevii, C. tropicalis, K. exigua*	[57]
Pitmud used in strong-flavor liquor	China	DGGE and NGS	2017	*Candida, Wickerhamomyces, Debaryomyces, Saccharomyces, Pichia*	[58]

Table 1. Recent studies on the diversity of yeasts present during the elaboration of fermented beverages.

3.2. Isolation of strains

Although the sources of yeasts can be quite diverse, isolates are preferably obtained from the initial must and along the fermentation process, intending that this system itself directs to select the best adapted strains. Two possibilities arise at this stage: (1) to isolate random species using a general solid media (PDA or YEPDA) or with an agent leading to add visual differentiation in colony morphologies (WL nutrient agar) or (2) to focus on a specific group using selective media (lysine media for nS or nutrient medium supplemented with sodium metabisulfite and ethanol for *Saccharomyces*) [3, 81]. If one of the purposes is to complement the community study, the first would be the best choice, but if the selection is directed to certain species, the second would be best. After isolating several strains, characterization and selection will take place.

3.3. Characterization (qualitative approach)

The desirable aspects of fermentative yeast are quite variable; preference is given to one or the other, depending on each case [75]. Some of the most important and various studies in which they were applied are summarized in **Table 2**. Among this characteristics, it can be considered essential to evaluate: resistance to high concentrations of sugar and ethanol, high fermentation performance (this may not apply when selecting nS yeasts), low production of sulfurous compounds, and volatile acidity and implantation aptitude [79–82]. This last trait is not always performed, even though it is almost imperative, as it requires the design and implementations of molecular methods, such as pulsed field gel electrophoresis (PFGE), mini- or microsatellite markers, qPCR, or others [65, 66]. Once the survival capacity of the strains is determined, the next interesting feature is the impact they exert on sensory aspects. Olfactory qualities of yeast are usually assessed by determining their enzymatic activity, including glucosidases, lyases, proteases, or reductases, which can release or produce active odorant compounds [75]. Also, visual aspect can be affected by yeasts through their enzymatic activity (hydroxycinnamate decarboxylase) or their ability to excrete pyruvate and acetaldehyde, which can lead to the formation of highly stable colorant compounds as pyranoantocyanins that help to improve the visual quality of red wines [77, 83]. Besides, it is quite useful to analyze the nitrogen requirements of yeasts, if they possess the killer phenotype and flocculation ability (particularly important in beer and sparkling wines). Furthermore, some features can only be qualitatively studied such as the presence or absence of certain enzymes, as well as which killer phenotype they possess. Some other are by definition quantitative, as fermentation performance, and several characteristics can be determined both ways, qualitatively and quantitatively, which is the case of tolerance feature.

The methods to evaluate resistant traits of yeast strains are usually based in the qualitative measure of the growth of the microorganism in a synthetic medium containing the limiting compound after a specific time of incubation [82, 84–86]. In the case of ethanol tolerance, there are a broad number of tests that could define this characteristic [87]: (a) the ability to grow in the presence of ethanol, (b) the degree of survival after exposure to a certain concentration, or (c) the maximal ethanol production capacity. Some of these methods are

Characteristic of interest	References
Tolerance to inhibitors (ethanol, SO_2, sugar, pH)	Nikolau et al. [59]; Fiore et al. [60]; Arrizon et al. [61]; Capece et al. [62]; Tristezza et al. [63]
Implantation aptitude	Lopes et al. [64]; Capece et al. [62]; Perrone et al. [65]; Alonso del real [66]
Fermentation vigor (efficiency, ethanol yield, speed)	Tristezza et al. [63]; Ribeiro et al. [67]
Low nitrogen requirements	Arrizon et al. [61]; Julien et al. [68]; Gardner et al. [69]
Enzymatic activity	Fiore et al. [60]; Capece et al. [62]; Capece et al. [70], Csoma et al. [71]; Romano et al. [72]
Color and aroma enhancement	Belda et al. [75]; Steensels et al. [76]; Morata et al. [77]; Morata et al. [78].
Production of specific metabolites (sulfurous, glycerol, fatty acids, other alcohols)	Nikolau et al. [59]; Capece et al. [62, 70]; Tristezza et al. [63]
Killer phenotype	Zagorc et al. [73].
Flocculation	Silva et al. [74]

Table 2. Main aspects considered to select fermentative yeasts.

qualitatively measured, and some other imply the set up and monitor of a fermentation, requiring large amounts of resources and hindering the possibility to evaluate a high number of strains. However, fermentation parameters such as ethanol yield, productivity, and maximum specific velocity of cell growth must be measured during the traditional process [22].

Additionally, the fermentative traits are often individually evaluated, while in fermentation several stress factors intervene together and increase in number and magnitude along the process. Testing more than one inhibitor in a qualitative evaluation complicates the selection based only in the presence or absence of growth; thus, a more objective method should be used to compare between strains.

4. Quantitative methods used for the selection of fermentative yeast

An alternative to assess the tolerance of yeast strains to limiting factors is to study the growth kinetics of the strains exposed to the inhibiting condition. This can be achieved by traditional methods (plate counts), implying more resources but being more reliable, or by means of optical density (OD) measurements of the yeast in the media to test. Microbial growth data obtained by absorbance measurements permits to obtain kinetic parameters, which can be transformed into quantitative and comparable variables such as (a) detection time, the time to reach the detection level of Bioscreen (Automated OD reader equipment) and its period includes the lag time; (b) maximum growth rate could be estimated with the slope of the tangent of exponential phase; and (c) maximum population density, asymptotic level of OD at the end of exponential phase (**Figure 1**).

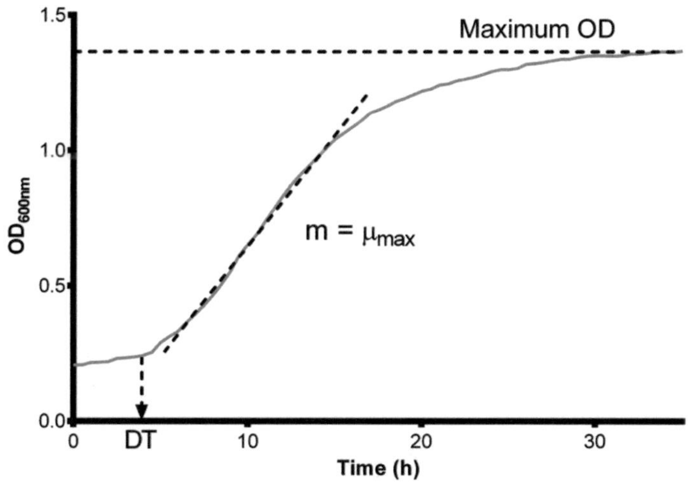

Figure 1. Kinetic parameters for microbiological growth using OD measurements. Detection time (DT), maximum growth rate (μmax), and maximum population density (maximum OD).

Measurement of microbial growth by using turbidity and cells' kinetic behavior description offer some benefits. A considerable amount of data could be generated faster and in real time. Nonetheless, this approach possesses some limitations such as the need to conduct the study in a high-transmittance liquid. Another factor to be considered is that cell concentration required to detect a shift in the measurement is at least 6 log CFU/mL.

4.1. Applications in food matrices

Analytical methods based on OD by assessing the final cell density have been used to evaluate the growth of foodborne pathogens in culture media, translucent liquid foods, and diluted food extracts [88–91]. Data from these studies have offered the possibility to obtain information to determine which food matrices and conditions could be inhibitory of pathogens growth. Also, this approach allows the comparison of strain variability of foodborne pathogens under several growth conditions, e.g., pH and salt concentration [92].

This method has allowed to rapidly evaluate the efficacy of antifungal compounds on the germination of fungal spores. The non-inhibitory concentration (NIC) and the minimum inhibitory concentration (MIC) for different environmental conditions can be calculated mathematically using the OD data [93]. MIC is considered the lowest concentration at which no growth is observed, while NIC is the lowest concentration at which any inhibitory effect is observed.

4.2. Potential applications for the selection of fermentative yeast

Quantitative measurement of the growth of microorganisms based in OD could be used to assess strain tolerance in fermentation-associated conditions. Some strategies and parameters have been proposed in base to OD data.

The MIC and NIC values are related to the susceptibility/tolerance of the microorganisms to stressful conditions. These parameters were proposed by Lambert and Pearson [94] to evaluate microorganisms not associated to fermentation processes, such as *Escherichia coli*, *Pseudomonas aeruginosa*, and *Staphylococcus aureus*. Arroyo-López et al. [18] used MIC and NIC to compare the growth of 29 strains of yeasts (*S. cerevisiae*, *Saccharomyces paradoxus*, *Saccharomyces bayanus*, *Saccharomyces kudriavzevii*, *Saccharomyces mikatae*, *Saccharomyces arboriculus*, *Saccharomyces cariocanus*, and some nS) in a medium containing ethanol. Álvarez-Pérez et al. [95] used the MIC among the criteria to select *S. cerevisiae* strains to produce *Picudo Rosé* wines with different aromatic profiles in Spain.

The area under the OD-time curve could be estimated, and it represents the population generated along the incubation period at a specific condition. The inhibitory effect could be estimated as a relative fraction of the area obtained under an optimal condition (control). **Figure 2** shows an example of this parameter, in which the growth of *S. cerevisiae* in synthetic medium is contrasted with the yeast behavior at the presence of ethanol and SO_2 [96].

When OD measurements are obtained with an automatized spectrophotometer, the detection time (DT) is reported for each tested condition. Low DT values are associated with the ability of microorganisms to adapt to a specific condition, and consequently, they could start their growth in short time.

Ortiz-Barrera et al. [97] reported DT for 90 nS strains grown in artificial medium containing 6% ethanol or 30 mg of total SO_2 (**Figure 3**). This parameter was used to select strains with the best ability to grow under stressful conditions (ethanol and SO_2) associated to wine (enclosed in the square in the figure).

In log growth phase, a close relationship between OD and cell concentration is apparent due to the fact that most of the cells are in an active multiplication process which in turn is used to estimate growth rate. In **Figure 4**, growth rate values of 12 *Saccharomyces* strains incubated in grape juice media with ethanol and SO_2 are shown (unpublished data).

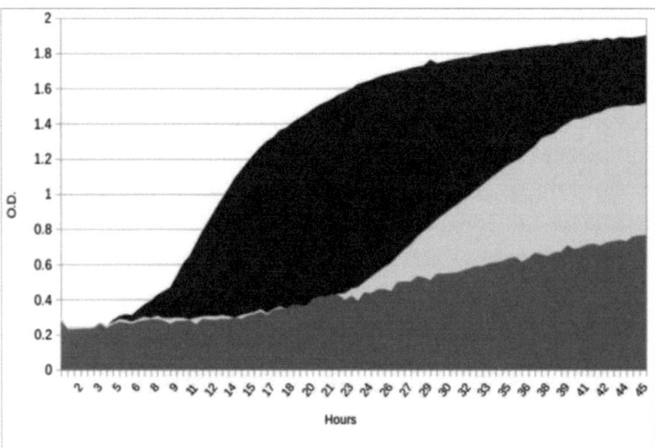

Inhibitor	Area (OD.h)	Fractional area
Ethanol	37.5	37.5/60.48=**0.62**
SO_2	23.5	23.5/60.48=**0.39**
Control	60.48	60.48/60.48=**1.00**

Figure 2. OD curves over time of *S. cerevisiae* growth in synthetic medium added with ethanol (12%), SO_2 (200 mg/L) and without inhibitors (control).

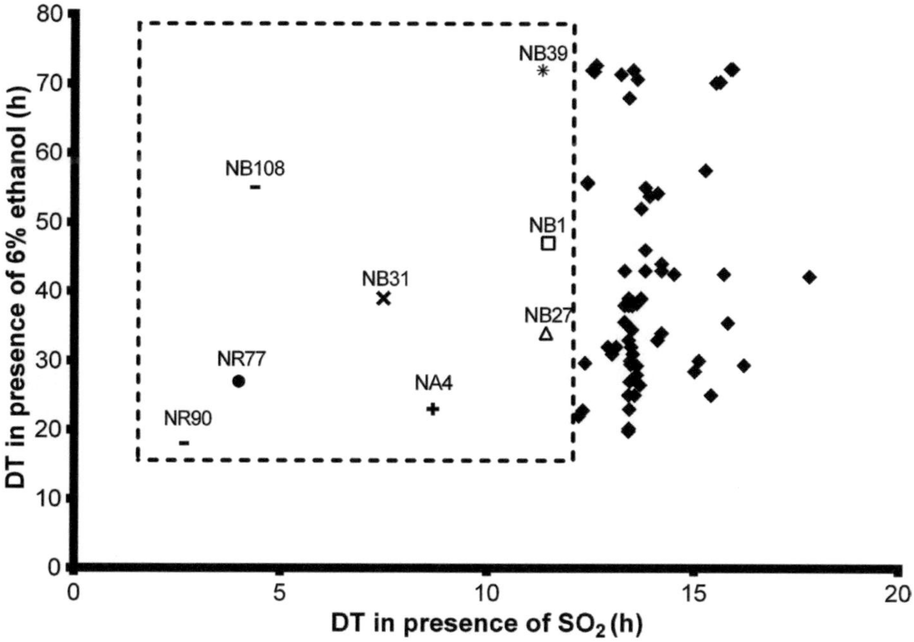

Figure 3. Detection times of 90 non-*Saccharomyces* strains in YPD (pH 3.5, 20°Bx) containing ethanol (6%) or SO_2 (30 mg·L^{-1}). Selected strains are shown in the inset.

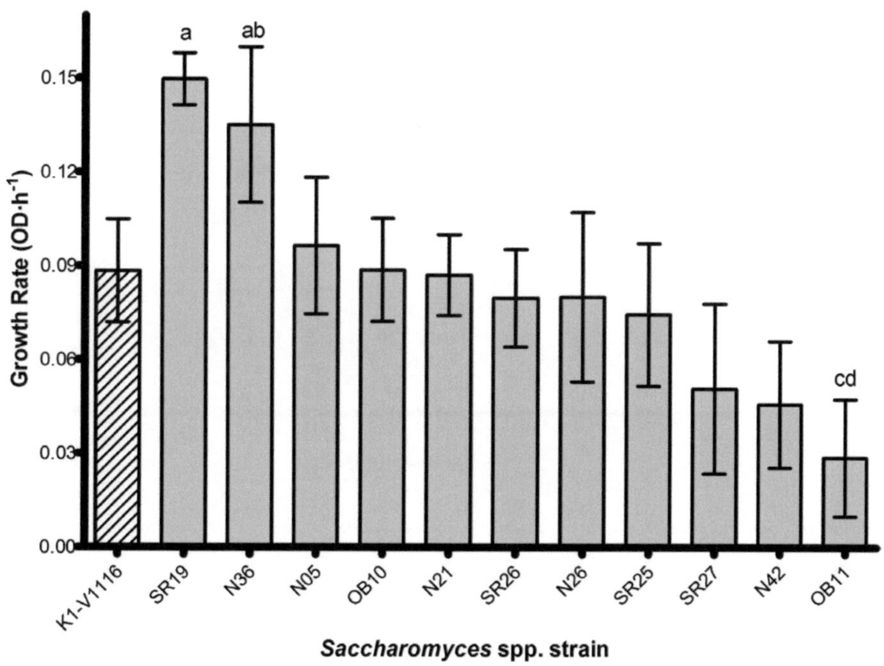

Figure 4. Growth rate of a commercial strain (stripped bar) and 12 *Saccharomyces* strains (solid bars) in grape juice media with 14% of ethanol, 50 mg/L of SO_2, and pH 3.5. Different letters denote statistical difference. Bars without letters correspond to ABCD.

Growth rate and DT are complementary parameters to describe the ability of yeast to adapt and grow in adverse conditions, similar to those found in fermentative beverages. Thus, both parameters (growth rate and DT) could be simultaneously used to select fermentative yeast strains (**Figure 5**).

Based on OD data, Salvadó et al. [98] estimated kinetic parameters of yeast strains (*S. cerevisiae* and nS) grown in a medium with ethanol (0–194.45 g/l) or incubated at different temperatures (4–46°C) by using Gompertz equation modified by Zwietering et al. [99]

$$y = D \times \exp\{-\exp[((\mu_{max})/D) \times (\lambda - t))+1]\} \quad (1)$$

where $y = \ln(OD_t/OD_0)$, $D = \ln(OD_{max}/OD_0)$ is the asymptotic maximum value reached, OD_0 is the initial OD, OD_t is the OD at time t, μmax is the maximum specific growth rate, and λ is the lag phase duration (h).

Based in μmax these authors generated a secondary model as a function of ethanol concentration and temperature of fermentation. Model prediction pointed to the temperature fermentation as the principal factor to promote the dominance of *S. cerevisiae* over other yeast genera during fermentation process. Only a few authors have applied this modeling strategy in yeast comparison [66, 100].

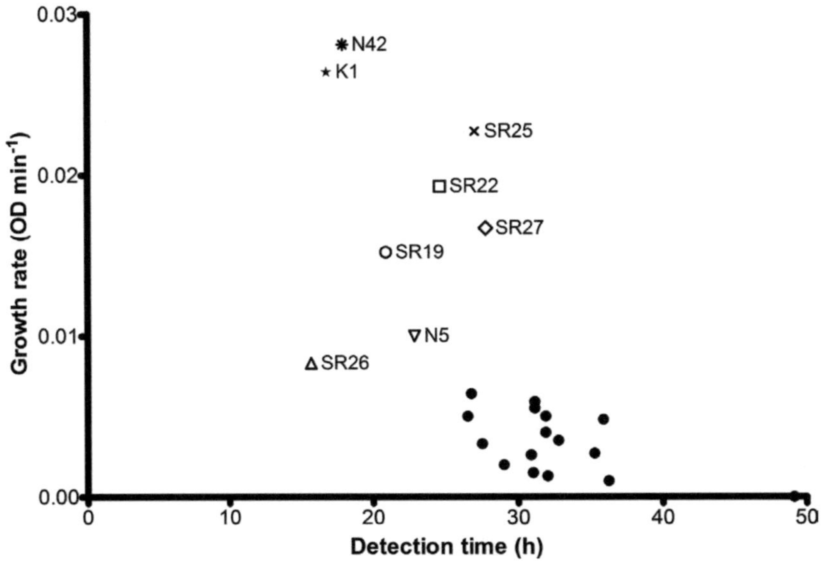

Figure 5. Detection times and growth rate of 29 strains of *Saccharomyces* grown in YPD medium with 12% of ethanol.

5. Conclusions

The vast variety of fermented beverages around the world and with them a great diversity of yeast present in each has encouraged the search for exceptional native yeast strains to be selected in order to improve the quality of the products and the fermentation process itself. Molecular techniques have become a good strategy to elucidate yeast diversity and composition during any fermentation process. However, methods to evaluate the tolerance of yeast strains, developed during the last decades, are usually based in the qualitative measure of the growth of the microorganism in a medium containing the limiting compound after a specific time of incubation. Studying strains growth in limiting media through optical density measurements permits to estimate quantitative and comparable parameters fast and inexpensively, providing an insight on the fitness of each strain to certain environment, lag phase duration, growth rate, and maximum population and then performing a rapid initial selection of the strains.

Author details

Dalia E. Miranda Castilleja, Jesús A. Aldrete Tapia, Sofía M. Arvizu Medrano, Montserrat Hernández Iturriaga, Lourdes Soto Muñoz and Ramón Á. Martínez Peniche*

*Address all correspondence to: alvar@uaq.mx

Microbial Food Safety Group, Food Research Department, Chemistry Faculty, Universidad Autónoma de Querétaro, Querétaro, Mexico

References

[1] Callejon RM, Clavijo A, Ortigueira P, Troncoso AM, Paneque P, Morales ML. Volatile and sensory profile of organic red wines produced by different selected autochthonous and commercial *Saccharomyces cerevisiae* strains. Analytica Chimica Acta. 2010;**660**(1-2):68-75. DOI: 10.1016/j.aca.2009.09.040

[2] Fleet GH. Yeast interactions and wine flavour. International Journal of Food Microbiology. 2003;**86**:11-22. DOI: 10.1016/S0168-1605(03)00245-9

[3] Rainieri S, Pretorius I. Selection and improvement of wine yeast. Annals of Microbiology. 2000;**50**:15-31. DOI: 10.1007/s13213-010-0098-0

[4] Jolly NP, Varela C, Pretorius IS. Not your ordinary yeast: Non-Saccharomyces yeasts in wine production uncovered. FEMS Yeast Research. 2014;**14**(2):215-237. DOI: 10.1111/1567-1364.12111

[5] Ciani M, Comitini F, Mannazzu I, Domizio P. Controlled mixed culture fermentation: A new perspective on the use of non-Saccharomyces yeasts in winemaking. FEMS Yeast Research. 2009;**10**(2):123-133. DOI: 10.1111/j.1567-1364.2009.00579.x-

[6] Capozzi V, Garofalo C, Chiriatti MA, Grieco F, Spano G. Microbial terroir and food innovation: The case of yeast biodiversity in wine. Microbiological Research. 2015;**181**:75-83. DOI:10.1016/j.micres.2015.10.005

[7] García M, Esteve-Zarzoso B, Arroyo T. Non-saccharomyces yeasts: Biotechnological role for wine production, grape and wine biotechnology. Morata A, editor. InTech, Rijeka, Croatia; 2016. DOI: 10.5772/64957

[8] Lachance M. Yeast communities in a natural tequila fermentation. Antonie Van Leeuwenhoek. 1995;**68**:151-160. DOI: 10.1007/BF00873100

[9] Escalante-Minakata P, Blaschek H, Barba de la Rosa A, Santos L, De Leon-Rodríguez A. Identification of yeast and bacteria involved in the mezcal fermentation of Agave Salmiana. Letters in Applied Microbiology. 2008;**46**:626-630. DOI: 10.1111/j.1472-765X.2008.02359.x

[10] Portugal C, Paron de Silva A, Bortoletto AM, Alcarde AR. How native yeasts may influence the chemical profile of the Brazilian spirit cachaça. Food Research International. 2016;**91**:18-25. DOI: 10.1016/j.foodres.2016.11.022

[11] Zuzuarregui A, del Olmo M. Analyses of stress resistance under laboratory conditions constitute a suitable criterion for wine yeast selection. Antonie Van Leeuwenhoek. 2004;**85**:271-280. DOI: 10.1023/B:ANTO.0000020162.21248.53

[12] Bauer FF, Pretorius IS. Yeast stress response and fermentation efficiency: How to survive the making of wine—A review. South African Journal of Enology and Viticulture. 2000;**21**:27-51

[13] Pretorius IS. Tailoring wine yeast for the new millennium: Novel approaches to the ancient art of winemaking. Yeast. 2000;**16**:675-729. DOI: 10.1002/1097-0061(20000615)16:8<675::AID-YEA585>3.0.CO;2-B

[14] Sanz ML, Martínez-Castro I. Carbohydrates. In: Moreno-Arribas V, Polo C, editors. Wine Chemistry and Biochemistry. New York: Springer Science +Business Media; 2009. pp. 231-248. DOI: 10.1007/978-0-387-74118-5

[15] Hohmann S. Osmotic stress signaling and osmoadaption in yeasts. Microbiology and Molecular Biology Reviews. 2002;**66**(2):300-372. DOI: 10.1128/MMBR.66.2.300-372.2002

[16] Slaninova I, Šesták S, Svoboda A, Farkaš V. Cell wall and cytoskeleton reorganization as the response to hyperosmotic shock in *Saccharomyces cerevisiae*. Archives of Microbiology. 2000;**173**:245-252. DOI: 10.1007/s002030000136

[17] Rodrigues F, Ludovico P, Leao C. Sugar Metabolism in yeast: An overview of aerobic and anaerobic glucose catabolism. In: Péter G, Rosa C, editors. The Yeast Handbook: Biodiversity and Ecophysiology of Yeasts. Heidelberg, Berlin: Springer Berlin; 2006. pp. 101-121

[18] Arroyo-López FN, Salvadó Z, Tronchoni J, Guillamón JM, Barrio E, Querol A. Susceptibility and resistance to ethanol in Saccharomyces strains isolated from wild and fermentative environments. Yeast. 2010;**27**:1005-1015. DOI: 10.1002/yea.1809

[19] Wang D, Xu Y, Hu J, Zhao G. Fermentation kinetics of different sugars by Apple yeast *Saccharomyces cerevisiae*. Journal of the Institute of Brewing. 2004;**110**(4):340-346. DOI: 10.1002/j.2050-0416.2004.tb00630.x

[20] Waleckx E, Gschaedler A, Colonna-Ceccaldi B, Monsan P. Hydrolysis of fructans from Agave tequilana Weber var. azul during the cooking step in a traditional tequila elaboration process. Food Chemistry. 2008;**108**:40-48. DOI: 10.1016/j.foodchem.2007.10.028

[21] Lappe-Oliveras P, Moreno-Terrazas R, Arrizón-Gaviño J, Herrera-Suaréz T, García-Mendoza A, Gschaedler-Mathis A. Yeasts associated with the production of Mexican alcoholic nondistilled and distilled Agave beverages. FEMS Yeast Research. 2008;**8**(7): 1037-1052. DOI: 10.1111/j.1567-1364.2008.00430.x

[22] Souza E, Rosa C, Morgano M, Serra G. Fermentation characteristics as criteria for selection of cachaça. World Journal of Microbiology and Biotechnology. 2004;**20**:19-24. DOI: 10.5713/ajas.14.0508

[23] Arrizon J, Gschaedler A. Increasing fermentation efficiency at high sugar concentrations by supplementing an additional source of nitrogen during the exponential phase of the tequila fermentation process. Canadian Journal of Microbiology. 2002;**48**:965-970. DOI: 10.1139/w02-093

[24] Espinosa E, de Billerbec G, Ardaillon D, Schuler A, François J, de Morais M. Influence of nitrogen supply on the production of higher alcohols/esters and expression of flavour-related genes in cachaça fermentation. Food Chemistry. 2013;**138**:701-708. DOI: 10.1016/j.foodchem.2012.10.147

[25] Lekkas C, Stewart G, Hill A, Taidi B, Hodgson J. Elucidation of the role of nitrogenous wort components in yeast fermentation. Journal of the Institute of Brewing. 2007;**113**(1): 3-8. DOI: 10.1002/j.2050-0416.2007.tb00249.x

[26] Bell S, Henschke P. Implications of nitrogen nutrition for grapes, fermentation and wine. Australian Journal of Grape and Wine Research. 2005;**11**:242-295. DOI: 10-1111/j.1755-0238.2005.tb00028.x

[27] Rankine B, editor. Making Good Wine. Melbourne, Australia: Sun Books ed; 1989

[28] Wells A, Osborne J. Production of SO_2 binding compounds and SO_2 by *Saccharomyces* during alcoholic fermentation and the impact on malolactic fermentation. South African Journal of Enology and Viticulture. 2011;**32**(2):267-279. DOI: 10.21548/32-2-1387

[29] Ciani M, Capece A, Comitini F, Canonico L, Siesto G, Romano P. Yeast interactions in inoculated wine fermentation. Frontiers in Microbiology. 2016;**7**:555. DOI: 10.3389/fmicb.2016.00555

[30] Arneborg N, Siegumfeldt H, Andersen GH, Nissen P, Daria VR, Rodrigo PG, et al. Interactive optical trapping shows that confinement is a determinant of growth in a mixed yeast culture. FEMS Microbiology Letters. 2005;**245**:155-159. DOI: 10.1016/j.femsle.2005.03.008

[31] Starmer WT, Ganter PF, Aberdeen V, Lachance MA, Phaff HJ. The ecological role of killer yeast in natural communities of yeasts. Canadian Journal of Microbiology. 1987;**33**:783-796. DOI: 10.1139/m87-134

[32] Kagan BL. Mode of action of yeast killer toxins: Channel formation in lipid bilayers. Nature. 1983;**302**:709-711. DOI: 10.1038/302709a0

[33] Schmitt MJ, Breinig F. The viral killer system in yeast: From molecular biology to application. FEMS Microbiology Reviews. 2002;**26**:257-276. DOI: 10.1111/j.1574-6976.2002.tb00614.x

[34] Lam FH, Ghaderi A, Fink GR, Stephanopoulos G. Engineering alcohol tolerance in yeast. Science. 2014;**346**(6205):71-75. DOI: 10.1126/science.1257859

[35] Alexandre H, Rousseaux I, Charpentier C. Relationship between ethanol tolerance, lipid composition and plasma membrane fluidity in *Saccharomyces cerevisiae* and *Kloeckera apiculata*. FEMS Microbiology Letters. 1994;**124**:17-22. DOI: 10.1111/j.1574-6968.1994.tb07255.x

[36] Meaden PG, Arneborg N, Guldfeldt LU, Siegumfeldt H, Jakobsen M. Endocytosis and vacuolar morphology in *Saccharomyces cerevisiae* are altered in response to ethanol stress or heat shock. Yeast. 1999;**15**:1211-1222. DOI: 10.1002/(SICI)1097-0061(19990915)15:12<1211::AID-YEA448>3.0.CO;2-H

[37] Salmon JM, Vincent O, Mauricio JC, Bely M, Barre P. Sugar transport inhibition and apparent loss of activity in *Saccharomyces cerevisiae* as a major limiting factor of enological fermentation. American Journal of Enology and Viticulture. 1993;**44**:56-64. DOI: 10.1139/w02-093

[38] Ibeas JL, Jiménez J. Mitochondrial DNA loss caused by ethanol in Saccharomyces flor yeast. Applied and Environmental Microbiology. 1997;**63**:7-12

[39] Hu XH, Wang MH, Tan T, Li JR, Yang H, Leach L, et al. Genetic dissection of ethanol tolerance in the budding yeast *Saccharomyces cerevisiae*. Genetics. 2007;**175**:1479-1487. DOI: 10.1534/genetics.106.065292

[40] Stanley D, Bandara A, Fraser S, Chambers PJ, Stanley GA. The ethanol stress response and ethanol tolerance of *Saccharomyces cerevisiae*. Journal of Applied Microbiology. 2010;**109**(1):13-24. DOI: 10.1111/j.1365-2672.2009.04657.x

[41] Alexandre H, Ansanay-Galeote V, Dequin S, Blondin B. Global gene expression during short-term ethanol stress *in Saccharomyces cerevisiae*. FEBS Letters. 2001;**498**:98-103. DOI: 10.1016/S0014-5793(01)02503-0

[42] Pampulha ME, Loureiro-Dias MC. Combined effect of acetic acid, pH and ethanol on intracellular pH of fermenting yeast. Applied Microbiology and Biotechnology. 1989;**31**:547-550. DOI: 10.1007/BF00270792

[43] Viegas CA, Rosa MF, Sa-Correia I, Novais JM. Inhibition of yeast growth by octanoic and decanoic acids produced during ethanolic fermentation. Applied and Environmental Microbiology. 1989;**55**(1):21-28

[44] Lafon-Lafourcade S, Geneix C, Ribéreau-Gayon P. Inhibition of alcoholic fermentation of grape must by fatty acids produced by yeasts and their elimination by yeast ghost. Applied and Environmental Microbiology. 1984;**47**(6):1246-1249

[45] Tamang JP, Watanabe K, Holzapfel WH. Review: Diversity of microorganisms in global fermented foods and beverages. Frontiers in Microbiology. 2016;**7**:377. DOI: 10.3389/fmicb.2016.00377

[46] Cocolin L, Ercolini D, editors. Molecular Techniques in the Microbial Ecology of Fermented Foods. New York: Springer; 2008. DOI: 10.1007/978-0-387-74520-6

[47] Ercolini D. High-throughput sequencing and metagenomics: Moving forward in the culture-independent analysis of food microbial ecology. Applied and Environmental Microbiology. 2013;**79**(10):3148-3155. DOI: 10.1128/AEM.00256-13

[48] Cocolin L, Alessandria V, Dolci P, Gorra R, Rantsio K. Culture independent methods to assess the diversity and dynamics of microbiota during food fermentation. International Journal of Food Microbiology. 2013;**167**:29-43. DOI: 10.1016/j.ijfoodmicro.2013.05.008

[49] De Filippis F, Parente E, Ercolini D. Metagenomics insights into food fermentations. Microbial Biotechnology. 2017;**10**(1):91-102. DOI: 10.1111/1751-7915.12421

[50] Miguel G, Collela C, De Almeida E, Dias D, Schwan R. Physicochemical and microbiological description of Caxiri- a cassava and corn alcoholic beverage. International Journal of Food Science and Technology. 2015;**50**:2537-2544. DOI: 10.1111/ijfs.12921

[51] Portillo M, Mas A. Analysis of microbial diversity and dynamics during wine fermentation of Grenache grape variety by high-throughput barcoding sequencing. LWT Food Science and Technology. 2015;**72**:317-321. DOI: 10.1016/j.lwt.2016.05.009

[52] Ramos C, De Sousa E, Ribeiro J, Almeida T, Santos C, Abegg M, Schwan R. Microbiological and chemical characteristics of tarubá, an indigenous beverage produced from solid cassava fermentation. Food Microbiology. 2015;**49**:182-188. DOI: 10.1016/j.fm.2015.02.005.

[53] Wang C, Mas A, Esteve-Zarzoso B. Interactions between *Hanseniaspora uvarum* and *Saccharomyces cerevisiae* during alcoholic fermentation. International Journal of Food Microbiology. 2015;**206**:67-74. DOI: 10.1016/j.ijfoodmicro.2015.04.022

[54] Xu-Cong L, Zhi-Chao C, Rui-Bo J, Zhi-Bin L, Wen Z, Shao-Jun C, et al. Microbial community structure and dynamics during the traditional brewing of Fuzhou Hong Qu glutinous rice wine as determined by culture-dependent and culture-independent techniques. Food Control. 2015;**57**:216-224. DOI: 10.1016/j.foodcont.2015.03.054

[55] Maturano YP, Mestre VM, Combina M, Toro ME, Vázquez F, Esteve-Zarzoso B. Culture-dependent and independent techniques to monitor yeast species during cold soak carried out at different temperatures in winemaking. International Journal of Food Microbiology. 2016;**237**:142-149. DOI: 10.1016/j.ijfoodmicro.2016.08.013

[56] Sankar S, Keot J, Das S, Sarma K, Barooah M. Metagenomics analysis of microbial communities associated with a traditional rice wine starter culture (Xaj-pitha) of Assam, India. Biotechnology. 2016;**6**:153. DOI: 10.1007/s13205-016-0471-1

[57] Santiago-Urbina J, Peña-Montes C, Nolasco-Cancino H, Ruiz-Terán F. PCR-DGGE analysis of the yeast population associated with natural fermentation of taberna. Journal of Microbiology, Biotechnology and Food Sciences. 2016;**6**:758-763. DOI: 10.15414/jmbfs.2016.6.2.758-763

[58] Liu M, Tang Y, Zhao K, Liu Y, Guo X, Ren D, et al. Determination of the fungal community of pit mud in fermentation cellars for Chinese strong-flavor liquor, using DGGE and Illumina MiSeq sequencing. Food Research International. 2017;**91**:80-87. DOI: 10.1016/j.foodres.2016.11.037

[59] Nikolaou E, Soufleros EH, Bouloumpasi E, Tzanetakis N. Selection of indigenous *Saccharomyces cerevisiae* strains according to their oenological characteristics and vinification results. Food Microbiology. 2006;**23**(2):205-211. DOI: 10.1016/j.fm.2005.03.004

[60] Fiore C, Arrizon J, Gschaedler A, Flores J, Romano P. Comparison between yeast from grape and agave must for trades of technological interest. World Journal of Microbiology and Biotechnology. 2005;**21**(3):1141-1147

[61] Arrizon J, Fiore C, Acosta G, Romano P, Gschaedler A. Fermentation behavior and volatile compound production by agave and grape must yeasts in high sugar Agave tequilana and grape must fermentations. Antonie Van Leeuwenhoek. 2006;**89**:181-189. DOI: 10.1007/s10482-005-9022-1

[62] Capece A, Romaniello R, Siesto G, Pietrafesa R, Massari C, Poeta C. et al. Selection of indigenous *Saccharomyces cerevisiae* strains for Nero d'Avola wine and evaluation of selected starter implantation in pilot fermentation. International Journal of Food Microbiology. 2010;**144**:187-192. DOI: 10.1016/j.ijfoodmicro.2010.09.009

[63] Tristezza M, Fantastico L, Vetrano C, Bleve G, Corallo D, Grieco F, et al. Molecular and technological characterization of *Saccharomyces cerevisiae* strains isolated from natural fermentation of Susumaniello grape must in Apulia, Southern Italy. International Journal of Microbiology. 2014;**2014**:897427. DOI: 10.1155/2014/897428

[64] Lopes CA, Rodríguez ME, Sangorrín M, Querol A, Caballero A. Patagonian wines: Implantation of an indigenous strain of *Saccharomyces cerevisiae* in fermentations conducted in traditional and modern cellars. Journal of Industrial Microbiology & Biotechnology. 2007;**34**(2):139. DOI: 10.1007/s10295-006-0178-0

[65] Perrone B, Giacosa S, Rolle L, Cocolin L, Rantsiou K. Investigation of the dominance behavior of *Saccharomyces cerevisiae* strains during wine fermentation. International Journal of Food Microbiology. 2013;**165**:156-162. DOI: 10.1016/j.ijfoodmicro.2013.04.023

[66] Alonso-del-Real J, Lairón-Peris M, Barrio E, Querol A. Effect of temperature on the prevalence of Saccharomyces non cerevisiae species against a *S. cerevisiae* wine strain in wine fermentation: Competition, physiological fitness, and influence in final wine composition. Frontiers in Microbiology. 2017;**8**:150. DOI: 10.3389/fmicb.2017.00150

[67] Ribeiro LS, Duarte WF, Dias DR, Schwan RF. Fermented sugarcane and pineapple beverage produced using *Saccharomyces cerevisiae* and non-Saccharomyces yeast. Journal of the Institute of Brewing. 2015;**121**:262-272. DOI: 10.1002/jib.218

[68] Julien A, Roustan JL, Dulau L, Sablayrolles JM. Comparison of nitrogen and oxygen demands of enological yeasts: Technological consequences. American Journal of Enology and Viticulture. 2000;**51**:215-222

[69] Gardner JM, Poole K, Jiranek V. Practical significance of relative assimilable nitrogen requirements of yeast: A preliminary study of fermentation performance and liberation of H2S. Australian Journal of Grape and Wine Research. 2002;**8**:175-179. DOI: 10.1111/j.1755-0238.2002.tb00253.x

[70] Capece A, Romaniello R, Pietrafesa R, Romano P. Indigenous *Saccharomyces cerevisiae* yeasts as a source of biodiversity for the selection of starters for specific fermentations. BIO Web of Conferences.. 2014;**3**:02003. DOI: 10.1051/bioconf/20140302003

[71] Csoma H, Zakany N, Capece A, Romano P, Sipiczki M. Biological diversity of Saccharomyces yeasts of spontaneously fermenting wines in four wine regions: Comparative genotypic and phenotypic analysis. International Journal of Food Microbiology. 2010;**140**:239-248. DOI: 10.1016/j.ijfoodmicro.2010.03.024

[72] Romano P, Fiore C, Paraggio M, Caruso M, Capece A. Function of yeast species and strains in wine flavour. International Journal of Food Microbiology. 2003;**86**:169-180. DOI: 10.1016/S0168-1605(03)00290-3

[73] Zagorc T, Maraz A, Cadez N, Jemec KP, Peter G. Indigenous wine killer yeasts and their application as a starter culture in wine fermentation. Food Microbiology. 2001;**18**:441-445. DOI: 10.1006/fmic.2001.0422

[74] Silva CLC, Rosa CA, Oliveira ES. Studies on the kinetic parameters for alcoholic fermentation by flocculent *Saccharomyces cerevisiae* strains and non-hydrogen sulfide-producing strains. World Journal of Microbiology and Biotechnology. 2006;**22**(8):857-863. DOI: 10.1007/s11274-005-9115-z

[75] Belda I, Ruiz J, Alastruey-Izquierdo A, Navascués E, Marquina D, Santos A. Unraveling the enzymatic basis of wine "Flavorome": A phylo-functional study of wine related yeast species. Frontiers in Microbiology. 2016;**7**:12. DOI: 10.3389/fmicb.2016.00012

[76] Steensels J, Meersman E, Snoek T, Saels V, Verstrepen J. Large-scale selection and breeding to generate industrial yeasts with superior aroma production. Applied and Environmental Microbiology. 2014;**80**(22):6965-6975. DOI: 10.1128/AEM.02235-14

[77] Morata A, González C, Suárez LA. Formation of vinylphenolic pyranoanthocyanins by selected yeasts fermenting red grape musts supplemented with hydroxycinnamic acids. International Journal of Food Microbiology. 2007;**116**(1):144-152. DOI: 10.1016/j.ijfoodmicro.2006.12.032

[78] Morata A, Gómez-Cordovés MC, Suberviola J, Bartolomé B, Colomo B, Suárez LA. Adsorption of Anthocyanins by yeast cell walls during the fermentation of red wines. Journal of Agricultural and Food Chemistry. 2003;**51**(14):4084-4088. DOI: 10.1021/jf021134u

[79] Borrull A, Poblet M, Rozès N. New insights into the capacity of commercial wine yeasts to grow on sparkling wine media. Factor screening for improving wine yeast selection. Food Microbiology. 2015;**48**:41-48. DOI: 10.1016/j.fm.2014.12.006

[80] Walker G, Van Dijck P. Physiological and molecular responses of yeast to the environment. In: Querol A, Fleet G, editors. The Yeast Handbook: Yeast in Food and Beverages. Berlin: Springer-Verlag; 2006. ISBN 3-540-28388-9

[81] Manzanares P, Vallés S, Viana F. Non-saccharomyces yeasts in the winemaking process. In: Carrascosa A, Muñoz R, González R, editors. Molecular Wine Microbiology. Academic Press. London, United Kingdom. 2011. ISBN: 978-0-12-375021-1

[82] Regodón J, Perez F, Valdes M, De Miguel C, Ramírez M. A simple and effective procedure for selection of wine yeast strains. Food Microbiology. 1997;**14**:247-254. DOI: 10.1006/fmic.1996.0091

[83] Morata A, Gómez-Cordovés MC, Colomo B, Suárez JA. Pyruvic acid and acetaldehyde production by different strains of *Saccharomyces cerevisiae*: Relationship with Vitisin A and B formation in red wines. Journal of Agricultural and Food Chemistry. 2003;**51**(25):7402-7409. DOI: 10.1021/jf0304167

[84] Mauriello G, Capece A, D'Auria M, Garde-Cerdán T, Romano P. SPME-GC method as a tool to differentiate VOC profiles in *Saccharomyces cerevisiae* wine yeasts. Food Microbiology. 2009;**26**(3):246-252. DOI: 10.1016/j.fm.2009.01.003

[85] Lee YJ, Choi YR, Lee SY, Park JT, Shim JH, Park KH, et al. Screening wild yeast strains for alcohol fermentation from various fruits. Mycobiology. 2011;**39**(1):33-39. DOI: 10.4489/MYCO.2011.39.1.033

[86] de Souza AP, Vicente MA, Klein RC, Fietto LG, Coutrim MX, de Cássia Franco Afonso RJ, et al. Strategies to select yeast starters cultures for production of flavor compounds in cachaça fermentations. Antonie Van Leeuwenhoek. 2012;**101**(2):379-392. DOI: 10.1007/s10482-011-9643-5

[87] Snoek T, Verstrepen K, Voordeckers K. How do yeast cells become tolerant to high ethanol concentrations? Current Genetics. 2016;**62**:475-480. DOI: 10.1007/s00294-015-0561-3

[88] Nilsson RE, Ross T, Bowman JP. Variability in biofilm production by *Listeria monocytogenes* correlated to strain origin and growth conditions. International Journal of Food Microbiology. 2011;**150**:14-24. DOI: 10.1016/j.ijfoodmicro.2011.07.012

[89] Ferreira V, Wiedmann M, Teixeira P, Stasiewicz MJ. *Listeria monocytogenes* persistence in food-associated environments: Epidemiology, strain characteristics, and implications for public health. Journal of Food Protection. 2014;**77**:150-170. DOI: 10.4315/0362-028X.JFP-13-150

[90] Posada-Izquierdo G, Del Rosal S, Valero A, Zurera G, Sant'Ana AS, Alvarenga V, Pérez-Rodríguez F. Assessing the growth of *Escherichia coli* O157:H7 and *Salmonella* in spinach, lettuce, parsley and chard extracts at different storage temperatures. Journal of Applied Microbiology. 2006;**120**(6):1701-1710. DOI: 10.1111/jam.13122

[91] Ortiz S, López V, Martínez-Suárez JV. The influence of subminimal inhibitory concentrations of benzalkonium chloride on biofilm formation by *Listeria monocytogenes*. International Journal of Food Microbiology. 2014;**189**:106-112

[92] Lianou A, Koutsoumanis K. Effect on the growth environment on the strain variability of *Salmonella enterica* kinetic behavior. Food Microbiology. 2011;**28**(4):828-837. DOI: 10.1016/j.fm.2010.04.006

[93] Medina A, Lambert RJ, Magan N. Rapid throughput analysis of filamentous fungal growth using turbidimetric measurements with the Bioscreen C: A tool for screening antifungal compounds. Fungal Biology. 2006;**116**(1):161-169. DOI: 10.1016/j.funbio.2011.11.001

[94] Lambert RJW, Pearson J. Susceptibility testing: Accurate and reproducible minimum inhibitory concentration (MIC) and non-inhibitory concentration (NIC) values. Journal of Applied Microbiology. 2000;**88**:784-790. DOI: 10.1046/j.1365-2672.2000.01017.x

[95] Álvarez-Pérez JM, Álvarez-Rodríguez M, Campo E, Sáenz de Miera LE, Ferreira V, Hernández-Orte P, et al. Selection of *Saccharomyces cerevisiae* strains applied to the production of Prieto Picudo Rosé wines with a different aromatic profile. South African Journal of Enology and Viticulture. 2014;**35**:242-256. DOI: 10.21548/35-2-1013

[96] Miranda-Castilleja DE, Ortiz-Barrera E, Arvizu-Medrano SM, Pacheco R, Aldrete-Tapia JA, Martínez-Peniche RA. Aislamiento, selección e identificación de levaduras *Saccharomyces* spp. Nativas de viñedos en Querétaro, México. Agrociencia. 2015;**49**: 759-773

[97] Ortiz-Barrera E, Miranda-Castilleja DE, Arvizu-Medrano SM, Pacheco-Aguilar JR, Aldrete-Tapia JA, Hernández-Iturriaga M, Martínez-Peniche RA. Enological potential of native non-*Saccharomyces* yeasts from vineyards established in Querétaro, Mexico. Revista Chapingo Serie Horticultura. 2015;**21**(2):169-183. DOI: 10.5154/r.rchsh.2015.01.001

[98] Salvadó Z, Arroyo-López FN, Barrio E, Querol A, Guillamón JM. Quantifying the individual effects of ethanol and temperature on the fitness advantage of *Saccharomyces cerevisiae*. Food Microbiology. 2011;**28**:1155-1161. DOI: 10.1016/j.fm.2011.03.008

[99] Zwietering MH, Jongenburger I, Rombouts FM, van't Riet K. Modeling of the bacterial growth curve. Applied and Environmental Microbiology. 1990;**56**(6):1875-1881

[100] García-Ríos E, Gutiérrez A, Salvadó Z, Arroyo-López FN, Guillamon JL. The fitness advantage of commercial wine yeasts in relation to the nitrogen concentration, temperature, and ethanol content under microvinification conditions. Applied and Environmental Microbiology. 2014;**80**: 704-713. DOI: 10.1128/AEM.03405-13

Chapter 4

Wine Spoilage Yeasts: Control Strategy

Carlos Escott, Iris Loira, Antonio Morata,
María Antonia Bañuelos and
José Antonio Suárez-Lepe

Additional information is available at the end of the chapter

http://dx.doi.org/10.5772/intechopen.69942

Abstract

Traditionally in winemaking, sulphur dioxide (SO_2) is chemically the most widely used for microflora control as antimicrobial preservative. Other tested compounds for selective yeast control are sorbic and benzoic acids. Herein, we discuss the effectiveness and the application of traditional and novel treatments and biotechnologies for chemical and biological control of wine spoilage yeasts. The versatility of the killer toxins and the antimicrobial properties of natural compounds such as carvacrol, essential oils and bioactive peptides will be considered. Some of the wine spoilage yeasts that are intended to control belong to the genera *Zygosaccharomyces, Saccharomycodes* and *Dekkera/Brettanomyces*, but also the non-*Saccharomyces* yeasts species dominating the first phase of fermentation (*Hanseniaspora uvarum, Hansenula anomala, Metschnikowia pulcherrima, Wickerhamomyces anomalus*) and some others, such as *Schizosaccharomyces pombe*, depending on the kind of wine to be produced.

Keywords: spoilage yeasts, essential oils, bioactive peptides, winemaking, wine, emerging biotechnologies, killer toxins, monitoring techniques

1. Introduction

Ever since the wild yeast colonies are controlled in grape musts, the fermentation has produced wine with differentiated organoleptic attributes. The control over yeast colonies during the entire winemaking process has given the winemaker the possibility of moulding these characteristics towards producing wine with better quality parameters.

Even though yeasts are responsible for transforming grape must into wine through fermentation, there are yeasts capable of spoiling it (**Figure 1**). Spoiling yeasts are, in most cases,

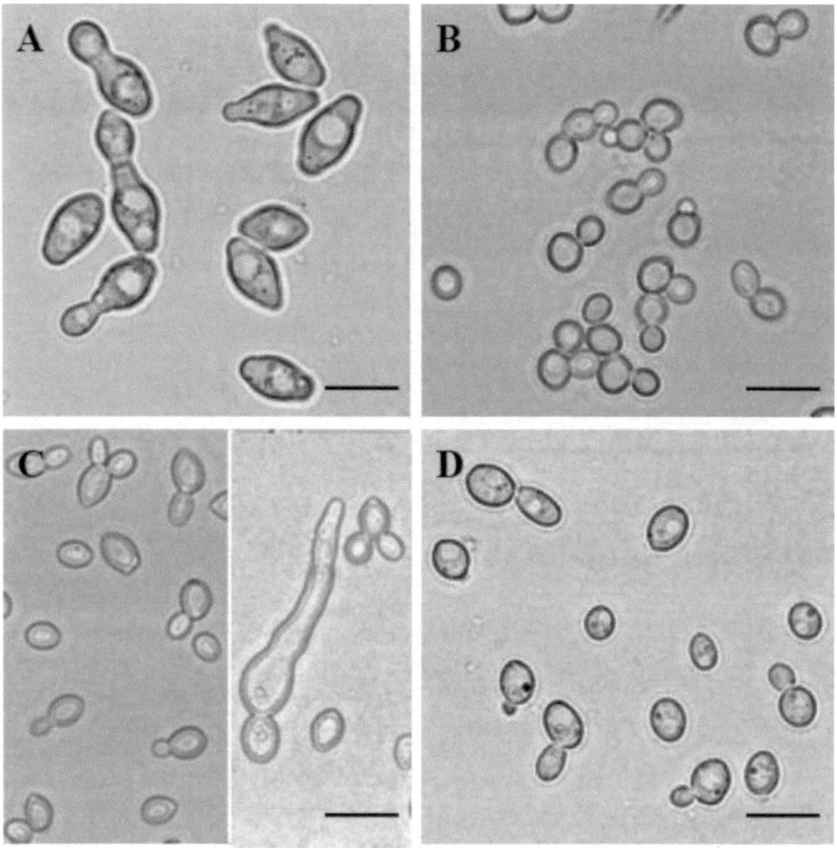

Figure 1. Optical microscopic images of some spoilage yeasts found in wine. (A) *S. ludwigii*; (B) *Z. rouxii*; (C) *Dekkera* spp.; (D) *P. anomala*. Bar scale: 10 μm.

resistant to harsh conditions such as high ethanol concentration, relatively low pH and lethal concentration of sulphites (SO_2) or dimethyl dicarbonate (DMDC) used as antimicrobial agents.

Some yeast genera can be considered as spoilage microorganisms due to their undesirable implantation in food in which they can cause nutritional and sensory quality degradation and consequently lead to major economic losses. Even their implication in relation to public health is currently under suspicion [1]. **Table 1** summarizes some yeast genera known as wine spoilage that also spoils certain food products, particularly specifying the main compounds affecting quality and the effect produced together with potential health hazard.

The uncontrolled use and misuse of antibiotics has caused increasing resistance in a broad group of pathogenic microorganisms, including food-borne pathogens, which, in addition to resisting the effect of antibiotics, are able to survive the processes of food preservation [16].

In this chapter, different technologies and treatments for the control of spoilage yeasts have been revised. These techniques were split into early and emerging technologies in accordance

Yeast species	Food product	Spoilage compounds	Effect observed	Health hazard
Zygosaccharomyces rouxii	Products with 50% sugar [2, 3]	Alcohol, esters [3]	Gas production: bubbling and package expansion [3]	
	Sweet wines [4]		Refermentation and CO_2 production [4]	
	Mould-ripened soft cheeses [5]			
	Fruit juices, sauces, carbonated soft drinks, salad dressings, ketchup [3, 6]	Alcohol, esters [3]	Gas production: bubbling and package expansion [3]	
Brettanomyces bruxellensis	Bulk, barrel matured and bottled wines [7, 8]	4-ethylphenol, 4-ethylguaiacol, acetic acid [7] Tetrahydropyridines [8]	Off aromas, cloudiness formation in sparkling wine, mousy aroma [7] Unpleasant mousy and medicinal taints [8]	
Pichia anomala	Dairy and baking products, beer, high salt environments and silage [9]			No restriction on handling and no risk to healthy persons [9]
	Lactic acid-rich products [10]			
	Wine, winemaking [7, 8]	Ethyl acetate [7] Acetaldehyde, esters, acetic acid [8]	Oxidation of ethanol [8]	
Pichia guilliermondii	Grape juice [7]	4-ethylphenol [7]	Off-odours barnyard-like or horsey [7]	
	Soft drinks [6]			
Pichia membranaefaciens	Mould-ripened soft cheeses [5]	Acetaldehyde [11]	Chalky film layer [11]	
Saccharomycodes ludwigii	Food products with SO_2 as antiseptic [2]			
	Bottled wines [7]		Spoilage by sediment or cloudiness formation [7]	
	Wine [12]	High acetoin level [12]	Flocculent sediment [11]	
Candida tropicalis	Fresh fruits: orange (*Citrus sinensis*) and pineapple (*Ananas comosus*) [13]			Candidiasis has not been transmitted by food products [5]
	Grated raw carrots [14]	CO_2 [14]	Increase in exudate and softening [14]	

Yeast species	Food product	Spoilage compounds	Effect observed	Health hazard
Hanseniaspora/ Kloeckera	Fresh must [15]			
	Various storage products [2]	Low ethanol concentration [2]	Undesirable fermentation products [2]	
	Must under fermentation [7]	Acetate production [7]	Aroma modification at early fermentation stage in winemaking [7]	

Table 1. Some of the most common yeasts often found in grape, musts and wines that can be considered spoilage yeast species in a wide range of food products.

with the novelty of their application in winemaking industry. A brief review of monitoring techniques as a tool for improving quality control in the winery is also included.

2. Technologies for spoilage yeast control

2.1. Early technologies

In the food industry, the control of the spoilage and pathogenic microorganisms was traditionally carried out by means of using thermal processes, to ensure the partial or total elimination of the microflora present [17, 18]. Together with an aseptic and hermetic packaging, it was possible to effectively extend the shelf life of the food products ensuring at the same time its microbiological safety [19, 20]. The main drawback of this kind of inactivation processes is the damage in organoleptic quality due to high processing temperatures. Another traditional way to fight against unwanted microorganisms is the addition of natural or chemically synthesized preservatives, such as organic acids (ascorbic, citric, benzoic, sorbic, etc.) and salts (potassium sorbate, sodium benzoate, sodium metabisulphite, etc.) [21, 22]. It is also possible to limit undesirable microbial development by modifying certain environmental parameters (temperature, pH, water activity, nutrient availability, toxic compounds, etc.) during the production process in order to hinder its growth. In the field of oenology, the ethanol tolerance is believed to be one of the main factors limiting yeast growth [23]. In addition, some of the antimicrobials most commonly used in winemaking are sulphur dioxide (SO_2), dimethyl dicarbonate and sorbic acid.

Yeast species resistant to one preservative also tend to be resistant to others with similar chemical composition. Such is the case of benzoic acid, sorbic acid and sulphur dioxide [24]. Also, sorbic acid resistance has demonstrated to be highly correlated to ethanol resistance [25]. In general, yeast resistance to preservatives seems to be strain dependent and also dependent on the physiological characteristics of the cells [6, 26].

Sulphur dioxide (SO_2) is the chemical additive mainly used in wineries as antioxidant and preservative to control bacteria, moulds and spoilage yeasts [11, 27] considering that its antiseptic property depends on the pH of the media [28]. However, in the last decades, its use is being reconsidered due to increasing allergic concerns. Researchers are looking for alternative methods to reduce the doses commonly added to grape juice and wine [29, 30].

High doses of SO_2 are needed to control the growth of *Dekkera bruxellensis* in red wine. Barata et al. [23] suggested an average value of 1 mg/L molecular sulphur dioxide to prevent *D. bruxellensis* development during red wine maturation in oak barrels. Similarly, *Saccharomycodes ludwigii* and *Zygosaccharomyces bailii* have a strong resistance to SO_2 and, in addition, *Z. bailii* also to organic acids [24].

Due to long-term exposure to SO_2, some wine yeasts have developed certain defence mechanisms to fight against this antimicrobial [28]. The ability of *Brettanomyces bruxellensis* to enter a viable but not culturable (VBNC) state as a survival strategy induced by the presence of SO_2 has been proved [31]. Molecular SO_2 levels as low as 0.2–0.4 mg/L can induce the VBNC cells in different *B. bruxellensis* strains (tested in a synthetic wine medium) [32]. However, the metabolism of this spoilage yeast is maintained, so synthesis and release of volatile phenols continues even under this non-reproductive state [32]. Benito et al. [33] claimed that 20 mg/L of free SO_2 is the minimum concentration required for inhibiting the enzymatic activity of *B. bruxellensis* at pH 3.5.

Dimethyl dicarbonate (DMDC) is a dimethyl ester of dicarbonic acid used as cold sterilant in food industry. It is legally authorized in the USA [34], Australia [35] and Europe [36] as chemical preservative for ensuring the microbiological stability of wines. DMDC antimicrobial effectiveness is maximal at low pH, high ethanol content and low microbial population [37]. At permissible usage levels, DMDC is more effective against yeast than bacteria or moulds [38]. The main mechanism of action of DMDC is to inactivate cellular enzymes such as glyceraldehyde-3-phosphate dehydrogenase and alcohol dehydrogenase through reaction with nucleophilic groups (imidazoles, amines and thiols) [39]. DMDC can be used to prevent spoilage yeasts growth in wines [40] as well as to stop alcoholic fermentation in the production of sweet wines [41] or to disinfect musts by removing native flora present [42].

In relation to their specific effect on spoilage yeasts, *Z. bailii, S. ludwigii* and *Lachancea thermotolerans* were found to be very sensitive to DMDC; complete cell death (initial population of 10^6 CFU/mL) can be achieved with a dose of 100–200 mg/L DMDC when added to red wine (12% v/v and pH 3.5) [43]. Later experiments carried by Zuehlke et al. [44] confirmed the toxicity of DMDC against *Z. bailii*. On the contrary, *Schizosaccharomyces pombe* and *Saccharomyces cerevisiae* can survive in the presence of higher concentrations of DMDC (minimum inhibitory concentration (MIC) of ≥300 mg/L). Moreover, the addition of DMDC in wines at the maximum dose legally permitted in Europe (200 mg/L; [36]) was proved to be ineffective against lactic acid and acetic acid bacteria [43]. However, resistance to DMDC highly depends on the media composition, since the amount needed to inhibit *Z. bailii* and *S. ludwigii* (10^6 cells/mL) when DMDC is added to grape must is higher, around 400 mg/L [42]. Based on the different yeast sensitivity towards DMDC, the use of this compound could be suggested as a technology to favour the microbial development of certain yeast species during the initial stages of alcoholic fermentation. The dominance of the selected yeast can be ensured if the inoculum is added 12 h after grape must treatment with DMDC [42].

Its use before the end of fermentations is not recommended since the antimicrobial effect of DMDC can be also exerted against the fermenting species such as *S. cerevisiae* and *Oenococcus oeni* [40]. These same authors stated that the action of DMDC added during vinification has a transitory nature, so the preservation effect does not last with time and a fractional dosage might be needed, especially if the goal is to protect the wine during barrel maturation [45].

There also exists a synergistic activity to increase the inactivation effect against wine yeast and bacteria between DMDC and sulphur dioxide in both potassium and sodium metabisulphite salts [40, 43]. In this regard, the use of DMDC allows a significant reduction of sulphur content in grape juice and semi-sweet wines [46].

Regarding health issues, DMDC does not represent any threat since it is naturally, rapidly and completely hydrolysed to methanol and carbon dioxide in aqueous solutions [46]. Notwithstanding, a disadvantage of the use of DMDC in winemaking is its potential toxicity during handling [39].

With relation to the use of weak acids, benzoic and sorbic acids are able to control spoilage yeasts in wines when used in certain concentrations. The lipophilic character of benzoic and sorbic acids allows their diffusion in its undissociated form through cell membranes [24]. However, so far, benzoic acid is not authorized for use in wine [47].

Z. bailii is a wine spoilage yeast species highly resistant to commonly used preservatives such as benzoic and sorbic acids [48]. Similarly, *Zygosaccharomyces rouxii* is an osmophilic yeast responsible for spoilage in high sugar content beverages. Among the chemical compounds with antimicrobial properties used in food products, potassium sorbate, sodium benzoate, dimethyl dicarbonate and vanillin showed good results to control the growth of *Z. rouxii* in concentrated grape juice [26]. After applying a mathematical model, a maximum inhibitory effect with average growth reduction of 40% was estimated using each of the four preservatives alone with the following doses: potassium sorbate (>4.7 mM), sodium benzoate (>10.4 mM), dimethyl dicarbonate (>1 mM) and vanillin (>30 mM).

B. bruxellensis has also a strong resistance to benzoic and sorbic acids. The required sorbic acid dose to arrest its growth and its enzymatic activity exceeds long the legal limit in wine (250 mg/L) [33]. And as it was previously mentioned, benzoic acid is still forbidden as wine additive. According to Loureiro and Malfeito-Ferreira [49], *B. bruxellensis* can stand up to 950 mg/L of sorbic acid at pH 3.5.

Among the physicochemical parameters that can be handled in the cellar, the storage temperature and the level of ethanol in wine together can exert a synergistic limiting effect on the growth of the wine spoilage yeast *B. bruxellensis*. The optimal combination range of both parameters to halt the proliferation of these microorganisms is above 14% v/v ethanol content and below 12°C storage temperature [50]. Similarly, trials in synthetic media revealed that *D. bruxellensis* proliferation can be hindered by ethanol concentrations above 14.5–15.0% (v/v) [23]. According to Couto et al. [51], the temperature-time binomial necessary for the thermal destruction of *Dekkera/Brettanomyces* yeast cells largely depends on the ethanol and phenolic contents, particularly ferulic acid, of the wine. Significant reductions in population of contaminated wine samples (initial inoculum of approximately 1×10^7 CFU mL^{-1}) can be achieved with temperatures above 35°C. It is worth mentioning that *B. bruxellensis* behaves differently in white and red wines, showing less development in white wines [23]. In addition, SO_2 requirements may be reduced with low storage temperatures, as the optimal growth temperature for *Brettanomyces* is 25–28°C [52]. These authors provided a limiting range of below 15°C and above 0.4 mg/L molecular SO_2 for controlling spoilage by *Brettanomyces* in stored red wines, although complete eradication cannot be ensured with these conditions.

2.2. Emerging technologies

2.2.1. High hydrostatic pressure

The high hydrostatic pressure (HHP) treatment is a non-thermal process used to inhibit pathogenic microorganisms including spoilage microbes as well as enzymes [53]. According to these authors, the food products that are treated do not change nutritional or modify their sensory quality. The HHP uses pressure-transmitting liquids (water, ethanol solutions, sodium benzoate solutions, etc.) to homogeneously transfer pressure to the food sample [54]; the process may be batch, semi-continuous or continuous. The use of HHP is also expected to increase the shelf life [55] of a wide range of food products able to be treated including meat, eggs, vegetables, seafood [53], fruits such as sweet cherries [56] or mango pulp [57], juices [55, 58] and Serrano ham [59], among others.

In the winemaking industry, the HHPs have been used to reduce or inhibit the presence of spoilage yeasts such as *B. bruxellensis*. In this matter, Van Wyk and Silva [60] have used 200 MPa for 3 min achieving a reduction of up to 5.8 log but higher than 6 log when increasing the pressure to 400 MPa during a much shorter period of time (5 seg). These parameters were also evaluated by González-Arenzana et al. [61] but also considering pH and amount of ethanol in wines; high ethanol and high pH needed less pressure (100 MPa) to inactivate *B. bruxellensis* cells, while at lower ethanol and lower pH the pressure used was higher (200 MPa). The HHP could also be used as a treatment for grapes in order to reduce wild yeast populations in order to be able to use yeast starters [62]; HHP has shown to increase pigment extraction and ethanol and methanol increase in wines after treatment. In the case of beer pasteurization, the use of 600 MPa allows beer with different ethanol content to have more than 7 log reductions of *S. cerevisiae* within 5 s [63].

2.2.2. Pulsed light irradiation

The pulsed light irradiation is a technique used to inactivate microorganisms in food products. The pulses are stored electricity in UV lamps that do not contain mercury but xenon [64], released in very short periods of time (fractions of millionths or thousandths of a second) that are rich in UV-C light (from 200 to 280 nm in the spectrum) [65]. This sterilization technique can cause DNA damage in the same way that continuous UV light does, nonetheless irradiation with pulsed light may also induce distortion in cell membranes and vacuoles [66]; therefore, the pulsed light irradiation may be considered an improved version of the UV treatment [64].

Pulsed light has been evaluated in the last decade as an efficient technique against foodborne spoilage microorganisms in commercial and fresh fruit juices using pulsed light against *S. cerevisiae* and pathogenic bacteria [67], or apple juices [68] where the combination of pulsed light with ultrasound has been evaluated as an alternative to inactivate *S. cerevisiae*: fresh strawberries [69] with reductions of 1 log colonies of yeast and microfungi lengthening shelf life 2 days; cured meat products [70] against pathogenic bacteria. Nevertheless, there are studies on the use of pulsed light on several food products; in wine or in the winemaking industry the evaluation of this technique appears to be scarce. The use of pulsed light might not be limited to just inactivating spoilage yeast in grapes and musts but also as a technique to potentially improve polyphenols extraction during maceration, thus increasing pigment compounds in wine.

2.2.3. *Electric pulses*

The electric pulses (EPs) are a method used for food preservation that avoids the use of chemical compounds as well as the use of thermal treatments [71]. Thus, it can be considered a cold temperature treatment that may preserve foodborne properties.

Hülsheger et al. [72] demonstrated that EPs have different effect on microorganisms. Yeasts and Gram-positive bacteria are more resistant than Gram-negative bacteria to the disruption caused by cellular structures by low-energy pulses while most of the microorganisms (>99%) die with high-energy pulses. The colonies were also more susceptible to die at logarithmic growth phase than when in steady state. Besides the growth phase, Gáskovà et al. [73] have shown that there are other factors involved in the efficiency of electric pulses for killing yeasts; such factors include the amplitude and the duration of the pulses, the size of the yeast cells and the temperature and conductivity of the media where the yeasts grow.

Electric pulses have been used lately in the production of fruit juices for its efficacy in microbial reduction and for keeping their sensorial and nutritional properties [74]. The use of EP to eliminate spoilage yeasts such as *S. cerevisiae* in orange juice requires less intensity pulses than those to inactivate other pathogens microorganisms such as the bacteria *Escherichia coli* or *Listeria innocua* [75].

The effect of EP could be combined with the use of other antimicrobial technology like the lyticase digestion; the combined effect of both techniques may be a biotechnological application for spoilage yeast control since the use of mild EP (2–4.5 kV/cm) affects the cell-wall porosity and therefore lower doses of lyticase are needed to effectively inactive *S. carlsbergensis, Kluyveromyces lactis* and *K. marxianus* [76].

Biotechnology applications of the electric pulses or the electroporation, other than spoilage yeasts control, may include genetic transformation of cells by incorporation of foreign DNA, extraction of intracellular metabolites and biomass drying [71].

2.2.4. *Natural extracts*

The use of antimicrobials extracted from nature such as chitosan, essential oils (active ingredients: eugenol, allicin, carvacrol, thymol and limonene), spices or nisin is very widespread in food preservation [21], but the limits of its antifungal activity and their potential applications as winemaking additives remain to be further explored.

Chitosan (β-1,4-D-glucosamine) is a linear heteropolysaccharide obtained by deacetylation of chitin. The high density of amino groups present in its structure makes chitosan a bioactive polymer [77]. This polysaccharide with antimicrobial activity is found in nature being part of the exoskeletons of arthropods, diatoms and algae, and the cell wall of some fungi (particularly zygomycetes) [78]. In a study carried out by Rojo et al. [26], where a dose of 300 mg/L of chitosan was added to a high sugar culture media in order to assess its inhibition properties to control the wine spoilage yeast *Z. rouxii*, it was found that with such a high dose of chitosan, tripling the allowable dose in wines (100 mg/L), no inhibition effect exists. Previously, *Z. rouxii* had been already reported as sensitive yeast to chitosan (completely inactivated by 100–400 mg/L), but this result was only based on the study of one strain isolated from a spoiled carbonated beverage [79]. Similarly, *S. ludwigii* has been identified as yeast very resistant to chitosan (5 g/L of chitosan was required to inactivate *S. ludwigii*). The minimum inhibitory concentration for *Brettanomyces/Dekkera* ranges from 0.2 to 0.5 mg/mL of chitosan, depending on the molecular weight [80]. Low-molecular-weight chitosan has higher antifungal properties. With respect to impact on sensory quality, the addition of chitosan may adversely affect the colour of the wine, mainly through a reduction of the colour intensity [80]. Moreover, Gómez-Rivas et al. [77] observed that in mixed fermentations *B. bruxellensis* and *B. intermedius* are more sensitive to chitosan than *S. cerevisiae*. Taillandier et al. [81] have also tested the use of chitosan of fungal origin to control *B. bruxellensis* in winemaking and concluded that its mechanism of action is a complex combination of both physical adsorption on cell wall and the biological interaction by changing permeability of cell membrane, together with an observed effect on the growing rate and the physiological state of *B. bruxellensis*. Petrova et al. [82] also reported on the efficacy of chitosan as strategy to control *B. bruxellensis* development in red wines, with an average reduction of 3 logs in the culturability. It has been proved that when applied in the vineyard chitosan can act as an inducer of the plant defence system, mainly by enhancing the polyphenolic phytoalexins synthesis, and therefore it represents a useful protection treatment against powdery mildew infection [83]. Chitosan can be also used as natamycin carrier to protect the cheese surface from fungal contamination and thus extend its shelf life [84]. It is also possible to use microencapsulation techniques with chitosan to dose oily compounds, such as essential oils, with the additional advantage of protecting the active ingredient (e.g. limonene) and thus preserving its antimicrobial properties [85].

S. cerevisiae has demonstrated strong resistant to chitosan, doses above 2 g/L are needed to fully inhibit it. Conversely, *B. bruxellensis* seems to be very susceptible to this natural extract, 0.2 g/L can effectively inactivate it. *Hanseniaspora uvarum* and *Z. bailii*, other spoilage wine yeasts, can be also inhibited by chitosan with 0.4 g/L [86]. However, all these results correspond to growth inhibition assays in laboratory media. As expected, these authors observed that the antimicrobial effect of chitosan is less under winemaking conditions. Regarding the performance of alcoholic fermentation, a slight delay in the beginning of the fermentation was observed, but without major repercussions [86].

Among the essential oils extracted from plants, those from cinnamon, clove, garlic, onion, oregano and thyme are the inhibitoriest against food spoilage yeasts [87]. Antimicrobial activity of these compounds is well documented [88].

Eugenol is the main component of clove oil (≈85%). So far, eugenol has proven to be effective as antibacterial and antifungal in several food products [89]. Both eugenol and thymol act on the membrane and cell wall, causing cell lysis [90]. These same authors suggested the use of eugenol and thymol as lysing agents to extract the genomic DNA from yeast cell instead of using zymolyase and sodium dodecyl sulphate (SDS). However, according to Kubo et al. [91] eugenol alone may not respond as effective antimicrobial against *Z. bailii* and *S. cerevisiae*, since the concentration required for ensuring the fungicidal effect is quite high (800 μg/mL).

Carvacrol and thymol are phenolic compounds being part of the essential oil from oregano and thyme [92]. According to Chavan and Tupe [93], low concentrations of carvacrol and thymol (≤64 mg/L; used independently) are effective in limiting the growth of several wine spoilage yeasts, including *Metschnikowia pulcherrima, Z. rouxii, S. pombe, Debaryomyces hansenii* and *D. bruxellensis*. Results for minimum inhibitory concentration values were comparable or even better than those of potassium metabisulphite. The antifungal mechanism of these two essential oils' active ingredients is based on the membrane damage and leakage of cytoplasmic content, thus increasing permeability.

As for allicin (diallyl thiosulphinate), it is the main bioactive component of garlic extract [94]. It is well known for its antioxidant, antibacterial and antifungal activities [95] and also for its anticancer activity [96, 97]. The main mechanism involved in the antimicrobial effect of allicin is based on its rapid reaction with thiol-containing proteins [94]. Thus, allicin can regulate the activity of enzymes containing very reactive or unshielded SH groups. With relation to its potent antifungal properties, allicin is effective against a wide range of yeasts, among them *S. pombe, S. cerevisiae* [98], *Aspergillus* spp. [99] and *Candida* spp. [100]. It also affects the synthesis of mycotoxins and inhibits the germination of spores as well as the growth of the hyphae [100]. Yoshida et al. [98] set the minimum inhibitory concentration (MIC) of allicin in 5 mg/mL for *S. pombe* and 10 mg/mL for *S. cerevisiae*. Moreover, chitosan and garlic extract, used independently, have also shown promising results as treatments for the control of fungal diseases of the vine due to pruning [101]. In the medical field, allicin and its derivatives have been proposed as antifungal prophylactic due to their strong efficacy in inhibiting microbial development [102]. However, further studies are still required to confirm the potential therapeutic use of allicin [103].

Despite their interesting antimicrobial properties, there still exist some limitations to the use of some of these natural extracts such as their solubility in water, their susceptibility to oxidation (part of its effectiveness is lost) and the impact on sensory attributes [21]. A possible solution to the low water solubility of some natural extracts is their transformation into nanoemulsions [104, 105] or the use of nanoencapsulation techniques to protect their antimicrobial properties [106].

The inhibition activity on the growth of some wine spoilage yeasts using different antimicrobial compounds was measured by agar diffusion tests (**Figure 2**). The dark areas around the disks suggest the inhibitory effect in growing colonies, thus demonstrating their effectiveness. This assay can be used to assess the susceptibility of the yeast species to each of the antimicrobials, taking into account the limitations due to the diffusion properties of the antifungal compounds in solid media.

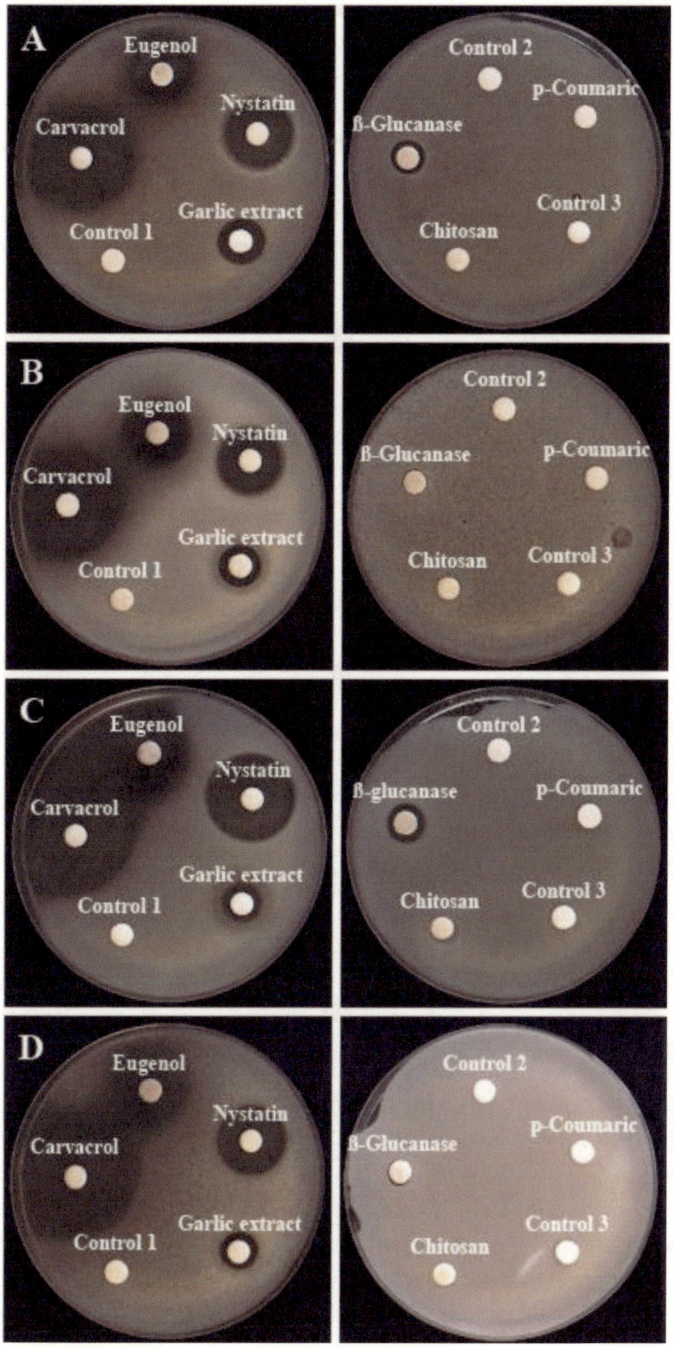

Figure 2. Agar diffusion tests for antimicrobial screening. Antimicrobials tested: eugenol, carvacrol, nystatin (antibiotic), garlic extract (allicin), chitosan, β-glucanase and *p*-coumaric acid. Controls used for the preparation of the antimicrobials: control 1-water, control 2-ethanol and control 3-acetic acid. Yeast species tested: (A) *S. ludwigii*, (B) *D. bruxellensis*, (C) *P. anomala* and (D) *Z. rouxii*.

Despite their interesting antimicrobial properties, there still exist some limitations to the use of some of these natural extracts such as their solubility in water and the impact on sensory attributes [21].

2.2.5. Antimicrobial peptides

Antimicrobial peptides are important molecules naturally occurring in the immune system [107] and may have different amino acids conformation. Their importance as antimicrobial agents increased when bacteria and other pathogens became more resistant to drugs [108]. Among the most studied peptides are tritrpticin and indolicidin, related cathelicidins [109] rich in Arg and Trp residues [110] as well as the lactoferricin. Other peptides having microbicidal activity are cecropins, defensins, magainins, melittin and alamethicin [111].

Tritrpticin is a positively charged peptide with tryptophan (Trp) residues [112] that has antifungal activity besides having antibacterial activity against Gram-positive and Gram-negative bacteria [113, 114]; it is naturally found in granules of bovine and porcine neutrophils [115] as well as in insects [107, 116, 117]. The electrostatic interaction between the cationic residue and the negative charge of phospholipids in the membrane has been recognized as an important factor in the microbicidal action found in these biomolecules [118].

Indolicidin is another peptide with antimicrobial activity against both fungi and bacteria (Gram-positive and Gram-negative) [119]. Similar to tritrpticin, indolicidin is a tryptophan (Trp)-rich peptide [120] that also contains proline (Pro) residues in its configuration [121]; its structure resembles that of detergent micelles and of phospholipid vesicles [122]. It has been isolated from cytoplasmic granules of bovine neutrophils [123] but it is also present in humans and other mammals as well as in some primitive vertebrates [124]. It acts directly on the lipidic bilayer [125] causing disruption in the cell membrane.

Lactoferricin B (Lfcin B) is a peptide obtained during the gastric digestion of the bovine protein lactoferrin [126]. Despite the fact that other mammals including humans also produce lactoferricins, the Lfcin B has higher antimicrobial potency [127]. Lfcin B shows different properties including antibacterial, antifungal, antiviral, antitumour, anti-inflammatory and immunoregulatory [127].

Antimicrobial peptides are important for the control of spoilage yeasts in food products. Lfcin B derivatives reduce spoilage yeasts such as *D. bruxellensis* [128] and *Z. bailii* and *Z. bisporus* [129] in wine production. Peptides are used to control the most common spoilage yeasts in dairy products such as *K. marxianus, D. hansenii, Candida* spp. and other spoilage yeasts such as *Zygosaccharomyces microellipsoides, Rhodotorula mucilaginosa, Yarrowia lipolytica, Torulospora* and *Pichia* [130]. Mozzarella cheese packaged with films of chitosan with up to 60% lysozyme has antimicrobial activity against yeast *Pseudomonas fluorescens* and other bacteria and moulds [131].

In medical applications, indolicidin has antimicrobial activity against different pathogens such as yeasts, viruses, bacteria, fungi and protozoa [132]. Indolicidin has been tested against candidiasis produced by fungi *Candida albicans* [133], and nosocomial pathogen yeasts *Candida krusei* and *Candida parapsilosis* [134] responsible for bloodstream infections

[109]. It is also known that indolicidin directly binds DNA forming a complex with the catalytic HIV-1 integrase [124] producing inhibitory activity in it with an implication on the HIV virus replication cycle [135]. Lfcin B has fungicidal activity against the pathogenic yeast *Candida tropicalis* and fungicidal properties against the fungi *Candida neoformans* and *C. albicans* [136].

A drawback in the use of some antimicrobial peptides like tritrpticin and indolicidin for treatment of infectious diseases produced by pathogen agents is the hemolytic activity observed against blood hematocytes [113, 115, 132]; therefore, it is important to use alternative antimicrobial peptides with less toxicity in immunocompromised patients [133].

2.2.6. Killer toxins and β-glucanases

Killer toxins are known as pore proteins produced by yeast metabolism and are able to kill sensitive yeast cells by forming cell wall or cell membrane disruptions [137]. There are different killer toxins produced by either *Saccharomyces* and non-*Saccharomyces* yeasts used as antimicrobials.

Among the toxins produced by non-*Saccharomyces*, Tdkt is produced by the species *Torulaspora delbrueckii* with β-glucanase and chitinase activity against yeasts *B. bruxellensis, Pichia guilliermondii, Peronospora manshurica* and *Pichia membranaefaciens* [138]; toxin Kwkt is obtained from the species *Kluyveromyces wickerhamii* and toxin Pikt is produced by the species *Wickerhamomyces anomalus* (formerly *Pichia anomala*), both are used against spoilage yeasts *Brettanomyces/Dekkera* [139]; toxin Kpkt is produced by the species *Tetrapisispora phaffii* and it could be used to control spoilage yeasts in winemaking [140] since it has proven fungicidal activity against yeast *Hanseniaspora/Kloeckera* [141]. This last yeast is also used as killer yeast against some fungal pathogens in plants and also to fight infections produced by fungi in animals and humans due to its high β-1,3-glucanase activity producing disruptions in the cell wall [142].

S. cerevisiae has also shown the ability to produce killer toxins; four different killer yeasts from this species produce killer toxins K1, K2, K28 and Klus. These yeasts are able to produce the toxins during spontaneous wine fermentation as the population of other yeasts decreases [143]. The killer toxin K1, for example, having the glycosylated subunits α and β, is released out of the cytoplasm [144]. The killer toxin forms a channel in the membrane of target cells through which ions are conducted to the exterior producing cellular death. The activity observed in some killer toxins is due to β-glucanases, which has been used to produce a synthetic β-glucanases preparation as antibacterial material against yeasts *D. bruxellensis* and *Z. bailii* [145]. The effect of β-glucanases on the cell integrity can be observed in **Figure 3** where cells of the species *S. ludwigii* have been disrupted after the addition of two different β-glucanases.

Even though both compounds have same disruptive inhibition mechanism, the activity of the β-glucanases is not the same. Such an activity is compared in **Figure 4** where YPD liquid growing media have been inoculated with four different wine spoilage yeasts: *S. ludwigii, D. bruxellensis, P. anomala* and *Z. rouxii*. The media where β-glucanase II has been used did not show any growth indicating total effectiveness of the treatment, while the media where

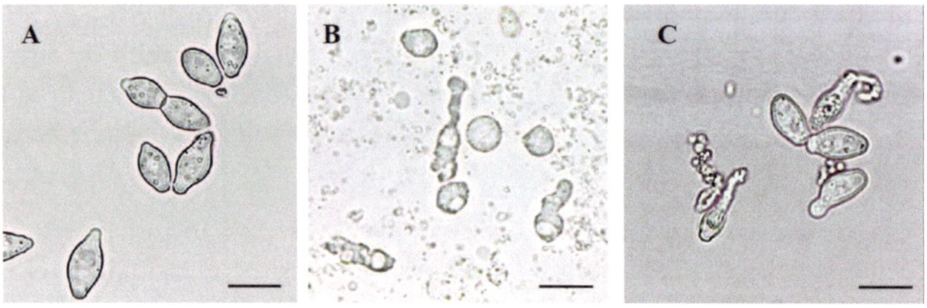

Figure 3. Effect of β-glucanases on the yeast's cell wall after the addition of β-glucanase I (B) and β-glucanase II (C) to *S. ludwigii* (A). Bar scale 10 μm.

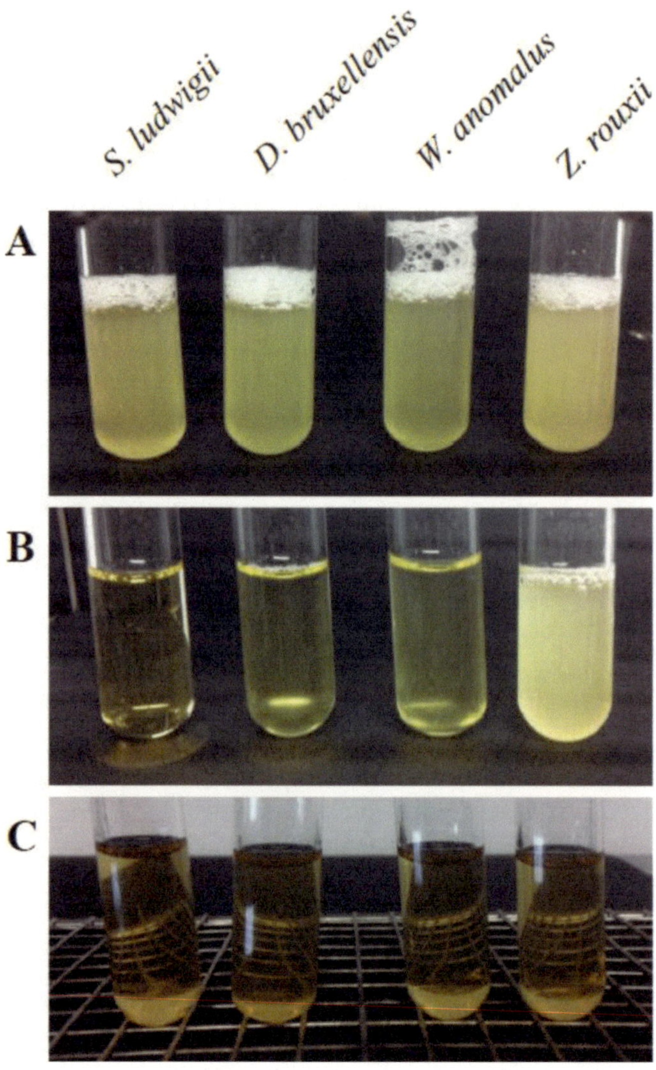

Figure 4. Comparison of the effect of two β-glucanases in YPD liquid growing media with same optical density (OD) of *S. ludwigii, D. bruxellensis, P. anomala* and *Z. rouxii*. The trials are grouped in (A) control, (B) β-glucanase I and (C) β-glucanase II.

β-glucanase I has been added showed CO_2 production and turbidity formation in function of the sensibility of each yeast species. Wine spoilage yeast *S. ludwigii* appears sensitive to both β-glucanases, as no foam has been formed after the addition of either compound.

In the case of β-glucanase, the inhibition zone observed in the agar plates suggests a low diffusion ability and thus limited antimicrobial activity; however, when grown in liquid media in the presence of same β-glucanase all the spoilage yeasts showed sensitivity (see **Figure 2**).

The yeasts could be neutral or sensitive to toxins produced by killer yeasts. In this matter, [146] have shown that the yeast species *B. bruxellensis, B. anomalus, D. anomala, B. custerisanus, H. uvarum, S. cerevisiae and Z. bailii* have either sensitive or neutral nature towards killer toxins produced by the species *K. wickerhamii* and *C. pyralidae*. Santos et al. [147] have also described the resistance of the species *S. cerevisiae* to the toxin CYC 1410 produced by the species *Ustilago maydis* during wine production; therefore, this procedure could inhibit the implantation of *B. bruxellensis* spoilage yeast. Cytoplasmatic β-glucanases have also shown, on the other hand, interesting contribution to wine-sensory properties as polysaccharides are released from the autolysis of non-*Saccharomyces* yeast cells during wine production [148]. Polysaccharides can improve mouth-feel properties of wines by changing their viscosity.

A summary of certain emerging antifungal compounds is presented in **Figure 5**.

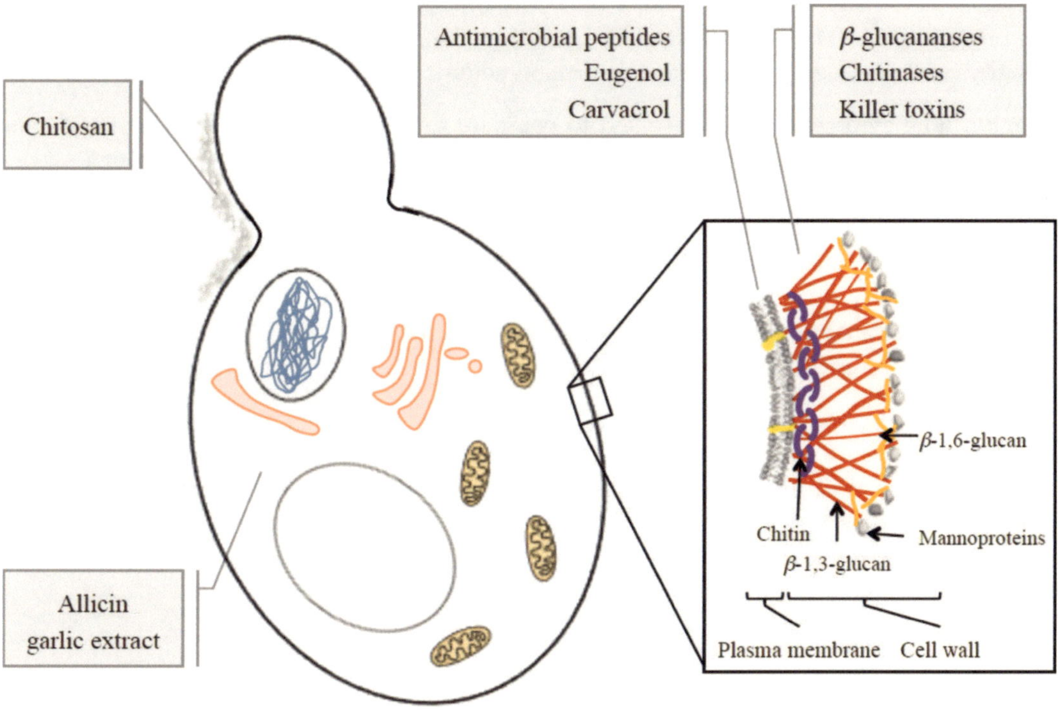

Figure 5. Schematic representation of the main mechanisms involved in the antimicrobial effect of different compounds on a yeast cell. The arrows indicate the target of each compound in the yeast structure: Plasma membrane (antimicrobial peptides, eugenol and carvacrol), cell wall (β-glucanases, chitinases and killer toxins), endoplasmatic enzymes (allicin) and physical adsorption on cell wall (chitosan).

3. Yeast populations monitoring techniques

Parallel to the use of technologies that reduce the colonization of spoilage yeasts in food products, there are technologies able to monitor the yeasts and bacteria populations and their nature in the alcoholic and malolactic fermentations during wine production. One of these techniques is the flow cytometry (FCM) [149]; other techniques are fluorescence-activated cell sorting (FACS), quantum dots (QDs) and emerging monitoring technologies such as mass cytometry (Cy-TOF), imaging flow cytometry and, recently, spectral cytometry.

FCM, used for counts of microorganism populations in wine, allowed the simultaneous determination of yeasts for as low as 10^3 cells/mL and, despite its size, populations of malolactic bacteria higher than 10^4 cells/mL [150]. FCM is able to show real-time situation of microorganisms in different matrices although it is considered complicated implementation due to the cost of reagents and the need of recruiting trained staff. FCM that relies on the use of complementary fluorescence dying to selectively determine specific type of microorganisms [151] through, for example, the metabolic enzyme activity or antigen expression is known as fluorescence-activated cell sorting (FACS). This technology is useful in monitoring changes in yeast cellular organelles' biogenesis such as the mitochondria [152].

The QD is a technology based on the production of semiconductor nanoparticles made from cadmium salts crystals with chloride (Cl), tellurium (Te) and sulphide (S) by target microbial cells such as spoilage yeast [153]. These nanoparticles are used as biomarkers of nucleic acids and proteins and can be detected visually when they are excited by a light source. QDs are photostable and they have a wide range of absorption while they have a narrow emission peak [154].

Imaging flow cytometry has been used to compare the mode of action in which different killer toxins affect cell structures. Comitini et al. [140] saw that the toxin Kpkt from *T. phaffii* had different mode of action from that of the toxin K1 from *S. cerevisiae*; all yeast cells dead to toxin K1 after well-known membrane disruption forms potassium channels, have nucleic acids stained with propidium iodide (PI) while dead cells exposed to Kpkt are not all stained with PI. These results suggested a difference in the way that both toxins have to kill yeast cells. Flow cytometry analyses showed that Kpkt causes disruption in the cell-wall integrity due to very specific β-glucanase activity [15].

The spectral cytometry was developed as to increase the accuracy and the precision of flow cytometry results by a higher resolution obtained with spectral analysis from more discrete bands of emission of multiple stained samples [155] as in the case of microbial counts in foodborne matrices stained with PI. Spectral measurements combined with flow cytometry technique allow obtaining fluorescence and RAMAN spectra analysis of large particles in chemical and biological processes [156].

4. Future trends

Several antimicrobial techniques have been developed to control the presence of spoilage yeasts in food products through the years. These techniques aim to diminish the negative economic impact of having contaminated/spoiled products as well as the potential health threat that this may represent. Some of these techniques are already in use by the winemaking industry and others may be explored by the different process of stages from vine to bottled wine.

The trend observed is that the winemaking industry is targeting the use of innocuous control techniques to avoid spoilage yeasts during the entire process in order to preserve varietal aromas from grapes, to protect and to extend the anthocyanins extraction yield and, in the best-case scenario, to improve the overall quality of wines.

Acknowledgements

This work was funded by the INNTER WINARGAL 2015 project (CDTI-MINECO).

Author details

Carlos Escott[1], Iris Loira[1], Antonio Morata[1]*, María Antonia Bañuelos[2] and José Antonio Suárez-Lepe[1]

*Address all correspondence to: antonio.morata@upm.es

1 Department of Chemistry and Food Technology, Universidad Politécnica de Madrid, Madrid, Spain

2 Department of Biotechnology and Plant Biology, Universidad Politécnica de Madrid, Madrid, Spain

References

[1] Fleet G. The commercial and community significance of yeasts in food and beverage production. Yeasts in Food and Beverages. Springer-Verlag, Berlin Heidelberg; 2006. pp. 1-12

[2] Palpacelli V, Ciani M, Rosini G. Activity of different 'killer' yeasts on strains of yeast species undesirable in the food industry. FEMS Microbiology Letters. 1991;**84**(1):75-78

[3] Rojo MC, López FA, Lerena M, Mercado L, Torres A, Combina M. Effects of pH and sugar concentration in *Zygosaccharomyces rouxii* growth and time for spoilage in concentrated grape juice at isothermal and non-isothermal conditions. Food Microbiology. 2014;**38**:143-150

[4] Alonso A, Belda I, Santos A, Navascués E, Marquina D. Advances in the control of the spoilage caused by *Zygosaccharomyces* species on sweet wines and concentrated grape musts. Food Control. 2015;**51**:129-134

[5] Fleet G. Spoilage yeasts. Critical Reviews in Biotechnology. 1992;**12**(1-2):1-44

[6] Martorell P, Stratford M, Steels H, Fernández-Espinar MT, Querol A. Physiological characterization of spoilage strains of *Zygosaccharomyces bailii* and *Zygosaccharomyces rouxii* isolated from high sugar environments. International Journal of Food Microbiology. 2007;**114**(2):234-242

[7] Loureiro V, Malfeito-Ferreira M. Spoilage yeasts in the wine industry. International Journal of Food Microbiology. 2003 Sep 1;**86**(1-2):23-50

[8] Fleet GH. Yeast interactions and wine flavour. International Journal of Food Microbiology. 2003;**86**(1):11-22

[9] Passoth V, Fredlund E, Druvefors UA, Schnurer J. Biotechnology, physiology and genetics of the yeast *Pichia anomala*. FEMS Yeast Research. 2006 Jan;**6**(1):3-13

[10] Schaffrath R, Breunig KD. Genetics and molecular physiology of the yeast *Kluyveromyces lactis*. Fungal Genetics and Biology. 2000;**30**(3):173-190

[11] Du Toit M, Pretorius IS. Microbial spoilage and preservation of wine: Using weapons from nature's own arsenal—A review. South African Journal of Enology and Viticulture. 2000;**21**(Special Issue):74-96

[12] Romano P, Suzzi G. Origin and production of acetoin during wine yeast fermentation. Applied and Environmental Microbiology. 1996 Feb;**62**(2):309-315

[13] Akinmusire O. Fungal species associated with the spoilage of some edible fruits in Maiduguri Northern Eastern Nigeria. Advances in Environmental Biology. 2011:157-162

[14] Babic I, Hilbert G, Nguyen-The C, Guiraud J. The yeast flora of stored ready-to-use carrots and their role in spoilage. International Journal of Food Science and Technology. 1992;**27**(4):473-484

[15] Ciani M, Fatichenti F. Killer toxin of *Kluyveromyces phaffii* DBVPG 6076 as a biopreservative agent to control apiculate wine yeasts. Applied Environmental Microbiology. 2001 Jul;**67**(7):3058-3063

[16] Gyawali R, Ibrahim SA. Natural products as antimicrobial agents. Food Control. 2014;**46**:412-429

[17] Török T, King A. Thermal inactivation kinetics of food-borne yeasts. Journal of Food Science. 1991;**56**(1):6-9

[18] Gaze J. Microbiological aspects of thermally processed foods. Journal of Applied Microbiology. 2005;**98**(6):1381-1386

[19] Hotchkiss J. Safety Considerations in Active Packaging. Active Food Packaging. Springer, US; 1995. pp. 238-255

[20] Brody AL, Bugusu B, Han JH, Sand CK, McHugh TH. Innovative food packaging solutions. Journal of Food Science. 2008;**73**(8):R107–R116

[21] Raybaudi-Massilia RM, Mosqueda-Melgar J, Soliva-Fortuny R, Martín-Belloso O. Control of pathogenic and spoilage microorganisms in fresh-cut fruits and fruit juices

by traditional and alternative natural antimicrobials. Comprehensive Reviews in Food Science and Food Safety. 2009;**8**(3):157-180

[22] Davidson PM, Taylor TM, Schmidt SE. Chemical Preservatives and Natural Antimicrobial Compounds. In: Food microbiology. American Society of Microbiology, Washington, DC. 2013. pp. 765-801

[23] Barata A, Caldeira J, Botelheiro R, Pagliara D, Malfeito-Ferreira M, Loureiro V. Survival patterns of *Dekkera bruxellensis* in wines and inhibitory effect of sulphur dioxide. International Journal of Food Microbiology. 2008;**121**(2):201-207

[24] Warth AD. Resistance of yeast species to benzoic and sorbic acids and to sulfur dioxide. Journal of Food Protection. 1985;**48**(7):564-569

[25] Stratford M, Anslow P. Evidence that sorbic acid does not inhibit yeast as a classic 'weak acid preservative'. Letters in Applied Microbiology. 1998;**27**(4):203-206

[26] Rojo MC, López FA, Lerena M, Mercado L, Torres A, Combina M. Evaluation of different chemical preservatives to control *Zygosaccharomyces rouxii* growth in high sugar culture media. Food Control. 2015;**50**:349-355

[27] Ough C, Crowell E. Use of sulfur dioxide in winemaking. Journal of Food Science. 1987;**52**(2):386-388

[28] Divol B, du Toit M, Duckitt E. Surviving in the presence of sulphur dioxide: Strategies developed by wine yeasts. Applied Microbiology and Biotechnology. 2012;**95**(3):601-613

[29] Santos MC, Nunes C, Saraiva JA, Coimbra MA. Chemical and physical methodologies for the replacement/reduction of sulfur dioxide use during winemaking: Review of their potentialities and limitations. European Food Research and Technology. 2012;**234**(1):1-12

[30] Guerrero RF, Cantos-Villar E. Demonstrating the efficiency of sulphur dioxide replacements in wine: A parameter review. Trends in Food Science and Technology. 2015,3;**42**(1):27-43

[31] Serpaggi V, Remize F, Recorbet G, Gaudot-Dumas E, Sequeira-Le Grand A, Alexandre H. Characterization of the "viable but nonculturable" (VBNC) state in the wine spoilage yeast *Brettanomyces*. Food Microbiology. 2012;**30**(2):438-447

[32] Agnolucci M, Rea F, Sbrana C, Cristani C, Fracassetti D, Tirelli A, et al. Sulphur dioxide affects culturability and volatile phenol production by *Brettanomyces/Dekkera bruxellensis*. International Journal of Food Microbiology. 2010;**143**(1):76-80

[33] Benito S, Palomero F, Morata A, Calderón F, Suárez-Lepe J. Factors affecting the hydroxy-cinnamate decarboxylase/vinylphenol reductase activity of *Dekkera/Brettanomyces*: Application for *Dekkera/Brettanomyces* control in red wine making. Journal of Food Science. 2009;**74**(1):M15–M22

[34] 27-CFR-24.246-2002 A. Code of Federal Regulations, Title 27, Chapter 1, Subchapter A, Part 24, Subpart L, Section 24.246. Materials Authorized for the Treatment of Wine and Juice. 2002. Available from: https://www.law.cornell.edu/cfr/text/27/24.246 [Accessed: 3 July 2017]

[35] FAR A474-2004 A. Final Assessment Report, Application A474, Winemaking. Food Standards Australia New Zealand; 2004. Available from: http://www.foodstandards.gov.au/code/applications/pages/applicationa474winemaking/index.aspx [Accessed: 3 July 2017]

[36] CR-EC 643-2006 A. Commission Regulation (EC) No 643/2006. Official Journal of the European Union, 28.4.2006, L115/6-L115/9. 27 April 2006. Available from: http://extwprlegs1.fao.org/docs/pdf/eur65203.pdf [Accessed: 3 July, 2017]

[37] Ough C, Langbehn L, Stafford P. Influence of pH and ethanol on the effectiveness of dimethyl dicarbonate in controlling yeast growth in model wine systems. American Journal of Enology and Viticulture. 1978;**29**(1):60-62

[38] Bizri JN, Wahem IA. Citric acid and antimicrobials affect microbiological stability and quality of tomato juice. Journal of Food Science. 1994;**59**(1):130-135

[39] Bartowsky E. Bacterial spoilage of wine and approaches to minimize it. Letters in Applied Microbiology. 2009;**48**(2):149-156

[40] Renouf V, Strehaiano P, Lonvaud-Funel A. Effectiveness of dimethyldicarbonate to prevent *Brettanomyces bruxellensis* growth in wine. Food Control. 2008;**19**(2):208-216

[41] Divol B, Strehaiano P, Lonvaud-Funel A. Effectiveness of dimethyldicarbonate to stop alcoholic fermentation in wine. Food Microbiology. 2005;**22**(2):169-178

[42] Delfini C, Gaia P, Schellino R, Strano M, Pagliara A, Ambrò S. Fermentability of grape must after inhibition with dimethyl dicarbonate (DMDC). Journal of Agriculture and Food Chemistry. 2002;**50**(20):5605-5611

[43] Costa A, Barata A, Malfeito-Ferreira M, Loureiro V. Evaluation of the inhibitory effect of dimethyl dicarbonate (DMDC) against wine microorganisms. Food Microbiology. 2008;**25**(2):422-427

[44] Zuehlke J, Glawe D, Edwards C. Efficacy of dimethyl dicarbonate against yeasts associated with Washington State grapes and wines. Journal of Food Processing and Preservation. 2014;**39**(6):1016-1026

[45] Oelofse A, Pretorius I, Du Toit M. Significance of *Brettanomyces* and *Dekkera* during winemaking: A synoptic review. South African Journal of Enology and Viticulture. 2008;**29**(2):128-144

[46] Threlfall R, Morris J. Using dimethyldicarbonate to minimize sulfur dioxide for prevention of fermentation from excessive yeast contamination in juice and semi-sweet wine. Journal of Food Science. 2002;67(7):2758-2762

[47] Resolution Oeno 6/2006. Method OIV-MA-AS313-20. Determination of sorbic, benzoic and salicylic acid content in wine by the use of high-performance liquid chromatography. Compendium of International Analysis of Methods – OIV. Available from: http://www.oiv.int/public/medias/2519/oiv-ma-as313-20.pdf [Accessed: 3 July 2017]

[48] Steels H, James S, Roberts I, Stratford M. Sorbic acid resistance: The inoculum effect. Yeast. 2000;**16**(13):1173-1183

[49] Loureiro V, Malfeito-Ferreira M. Spoilage activities of *Dekkera/Brettanomyces* spp. In: Blackburn C, editor. Food Spoilage Microorganisms. Cambridge, UK: Woodhead Publisher; 2006. pp. 354-398

[50] Oswald TA, Edwards CG. Interactions between storage temperature and ethanol that affect growth of *Brettanomyces bruxellensis* in Merlot wine. American Journal of Enology and Viticulture. 2017;**68**:188-194. ajev. 2017.16102

[51] Couto JA, Neves F, Campos F, Hogg T. Thermal inactivation of the wine spoilage yeasts *Dekkera/Brettanomyces*. International Journal of Food Microbiology. 2005,Oct 25;**104**(3):337-344

[52] Zuehlke JM, Edwards CG. Impact of sulfur dioxide and temperature on culturability and viability of *Brettanomyces bruxellensis* in wine. Journal of Food Protection®. 2013;**76**(12):2024-2030

[53] Wang C, Huang H, Hsu C, Yang BB. Recent advances in food processing using high hydrostatic pressure technology. Critical Reviews in Food Science and Nutrition. 2016;**56**(4):527-540

[54] Yaldagard M, Mortazavi SA, Tabatabaie F. The principles of ultra high pressure technology and its application in food processing/preservation: A review of microbiological and quality aspects. African Journal of Biotechnology. 2008;**7**(16):2739-2767

[55] Koutchma T, Popović V, Ros-Polski V, Popielarz A. Effects of ultraviolet light and high-pressure processing on quality and health-related constituents of fresh juice products. Comprehensive Reviews in Food Science and Food Safety. 2016;**15**(5):844-867

[56] del Árbol JT, Pulido RP, La Storia A, Burgos MJG, Lucas R, Ercolini D, et al. Microbial diversity in pitted sweet cherries (*Prunus avium* L.) as affected by high-hydrostatic pressure treatment. Food Research International. 2016;**89**:790-796

[57] Pulido RP, Burgos MJG, Gálvez A, Lucas R. Changes in bacterial diversity of refrigerated mango pulp before and after treatment by high hydrostatic pressure. LWT-Food Science and Technology. 2017;**78**:289-295

[58] Shahbaz HM, Yoo S, Seo B, Ghafoor K, Kim JU, Lee D, et al. Combination of TiO_2-UV photocatalysis and high hydrostatic pressure to inactivate bacterial pathogens and yeast in commercial apple juice. Food and Bioprocess Technology. 2016;**9**(1):182-190

[59] Martínez-Onandi N, Castioni A, San Martín E, Rivas-Cañedo A, Nuñez M, Torriani S, et al. Microbiota of high-pressure-processed Serrano ham investigated by culture-dependent and culture-independent methods. International Journal of Food Microbiology. 2017;**241**:298-307

[60] van Wyk S, Silva FV. High pressure inactivation of *Brettanomyces bruxellensis* in red wine. Food Microbiology. 2017;**63**:199-204

[61] González-Arenzana L, Sevenich R, Rauh C, López R, Knorr D, López-Alfaro I. Inactivation of *Brettanomyces bruxellensis* by high hydrostatic pressure technology. Food Control. 2016;**59**:188-195

[62] Morata A, Loira I, Vejarano R, Bañuelos MA, Sanz PD, Otero L, et al. Grape processing by high hydrostatic pressure: Effect on microbial populations, phenol extraction and wine quality. Food and Bioprocess Technology. 2015;**8**(2):277-286

[63] Milani EA, Silva FV. Nonthermal pasteurization of beer by high pressure processing: Modelling the inactivation of *Saccharomyces cerevisiae* ascospores in different alcohol beers. High Pressure Research. 2016;**36**(4):595-609

[64] Turtoi M. Inactivation of *Saccharomyces cerevisiae* using new non-thermal technologies. A review. Romanian Biotechnological Letters. 2014;**19**(1):8901-8909

[65] Gomez-Lopez VM, Ragaert P, Debevere J, Devlieghere F. Pulsed light for food decontamination: A review. Trends in Food Science and Technology. 2007;**18**(9):464-473

[66] Takeshita K, Shibato J, Sameshima T, Fukunaga S, Isobe S, Arihara K, et al. Damage of yeast cells induced by pulsed light irradiation. International Journal of Food Microbiology. 2003;**85**(1):151-158

[67] Ferrario M, Alzamora SM, Guerrero S. Inactivation kinetics of some microorganisms in apple, melon, orange and strawberry juices by high intensity light pulses. Journal of Food Engineering. 2013;**118**(3):302-311

[68] Ferrario M, Alzamora SM, Guerrero S. Study of the inactivation of spoilage microorganisms in apple juice by pulsed light and ultrasound. Food Microbiology. 2015;**46**:635-642

[69] Luksiene Z, Buchovec I, Viskelis P. Impact of high-power pulsed light on microbial contamination, health promoting components and shelf life of strawberries. Food Technology and Biotechnology. 2013;**51**(2):284

[70] Ganan M, Hierro E, Hospital XF, Barroso E, Fernández M. Use of pulsed light to increase the safety of ready-to-eat cured meat products. Food Control. 2013;**32**(2):512-517

[71] Kotnik T, Frey W, Sack M, Meglič SH, Peterka M, Miklavčič D. Electroporation-based applications in biotechnology. Trends in Biotechnology. 2015;**33**(8):480-488

[72] Hülsheger H, Potel J, Niemann E. Electric field effects on bacteria and yeast cells. Radiation and Environmental Biophysics. 1983;**22**(2):149-162

[73] Gášková D, Sigler K, Janderova B, Plášek J. Effect of high-voltage electric pulses on yeast cells: Factors influencing the killing efficiency. Bioelectrochemistry and Bioenergetics. 1996;**39**(2):195-202

[74] Buckow R, Ng S, Toepfl S. Pulsed electric field processing of orange juice: A review on microbial, enzymatic, nutritional, and sensory quality and stability. Comprehensive Reviews in Food Science and Food Safety. 2013;**12**(5):455-467

[75] Aronsson K, Rönner U, Borch E. Inactivation of *Escherichia coli, Listeria innocua* and *Saccharomyces cerevisiae* in relation to membrane permeabilization and subsequent leakage of intracellular compounds due to pulsed electric field processing. International Journal of Food Microbiology. 2005;**99**(1):19-32

[76] Ganeva V, Galutzov B, Teissie J. Evidence that pulsed electric field treatment enhances the cell wall porosity of yeast cells. Applied Biochemistry and Biotechnology. 2014;**172**(3):1540-1552

[77] Gómez-Rivas L, Escudero-Abarca BI, Aguilar-Uscanga MG, Hayward-Jones PM, Mendoza P, Ramírez M. Selective antimicrobial action of chitosan against spoilage yeasts in mixed culture fermentations. Journal of Industrial Microbiology and Biotechnology. 2004;**31**(1):16-22

[78] Raafat D, Sahl H. Chitosan and its antimicrobial potential—A critical literature survey. Microbial Biotechnology. 2009;**2**(2):186-201

[79] Roller S, Covill N. The antifungal properties of chitosan in laboratory media and apple juice. International Journal of Food Microbiology. 1999;**47**(1):67-77

[80] Ferreira D, Moreira D, Costa EM, Silva S, Pintado MM, Couto JA. The antimicrobial action of chitosan against the wine spoilage yeast *Brettanomyces/Dekkera*. Journal of Chitin and Chitosan Science. 2013;**1**(3):240-245

[81] Taillandier P, Joannis-Cassan C, Jentzer J, Gautier S, Sieczkowski N, Granes D, et al. Effect of a fungal chitosan preparation on *Brettanomyces bruxellensis*, a wine contaminant. Journal of Applied Microbiology. 2015;**118**(1):123-131

[82] Petrova B, Cartwright ZM, Edwards CG. Effectiveness of chitosan preparations against *Brettanomyces bruxellensis* grown in culture media and red wines. OENO One. 2016;**50**(1):49-56

[83] Iriti M, Vitalini S, Di Tommaso G, D–AMICO S, Borgo M, Faoro F. New chitosan formulation prevents grapevine powdery mildew infection and improves polyphenol content and free radical scavenging activity of grape and wine. Australian Journal of Grape and Wine Research. 2011;**17**(2):263-269

[84] Fajardo P, Martins J, Fuciños C, Pastrana L, Teixeira J, Vicente A. Evaluation of a chitosan-based edible film as carrier of natamycin to improve the storability of Saloio cheese. Journal of Food Engineering. 2010;**101**(4):349-356

[85] Souza JM, Caldas AL, Tohidi SD, Molina J, Souto AP, Fangueiro R, et al. Properties and controlled release of chitosan microencapsulated limonene oil. Revista Brasileira de Farmacognosia. 2014;**24**(6):691-698

[86] Elmacı SB, Gülgör G, Tokatlı M, Erten H, İşci A, Özçelik F. Effectiveness of chitosan against wine-related microorganisms. Antonie Van Leeuwenhoek. 2015;**107**(3):675-686

[87] Conner D, Beuchat L. Effects of essential oils from plants on growth of food spoilage yeasts. Journal of Food Science. 1984;**49**(2):429-434

[88] Curtis O, Shetty K, Cassagnol G, Peleg M. Comparison of the inhibitory and lethal effects of synthetic versions of plant metabolites (anethole, carvacrol, eugenol, and thymol) on a food spoilage yeast (*Debaromyces hansenii*). Food Biotechnology. 1996;**10**(1):55-73

[89] Burt S. Essential oils: Their antibacterial properties and potential applications in foods—A review. International Journal of Food Microbiology. 2004;**94**(3):223-253

[90] Bennis S, Chami F, Chami N, Bouchikhi T, Remmal A. Surface alteration of *Saccharomyces cerevisiae* induced by thymol and eugenol. Letters in Applied Microbiology. 2004;**38**(6): 454-458

[91] Kubo I, Cespedes CL. Antifungal activity of alkanols: Inhibition of growth of spoilage yeasts. Phytochemistry Reviews. 2013;**12**(4):961-977

[92] Sivropoulou A, Papanikolaou E, Nikolaou C, Kokkini S, Lanaras T, Arsenakis M. Antimicrobial and cytotoxic activities of Origanum essential oils. Journal of Agricultural and Food Chemistry. 1996;**44**(5):1202-1205

[93] Chavan PS, Tupe SG. Antifungal activity and mechanism of action of carvacrol and thymol against vineyard and wine spoilage yeasts. Food Control. 2014;**46**:115-120

[94] Rabinkov A, Miron T, Konstantinovski L, Wilchek M, Mirelman D, Weiner L. The mode of action of allicin: Trapping of radicals and interaction with thiol containing proteins. Biochimica et Biophysica Acta (BBA)-General Subjects. 1998;**1379**(2):233-244

[95] Chang T, Jang H, Lin W, Duan P. Antifungal activities of commercial rice wine extracts of Taiwanese *Allium fistulosum*. Advances in Microbiology. 2016;**6**(07):471

[96] Oommen S, Anto RJ, Srinivas G, Karunagaran D. Allicin (from garlic) induces caspase-mediated apoptosis in cancer cells. European Journal of Pharmacology. 2004;**485**(1):97-103

[97] Lee J, Gupta S, Huang J, Jayathilaka LP, Lee B. HPLC-MTT assay: Anticancer activity of aqueous garlic extract is from allicin. Analytical Biochemistry. 2013;**436**(2):187-189

[98] Yoshida H, Katsuzaki H, Ishikawa K, Fukuda H, Fujino T, Suzuki A. Antimicrobial activity of the thiosulfinates isolated from oil-macerated garlic extract. Bioscience, Biotechnology, and Biochemistry. 1999;**63**(3):591-594

[99] Shadkchan Y, Shemesh E, Mirelman D, Miron T, Rabinkov A, Wilchek M, et al. Efficacy of allicin, the reactive molecule of garlic, in inhibiting *Aspergillus* spp. in vitro, and in a murine model of disseminated aspergillosis. Journal of Antimicrobial Chemotherapy. 2004 May;**53**(5):832-836

[100] Ankri S, Mirelman D. Antimicrobial properties of allicin from garlic. Microbial Infections. 1999;**1**(2):125-129

[101] Cobos R, Mateos RM, Alvarez-Perez JM, Olego MA, Sevillano S, Gonzalez-Garcia S, et al. Effectiveness of natural antifungal compounds in controlling infection by grapevine trunk disease pathogens through pruning wounds. Applied Environmental Microbiology. 2015 Sep;**81**(18):6474-6483

[102] Davis SR. An overview of the antifungal properties of allicin and its breakdown products—The possibility of a safe and effective antifungal prophylactic. Mycoses. 2005;**48**(2):95-100

[103] Marchese A, Barbieri R, Sanches-Silva A, Daglia M, Nabavi SF, Jafari NJ, et al. Antifungal and antibacterial activities of allicin: A review. Trends in Food Science and Technology. 2016;**52**:49-56

[104] Chang Y, McLandsborough L, McClements DJ. Physical properties and antimicrobial efficacy of thyme oil nanoemulsions: Influence of ripening inhibitors. Journal of Agriculture and Food Chemistry. 2012;**60**(48):12056-12063

[105] Maté J, Periago PM, Palop A. When nanoemulsified, d-limonene reduces *Listeria monocytogenes* heat resistance about one hundred times. Food Control. 2016;**59**:824-828

[106] Blanco-Padilla A, Soto KM, Hernandez Iturriaga M, Mendoza S. Food antimicrobials nanocarriers. Scientific World Journal. 2014;**2014**:837215

[107] Fogaça AC, Lorenzini DM, Kaku LM, Esteves E, Bulet P, Daffre S. Cysteine-rich antimicrobial peptides of the cattle tick *Boophilus microplus*: Isolation, structural characterization and tissue expression profile. Developmental & Comparative Immunology. 2004;**28**(3):191-200

[108] Schibli DJ, Epand RF, Vogel HJ, Epand RM. Tryptophan-rich antimicrobial peptides: Comparative properties and membrane interactions. Biochemistry and Cell Biology. 2002;**80**(5):667-677

[109] Jung HJ, Park Y, Hahm K, Lee DG. Biological activity of Tat (47-58) peptide on human pathogenic fungi. Biochemical and Biophysical Research Communications. 2006;**345**(1):222-228

[110] Chan DI, Prenner EJ, Vogel HJ. Tryptophan-and arginine-rich antimicrobial peptides: Structures and mechanisms of action. Biochimica et Biophysica Acta (BBA)-Biomembranes. 2006;**1758**(9):1184-1202

[111] Zhao H, Mattila J, Holopainen JM, Kinnunen PK. Comparison of the membrane association of two antimicrobial peptides, magainin 2 and indolicidin. Biophysics Journal. 2001;**81**(5):2979-2991

[112] Schibli DJ, Hwang PM, Vogel HJ. Structure of the antimicrobial peptide tritrpticin bound to micelles: A distinct membrane-bound peptide fold. Biochemistry. 1999;**38**(51):16749-16755

[113] Salay LC, Procopio J, Oliveira E, Nakaie CR, Schreier S. Ion channel-like activity of the antimicrobial peptide tritrpticin in planar lipid bilayers. FEBS Letters. 2004;**565**(1-3):171-175

[114] Yang S, Shin SY, Kim Y, Kim Y, Hahm K, Kim JI. Conformation-dependent antibiotic activity of tritrpticin, a cathelicidin-derived antimicrobial peptide. Biochemical and Biophysical Research Communications. 2002;**296**(5):1044-1050

[115] Nguyen LT, de Boer L, Zaat SA, Vogel HJ. Investigating the cationic side chains of the antimicrobial peptide tritrpticin: Hydrogen bonding properties govern its membrane-disruptive activities. Biochimica et Biophysica Acta (BBA)-Biomembranes. 2011;**1808**(9):2297-2303

[116] Tang Y, Shi Y, Zhao W, Hao G, Le G. Discovery of a novel antimicrobial peptide using membrane binding-based approach. Food Control. 2009;**20**(2):149-156

[117] Tomie T, Ishibashi J, Furukawa S, Kobayashi S, Sawahata R, Asaoka A, et al. Scarabaecin, a novel cysteine-containing antifungal peptide from the rhinoceros beetle, Oryctes rhinoceros. Biochemical and Biophysical Research Communications.2003;**307**(2):261-266

[118] Yang S, Shin SY, Lee CW, Kim Y, Hahm K, Kim JI. Selective cytotoxicity following Arg-to-Lys substitution in tritrpticin adopting a unique amphipathic turn structure. FEBS Letters. 2003;**540**(1-3):229-233

[119] Subbalakshmi C, Sitaram N. Mechanism of antimicrobial action of indolicidin. FEMS Microbiology Letters. 1998 Mar 1;**160**(1):91-96

[120] Falla TJ, Karunaratne DN, Hancock RE. Mode of action of the antimicrobial peptide indolicidin. Journal of Biological Chemistry. 1996;**271**(32):19298-19303

[121] Ladokhin AS, Selsted ME, White SH. Bilayer interactions of indolicidin, a small antimicrobial peptide rich in tryptophan, proline, and basic amino acids. Biophysics Journal. 1997;**72**(2):794-805

[122] Rozek A, Friedrich CL, Hancock RE. Structure of the bovine antimicrobial peptide indolicidin bound to dodecylphosphocholine and sodium dodecyl sulfate micelles. Biochemistry. 2000;**39**(51):15765-15774

[123] Friedrich CL, Rozek A, Patrzykat A, Hancock RE. Structure and mechanism of action of an indolicidin peptide derivative with improved activity against gram-positive bacteria. Journal of Biological Chemistry. 2001 Jun 29;**276**(26):24015-24022

[124] Marchand C, Krajewski K, Lee HF, Antony S, Johnson AA, Amin R, et al. Covalent binding of the natural antimicrobial peptide indolicidin to DNA abasic sites. Nucleic Acids Research. 2006;**34**(18):5157-5165

[125] Lee DG, Kim HK, Am Kim S, Park Y, Park S, Jang S, et al. Fungicidal effect of indolicidin and its interaction with phospholipid membranes. Biochemical and Biophysical Research Communications. 2003;**305**(2):305-310

[126] Bellamy W, Takase M, Wakabayashi H, Kawase K, Tomita M. Antibacterial spectrum of lactoferricin B, a potent bactericidal peptide derived from the N-terminal region of bovine lactoferrin. Journal of Applied Bacteriology. 1992;**73**(6):472-479

[127] Vogel HJ, Schibli DJ, Jing W, Lohmeier-Vogel EM, Epand RF, Epand RM. Towards a structure-function analysis of bovine lactoferricin and related tryptophan-and arginine-containing peptides. Biochemistry and Cell Biology. 2002;**80**(1):49-63

[128] Enrique M, Marcos JF, Yuste M, Martínez M, Vallés S, Manzanares P. Inhibition of the wine spoilage yeast *Dekkera bruxellensis* by bovine lactoferrin-derived peptides. International Journal of Food Microbiology. 2008;**127**(3):229-234

[129] Enrique M, Marcos JF, Yuste M, Martínez M, Vallés S, Manzanares P. Antimicrobial action of synthetic peptides towards wine spoilage yeasts. International Journal of Food Microbiology. 2007;**118**(3):318-325

[130] Frank JF. Milk and Dairy Products. Food Microbiology: Fundamentals and Frontiers. 3rd ed. American Society of Microbiology, Washington, DC. 2007. pp. 141-155

[131] Duan J, Park S, Daeschel M, Zhao Y. Antimicrobial chitosan-lysozyme (CL) films and coatings for enhancing microbial safety of mozzarella cheese. Journal of Food Science. 2007;**72**(9):M355–M362

[132] Ryge T, Doisy X, Ifrah D, Olsen J, Hansen P. New indolicidin analogues with potent antibacterial activity. The Journal of Peptide Research. 2004;**64**(5):171-185

[133] Benincasa M, Scocchi M, Pacor S, Tossi A, Nobili D, Basaglia G, et al. Fungicidal activity of five cathelicidin peptides against clinically isolated yeasts. Journal of Antimicrobial Chemotherapy. 2006 Nov;**58**(5):950-959

[134] Sader HS, Fedler KA, Rennie RP, Stevens S, Jones RN. Omiganan pentahydrochloride (MBI 226), a topical 12-amino-acid cationic peptide: Spectrum of antimicrobial activity and measurements of bactericidal activity. Antimicrobial Agents and Chemotherapy. 2004 Aug;**48**(8):3112-3118

[135] Krajewski K, Marchand C, Long Y, Pommier Y, Roller PP. Synthesis and HIV-1 integrase inhibitory activity of dimeric and tetrameric analogs of indolicidin. Bioorganic & Medicinal Chemistry Letters. 2004;**14**(22):5595-5598

[136] Vorland LH, Ulvatne H, Andersen J, Haukland HH, Rekdal Ø, Svendsen JS, et al. Lactoferricin of bovine origin is more active than lactoferricins of human, murine and caprine origin. Scandinavian Journal of Infectious Diseases. 1998;**30**(5):513-517

[137] Bussey H. K1 killer toxin, a pore-forming protein from yeast. Molecular Microbiology. 1991;**5**(10):2339-2343

[138] Villalba ML, Sáez JS, del Monaco S, Lopes CA, Sangorrín MP. TdKT, a new killer toxin produced by *Torulaspora delbrueckii* effective against wine spoilage yeasts. International Journal of Food Microbiology. 2016;**217**:94-100.

[139] Oro L, Ciani M, Bizzaro D, Comitini F. Evaluation of damage induced by Kwkt and Pikt zymocins against *Brettanomyces/Dekkera* spoilage yeast, as compared to sulphur dioxide. Journal of Applied Microbiology. 2016;**121**:207-214.

[140] Comitini F, Di Pietro N, Zacchi L, Mannazzu I, Ciani M. *Kluyveromyces phaffii* killer toxin active against wine spoilage yeasts: Purification and characterization. Microbiology. 2004;**150**(8):2535-2541

[141] Oro L, Zara S, Fancellu F, Mannazzu I, Budroni M, Ciani M, et al. TpBGL2 codes for a *Tetrapisispora phaffii* killer toxin active against wine spoilage yeasts. FEMS Yeast Research. 2014 May;**14**(3):464-471

[142] Anonymous N. Tetrapisispora phaffii. 2011. Available from: https://www.ncbi.nlm.nih.gov/genome/?term=Tetrapisispora+phaffii [Accessed: July 3, 2017]

[143] Maqueda M, Zamora E, Alvarez ML, Ramirez M. Characterization, ecological distribution, and population dynamics of *Saccharomyces* sensu stricto killer yeasts in the spontaneous grape must fermentations of southwestern Spain. Applied and Environmental Microbiology. 2012 Feb;**78**(3):735-743

[144] Marquina D, Santos A, Peinado J. Biology of killer yeasts. International Microbiology. 2002;**5**(2):65-71

[145] Enrique M, Ibáñez A, Marcos J, Yuste M, Martinez M, Vallés S, et al. β-glucanases as a tool for the control of wine spoilage yeasts. Journal of Food Science. 2010;**75**(1):M41–M45

[146] Mehlomakulu NN, Setati ME, Divol B. Characterization of novel killer toxins secreted by wine-related non-*Saccharomyces* yeasts and their action on *Brettanomyces* spp. International Journal of Food Microbiology. 2014;**188**:83-91

[147] Santos A, Navascués E, Bravo E, Marquina D. *Ustilago maydis* killer toxin as a new tool for the biocontrol of the wine spoilage yeast *Brettanomyces bruxellensis*. International Journal of Food Microbiology. 2011;**145**(1):147-154

[148] García M, Esteve-Zarzoso B, Arroyo T. Non-*Saccharomyces* yeasts: Biotechnological role for wine production. In Grape and Wine Biotechnology. InTech. 2016

[149] Longin C, Petitgonnet C, Guilloux-Benatier M, Rousseaux S, Alexandre H. Application of flow cytometry to wine microorganisms. Food Microbiology. 2017April;**62**:221-231

[150] Malacrinò P, Zapparoli G, Torriani S, Dellaglio F. Rapid detection of viable yeasts and bacteria in wine by flow cytometry. Journal of Microbiology Methods. 2001;**45**(2):127-134

[151] Veal D, Deere D, Ferrari B, Piper J, Attfield P. Fluorescence staining and flow cytometry for monitoring microbial cells. Journal of Immunological Methods. 2000;**243**(1):191-210

[152] Walker GM. Yeasts. Desk Encyclopedia of Microbiology. 2nd ed. London: Elsevier/Academic Press; 2009

[153] Khiyami MA, Almoammar H, Awad YM, Alghuthaymi MA, Abd-Elsalam KA. Plant pathogen nanodiagnostic techniques: Forthcoming changes? Biotechnology & Biotechnological Equipment. 2014;**28**(5):775-785

[154] Lim DV, Simpson JM, Kearns EA, Kramer MF. Current and developing technologies for monitoring agents of bioterrorism and biowarfare. Clinical Microbiology Reviews. 2005 Oct;**18**(4):583-607

[155] Goddard G, Martin JC, Naivar M, Goodwin PM, Graves SW, Habbersett R, et al. Single particle high resolution spectral analysis flow cytometry. Cytometry Part A. 2006;**69**(8):842-851

[156] Watson DA, Gaskill DF, Brown LO, Doorn SK, Nolan JP. Spectral measurements of large particles by flow cytometry. Cytometry Part A. 2009;**75**(5):460-464

Chapter 5

Contribution of Yeast in Wine Aroma and Flavour

Minas Mina and Dimitrios Tsaltas

Additional information is available at the end of the chapter

http://dx.doi.org/10.5772/intechopen.70656

Abstract

Organoleptic characteristics of wine, especially, the spectrum that is defined as flavour and aroma, are the most important parameters for assessing the quality of wine. The origin of these characteristics comes for four main sources: grapes, vinification, maturation and ageing. The final concentrations of various odour-active components (OAC) are highly depended on the yeast during fermentation. The major OAC that are formed during fermentation are volatile substances like esters, higher alcohols and carbonyl compounds. Decoding the origin and contribution of these OAC, the modern winemaker can direct and manipulate the yeast during fermentation on his benefit. These compounds are originated from the secondary metabolism of the yeast, understanding the role of the key parameters during fermentation influencing the OAC formation like temperature, yeast assimilable nitrogen (YAN) and suspended solids is vital for the final organoleptic characteristics of wine.

Keywords: yeast, wine, aroma, flavour, fermentation, volatiles, esters, higher alcohols

1. Introduction

Wine is the alcoholic beverage which is the product of fermentation, usually, of fresh grape must. Wine consists mainly of 86.8% water and weight by volume concentration of the following: 11.2% ethanol, 0.5% acids (volatile and non-volatile), 1% trace components (sugars, anions, cations, etc.) and only a very small portion of 0.5% of volatiles contributing to the aroma of wine [1, 2], often described as odour-active compounds (OAC). These compounds are part of the olfactory fingerprint of each wine. The concentration and the ratio between various groups of OAC are unique not only to each wine but also to each terroir, to the style of vinification and maturation procedures. Yeast has a major role in the wine aroma formation and modulation, apart from the formation of alcohol. According to Fleet [3], the way that yeast influences the final aroma of wine can follow these six mechanisms:

i. Involve in the biocontrol of moulds on grape, which influences quality before harvest.

ii. Perform alcoholic fermentation of must sugars and transform juice into wine; the *de novo* biosynthesis of the flavour and aroma compounds.

iii. Enzymatic conversion of flavour neutral, grape components into odour-active compounds.

iv. Alterations of OAC profile through the yeast autolysis products.

v. Absorption of grape juice components.

vi. Spoilage of bulk wine throughout the storage period and even after packaging.

vii. Influence growth of other spoilage microorganism, for example, lactic acid bacteria, acetic acid bacteria.

The single most important mechanism, which can be manipulated by the winemaker, of the above list is that of fermentation. Due to the fact that during the procedure of fermentation the largest concentration the final OAC in wine is formed [3–5]. The input and manipulation of winemaker that can influence the final product, this is done through decision-making for the implementation of various vinifications practises and, like the fermentation temperature [6–10], inoculation [11–13], addition of yeast assimilable nitrogen (YAN) [14–17] and the initial total concentration of the suspended solids of must before inoculation [18–21].

Throughout the fermentation process, the environment, which the yeast is called to function, is under continuous changes [22]. During fermentation, a vast amount of heat is being produce by the yeast, although this is counter balanced, and is easily control, by the modern temperature control tanks; temperature is one of the main limiting factors, for yeast growth. Also due to the sugar transformation and the utilisation of oxygen and YAN, yeast should engage various mechanisms in a diverse unaffordable growth conditions, called 'stress', to the emerging environment alterations [22]:

i. Osmotic pressure alterations

ii. Limitation of essential nutrients

iii. Ethanol toxicity

iv. Production of by-products toxic to the cell

The continuous need for adaptation to this kind of environment emerges the need for the corresponding adaptation on various responses, in order to maintain the intracellular metabolic activity [22].

Fermentation of sugars by yeast can be divided into two stages: the primary and the secondary metabolism. By primary, we mean the metabolism that is essential for yeast growth and cell division, producing compounds like ethanol, glycerol, acetaldehyde and acetic acid. Secondary metabolism is non-essential for growth and produces small molecules. Through this secondary metabolism, yeast adaptation procedures can mainly influence the final wine aroma profile [23]. Namely, the FAC generating mechanisms in which the yeast is engaged are the following [3]:

i. Utilising grape juice constituents

ii. Producing ethanol and other solvents that help to extract flavour components from grape solids

iii. Producing enzymes that transform neutral grape compounds into flavour active compounds

iv. Producing many hundreds of flavour active, secondary metabolites (e.g. acids, alcohols, esters, polyols, aldehydes, ketones and volatile sulphur compounds)

v. Autolytic degradation of dead yeast cells

2. Wine aroma: origin and type

Overall, the aroma of wine can be distinct into primary, being the OAC derived directly from the fruit; characteristic to the grape variety, secondary aroma generated during the vinification/fermentation and lastly maturation and ageing procedures are responsible for the group of aroma characteristics that are described as tertiary [24, 25].

The major OAC are divided into four big groups: esters, aldehydes, alcohols and terpenes [25, 26]. In addition to these, there are also two other groups that are very characteristic for specific grape varieties, these are pyrazines (primary aroma) and sulphur compounds like polyfunctional sulphur compounds (4-mercapto-4-methyl-pentan-2-one, 3-mercaptophenol) and dimethyl sulphide DMS [27]. Each one of these groups plays a unique role in the perception of the aroma character of the final wine. Also all these groups have a diverse formation pathway.

2.1. Flavour active groups contributing in wine aroma

2.1.1. Esters formation and contribution

Esters are the group with the highest importance in wine and are usually the most predominant in the formation of the flavour character of the final product [22]. Esters are formed with the combination of alcohols and organic acid with the elimination of water [25]. In wine, two types of this group can be met: first, the one that is modulated enzymatically by the yeast enzyme pool and second, the one that is formed during ageing [2, 28]. The enzymatic biosynthesis of ester is catalysed mainly by two types of enzymes: esterases and lipases. The final profile of esters in wine depends on various parameters, many authors pointed the variety and quantities of esters, in **Table 1**, are the main esters found in commercial wine [29]. The group of ethyl acetate, isoamyl acetate, isobutyl acetate, ethyl caproate and 2-phenyl acetate are descripted as the most important esters affecting wine flavour [28, 30, 31].

The net concentration of ester in wine at any given time varies due to the fact that wine is a fairly complex matrix substrate, with a number of different compounds involve in various procedures [2]. This depends on the enzymatic activities of synthesis and ester hydrolysis. Maximum concentrations of esters during fermentation observe around 9–12% of ethanol [8, 32].

Compound	Sweet wines (mg/l)	Dry wines (mg/l)
Ethyl acetate	96.56 ± 39.75	85.00 ± 12.54
Isobutyl acetate	0.07 ± 0.02	0.07 ± 0.04
Ethyl butyrate	0.31 ± 0.09	0.41 ± 0.05
Isoamyl acetate	1.81 ± 0.91	2.37 ± 0.62
Ethyl hexanoate	0.87 ± 0.41	1.06 ± 0.19
Hexyl acetate	0.06 ± 0.04	0.14 ± 0.14
Ethyl lactate	13.5 ± 6.6	23.00 ± 18.88
Ethyl octanoate	1.57 ± 0.73	2.11 ± 0.49
Ethyl decanoate	0.65 ± 0.26	0.56 ± 0.06
Benzyl acetate	0.004 ± 0.004	0.003 ± 0.001
2-Phenylethyl acetate	0.23 ± 0.17	0.21 ± 0.05
Ethyl dodecanoate	0.079 ± 0.053	0.021 ± 0.007

Table 1. Range of ester contents in commercial white wines [29].

Esters formed during alcoholic fermentation, by enzymes; fall into two main categories. The ethyl esters of organic acids and the acetates of higher alcohols [23]. The ethyl esters comprise of an alcohol group (ethanol) and an acid group (small, medium-chain fatty acid) (**Figure 1**). The acetate esters are comprised of an acid group (acetate) and an alcohol group which is either ethanol or a higher alcohol derived from amino acid metabolism (**Figure 2**). The latter are responsible for the pleasant fruity aroma of wines [2].

Formation of these two groups of esters during alcoholic fermentation involves a series of various proteins and genes. Today six genes have been identified, with their corresponding protein, to be involved in either the synthesis or hydrolysis of esters in yeast cells. Namely, these are ATF1, Lg-ATF1, ATF2, EHT1, EEB1 and IAH1 [2, 5]. The first three are involved in the mechanism of alcohol acetyltransferase with the first to be the most studied and important in the total quantity of esters formed. EHT1 is involved in the ethanol hexanol-transferase mechanism for the synthesis and hydrolysis of medium-chain fatty acids ethyl esters. EEB1 is

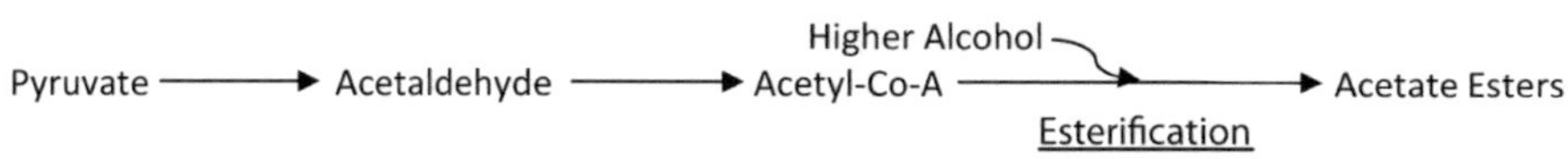

Figure 1. Formation of acetate esters by the esterification of acetyl-Co-A with a higher alcohol [23].

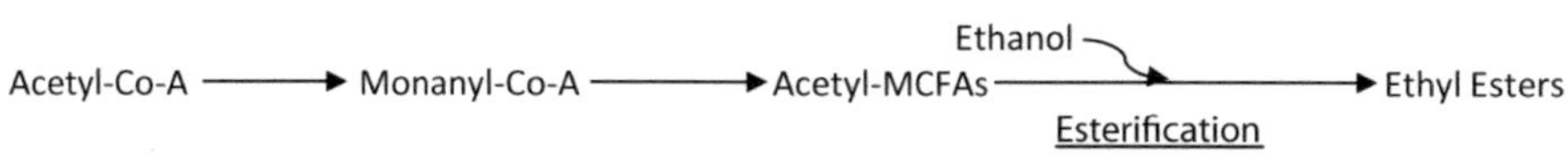

Figure 2. Formation of ethyl esters by the esterification of medium-chain fatty acid (MCFA) with a ethanol [23].

involved in the ethanol acyltrasferase and ethyl hydrolase mechanism. Lastly, the IAH1 is a gene involved in the mechanism of esterase, for the hydrolysis of acetate esters [2, 5].

The acetate ester synthesis is an energy requiring mechanism [31], the reasons why esters are formed during the fermentation procedure are not quite clear. One approach to this is that the synthesis is associated with the detoxification effect by the removal of fatty acids [31]. Even though the transfer of esters through the membrane is directly associated with the length of their chain that is varying from 100% small medium-chain to 8–17% for a long chain (ethyl decanoate) [2]. Another systematic approach for the formation is that of the maintenance in the balance of the acetyl-CoA and CoA-SH pool. It is proposed that these are formed as over-spill products from the fermentation through the sugar metabolism [31].

2.2. Higher alcohols or fusel alcohols origin and contribution

Higher alcohols refers to the group of alcohols that have more than two atoms of carbon on their molecule. This group of compounds with esters are the two biggest groups contributing to the aroma of wine. The vital step in the synthesis of these compounds is the formation of α-ketoacid [33]. Based on the origin of α-ketoacid fusel alcohols can be divided into two categories. The first has origin of α-ketoacid the amino acids and the second, the anabolic pathway of sugars [27]. From the first group, a list of higher alcohols is shown in **Table 2**, were as from the former 1-butanol and 1-pentanol are formed. Their contribution to wine aroma is consider positive when the concentration of these compounds is up to 300 mg/l, above this level the pungent odour is profound [25, 31]. The utilisation of nitrogen sources is strongly associated with the biosynthesis of these higher alcohols [17, 34, 35]. The nitrogen composition and nature (organic or mineral), of the must, are influencing the biosynthesis of these volatiles. It has been shown by many studies that the initial concentration and type of amino acids in the must in some cases is strongly associated with varietal aromas [5]. The observation that the increase in the concentration of certain amino acids led to the increase of the production of specific fusel alcohols let to the formulation of the Ehrlich pathway [33]. Also the well documented Ehrlich pathway intermediates can be found in bibliography, stating, the following **Table 2** of the 'substrate' amino acid and their corresponding fusel alcohol [5].

Amino acid	Higher alcohol
Leusine—Leu	3-Methylbutanol
Valine—Val	2-Methylpropanol
Isoleucine—Ile	2-Methylbutanol
Phenylalanine—Phe	2-Phenylethanol
Tyrosine—Tyr	2-(4-Hydroxylphenyl) ethanol
Tryptophan—Trp	2-(Indol-3-yl) ethanol
Methionine—Met	3-(methyl thio) propanol

Table 2. Flavour-producing amino acid catabolism via the Ehrlich pathway [5].

The biosynthesis in the Ehrlich pathway starts with the transamination of the amino acid producing a α-ketoacid [33]. Followed by the decarboxylation α-ketoacid to a fusel aldehydes [33]. Finally is the decisive step were the fusel aldehyde is reduced to fusel alcohol or oxidised to the corresponding fusel acid. This step is highly dependable on the growing condition during the utilisation of amino acids [27, 33]. In aerobic conditions, amino acids are converted predominantly to fusel acids where as in anaerobic the product of the pathway in almost entirely fusel alcohol. However, the procedure is far from simple, since a vast number of genes and their corresponding proteins are involve in every stage.

Four *S. cerevisiae* proteins have been implicated in the initial transamination step of the Ehrlich pathway (**Figure 3**). Twt1p (also known as Bat1p or Eca39p) is the mitochondrial branched-chain amino acid aminotransferase, and Twt2p (Bat2p or Eca40p) is the cytosolic isozyme. The mitochondrial isozyme is highly expressed in batch cultures during exponential growth and is repressed during stationary phase, while the cytosolic isozyme has the opposite expression pattern [33].

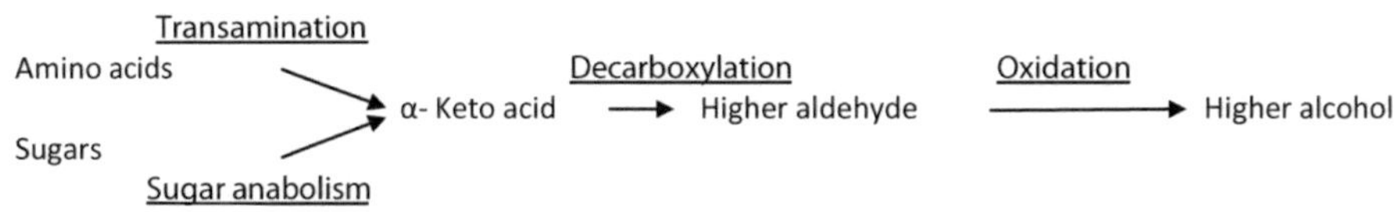

Figure 3. Formation of higher alcohols by the Ehrlich pathway and the anabolism of sugars.

The course of the formation of various higher alcohols was the study of different researchers. According to Fraile et al. [14, 36], the formation of different alcohols takes place at the end of the fermentation, when most of the amino acids have been consumed, whereas, according to Rapp and Versini [37], this synthesis occurs at the same time as ethanol production. A more recent study by Hernandez-Orte et al. [38] monitoring the course of formation of alcohols, among other volatiles, clearly shows that the formation of isoamyl, isobutanol and β-phenylethanol are generated throughout the entire alcoholic fermentation.

2.3. Carbonyl *c*ompounds origin and contribution

The two major compounds in this group are acetaldehyde and diacetyl. Acetaldehyde, which is one of the main metabolic intermediates in alcoholic fermentation, is the last precursor in the anaerobic pathway before ethanol. The pyruvate, end product of glycolysis, is converted to acetaldehyde by the pyruate decarboxylase enzymes, which is further converted to ethanol, by the dehydrogenase enzymes. Another source of acetaldehyde is the oxidation of ethanol during ageing or the activity of film forming yeast to the wine [25, 27] 'flor' effect.

Diacetyl is formed in small quantities in wine by the yeast. This can further metabolised to the corresponding end product of 2,3-butanediol or the intermediate, acetoin. Concentrations of up to 100 mg/L of acetaldehyde and 1–4 mg/L of diacetyl can be described as desirable and that these are contributing to the complexity of the aroma of wine [25, 27].

2.4. Thiols contribution and origin

Thiols are the group of alcohol compounds that oxygen was replaced by sulphur in the hydroxyl group. This compound group is very characteristic for wine aroma, especially for the Sauvignon blanc variety, and is mostly depended on yeast metabolism. The three thiols that have been identified in wine are 3-mercaptohexan-1-ol (3MH), 3-mercaptohexyl acetate (3MHA) and 4-mercapto-4-methylpentan-2-one (4MMP). These compounds apart from been characteristic are also very interesting for studying, exhibiting very low perception thresholds, that is, for 4MMP 3 ng/L, for 3MH 50–60 ng/L and for the acetylated form of 3MH the 3MHA is down to 2–4 ng/L [39]. In addition, the 3MH and 3MHA both have two enantiomers, as chiral molecules, R and S. The aroma the S-3MH and S-3MHA forms were described as passion fruit whereas the form R-3MH as grape fruit and R-3MHA as box tree [40].

Over the years, on Sauvignon blanc variety, two different molecule classes have been proposed as precursors of thiols, the amino acid-based compounds [41, 42] and the non-amino acid [43]. Most studies are focused on the first class of compounds and more specific to Cysteine-conjugates and Glutathione-conjugates. Interestingly, the amino compounds not only are the main precursors but also there was the evidence that glutathionylated precursors can be converted into cysteinylated precursors [44]. The non-amino acid precursor compounds were suggested to be the mesityl oxide and E-2-hexenal [43].

2.5. Terpenes: de novo synthesis and/or biotransformations by yeasts

Clearly terpenes are responsible for some of the most prominent, characteristic and important aromas in grapes and wines. It has been documented early on (1978) that beside grapes, yeasts are also capable of producing terpenes (citronellol, linalool and geraniol by *Kluyveromyces lactis* [45]. Enzymatic activity by yeasts is also possible in relation to liberation of terpenes from sugar molecules and β-glucosidase is well documented [46].

Saccharomyces cerevisiae shows some enzyme activity in different strains [47] but most studies demonstrate significantly higher enzyme production from non-*Saccharomyces* species [48–51]. Efforts to identify the most efficient non-*Saccharomyces* are showing the potential these yeasts have in modern wine making and mix cultures. *Torulaspora delbrueckii* and *Metschnikowia pulcherrima* are enhancing a very good aromatic profile if used in combination with *S. cerevisiae* [52–55]. Efforts to further our knowledge on the related pathways [56] as well as better exploiting the capacity of mixed cultures (*Saccharomyces* with non-*Saccharomyces*) are copious [57–62].

3. Fermentation conditions and influence to wine aroma

The winemaker through the process of vinifications has various parameters that can use in his/her benefit, in order to manipulate the outcome of the process. These parameters are the temperature and molecular oxygen availability during fermentation, maturation and ageing, the nitrogen source, for the growth and propagation of yeast, inoculation size and yeast strain of the starting culture [3, 31], as well as the nature and quantity of the solids derived from the grapes. All these play a decisive role in the vinification strategy and style that the winemaker wants to follow. Although for most of the OAC, the formation pathway and production promoting parameters are clear in some cases the knowledge behind synergies between these parameters are not quite apparent. The winemaker also has to deal with the changing environment of the fermentation, and more specific to the metabolism of sugar into ethanol and carbon dioxide [9, 22]. These two metabolites, but more essentially, ethanol build-up consecration plays a significant role for the physiology of the yeast. First, high concentrations of ethanol are related to the reduced water activity; this has a triggering effect to the production of various compounds to counter-balance this, but most importantly is the functional alteration of cell membrane that is influencing the uptake of various essential nutrients, important for the yeast survival and growth, including nitrogen compounds, YAN [9, 22]. Although fermentation is a well-known anaerobic pathway, carbon dioxide concentration also is influencing indirectly, the availability of proline utilisation. Specifically, the saturation of must with CO_2 is having as a consequence the elimination of dissolve molecular oxygen, which is needed by oxidase for the first step in proline degradation. This is precluding the utilisation of proline, which is the main amino acid in grape must.

3.1. Temperature effect on aroma formation

Temperature conditions are associated with all enzymatic reactions rate so forth the metabolism and growth of yeast among other microorganism. The temperature range between 15 and 25°C, during wine fermentation is considered favourable for yeast growth under winemaking conditions. The fact that aromatic profile can be modulated during fermentation was noticed very early, since temperature not only affects the volatile composition but also in the case of red wines the extraction of phenolic compounds from the skin and grape seeds [7, 10]. The most noticeable and well-known effect, to winemakers, of temperature is on the fermentation rate and completion which is defined by the total consumption of sugars. Fermentation at 28°C compared to one at 15°C was observed to be 2.5 times faster [7]. Temperature is influencing not only the production of FAC but also the concentration of primary metabolites like ethanol and glycerol, on which it seems temperature to have a reverse effect. In low temperature, the production of ethanol is counter to the glycerol production [7, 9]. From very early studies it was pointed out that the final concentration of esters, contributing to the fruity flavour of the wine, was favoured by low temperatures during fermentation [8]. Particularly esters associated with pleasant fruity aroma, like isoamyl acetate and n-hexyl-acetate, accumulated in higher consecrations at low temperatures. Whereas in high temperature fermentation higher accumulation of ester characterised as heavy odorants like ethyl-octanoate and ethyl decanoate, was observed. Higher final consecration of 2-phenylethyl-acetate was favoured in higher temperatures, by some authors is consider pleasant with rose like odour [6–8]. For thiols temperatures high as 20°C are more favour for their modulation, whereas low temperatures around 13°C show significant less modulation of the 3MH [63].

3.2. Nitrogen source (yeast assimilable nitrogen (YAN)) effect on aroma formation

Yeast assimilable nitrogen (YAN) concentration in grape must is a vital parameter not only completion of the fermentation but also for the production of volatile and nonvolatile metabolites [64, 65]. The depletion of YAN in grape juice during the early stages of fermentation is also triggering the entry to the stationary phase of yeast growth [35]. YAN, source in grape must is categorised into two types, the organic and the non-organic. The organic fraction, often referred to free amino (or amino acid) nitrogen FAN, is the total amount of the amino acids and some small peptides that can be utilised by the yeast. Ammonium nitrogen is the inorganic fraction. An initial concentration of 140 mg/L of YAN in the grape juice is considered to be the lowest threshold for the completion of an industrial fermentation, with low fermentation temperatures and low suspended solids [35]. Measurement of the initial YAN and supplementation of ammonium salts or mixtures of amino acids, to reach the lowest threshold of concentration, is a common practice in most of the wineries, as a prevention measurement to sluggish or incomplete fermentations. Supplementation of nitrogen during the early stages or even through the fermentation course not only results in high fermentation kinetics and yeast growth but also to the formation of various volatile and non-volatile compounds [66, 67]. Timing of the nutrition supplementation is also important since this is influencing the type of nitrogen intake by the yeast cell [35, 65]. Specifically ammonium ion has an inhibitory role in the uptake of amino acids, since in high concentration at the early stages of growth, the general amino acid permease (GAP) is not synthesised [15, 16]. This results in low uptake of amino acid during later stages of fermentation. Another parameter that has an inhibitory role in the amino acid uptake by the yeast is the CO_2. High pressure of CO_2 was observed to reduce the rate by which the amino acids are absorbed. Wines with high concentration levels more than 300 mg/L have showed high esters concentration and low concentrations of acids and higher alcohols [35]. The basic information regarding the initial concentration and ratio between organic and non-organic nitrogen can be obtained by a rather easy enzymatically or chemical method. Nature of the YAN, organic or not, plays also an important role in the outcome of the volatile profile [67]. Addition of amino acids in order to increase the YAN in low concentration grape juice, under the current regulation is forbidden. Use of amino acid enriched dry yeast preparation can provide the mean to serve this purpose; also these types of preparations are high in small peptides. Ratio between the two nitrogen sources is a good tool for the winemaker, to modify the aroma profile composition of the produced wine. It has been proven that the type of nitrogen supplementation resulted in quantitative differences for most of yeast metabolites related compounds, suggesting the importance of the supplementation decision-making process [35]. The concentration of acetates and medium-chain fatty acid esters, contributing to the fruity aroma, is favoured by the higher concentration of amino acids rather than ammonium concentrations. Also higher amino acid concentration is leading to higher concentration of fusel alcohols. High concentrations of ammonium as the sole nitrogen supplement, results in the increase of ethyl acetate and acetic acid [35, 64]. Other recent studies show that there is a close relation between the initial concentration and also most importantly profile of various amino acids for the production of certain aroma profile. Also the same study gives importance values to specific amino acids. Namely in the case of *S. cerevisiae,* they are leucine, isoleusine, valine, histidine, glutamine and proline under certain conditions. [35]. Whereas other researchers' show that, for the formation of volatile compounds: threonine, phenylalanine and aspartic acid are amino acids with the most important value. For thiols, high addition of assimilable nitrogen in the early stage of fermentation, in the form of ammonium (di-ammonium phosphate) seems to reduce the 3MH production [68]. In [56], researchers documented that the highest concentration of terpenes is obtained under conditions that stimulate glycolytic flux. Microaerobic and high assimilable nitrogen conditions, favour terpene accumulation.

3.3. Suspended solids and contribution to wine aroma

During the process of vinification and especially during the first stages of destemming and pressing it is inevitable the presence of grape solids in the must. These solids of various, origin, nature and size are generally referred to as 'sludge' [18]. The measurements by the winemakers assess the presence of these solids are the turbidity units (nephelometric turbidity units, NTU) and a wt% on suspended solids (total wet suspended solids, TWSS or dry TDSS% (w/w) [18]. The ease to measure NTU makes this measurement, the most widely used and accepted method of reference in wineries. A limitation of NTU measurement is that is not in a direct relation with the suspended solids quantity since this is a nephelometric measurement and is being influenced by the size and shape of the particles and refractive index of the medium [18]. Also NTU does not give us information regarding the composition of the suspended solids. Must suspended solids can influence white wines aroma profile in many different ways, directly and indirectly. Suspended solids are consider a good nutrient source, specially for amino acids, the role of which is consider to be the most crucial of all must substrates for the formation of volatile compounds [18–21, 69]. Another direct role of suspended solids is the high content of oxidative enzymes, which is enhanced by various grape moulds contaminations. Also some evidence suggests that grape tissues contain esterase, a limitating factor for the accumulation and final concentration of esters at the final product. Presence of SS on fermentation apart from the direct role on chemical composition, have also an important indirect role, that of the nucleation of the CO_2 [18] and the further release it to atmosphere. High accumulation of CO_2 produces higher concentrations of acetic acid by limiting the long chain fatty acid synthesis. Also over oversaturation of CO_2, is affecting the transport and utilisation by the yeast of amino acids [18].

3.4. Inoculation rate and contribution to wine aroma

Yeast starting cultures are extensively used by many wineries as a mean to control the course of fermentation avoiding slow or sluggish fermentations [13]. Now available on market are a large number of dry yeast cultures ready to use. Primarily the need to use is the easy completion of fermentation, without any technological folds like reduction smell SO_2 and high concentration of volatile acidity. The size though of the inoculation it seems to play a catalytic role in the overall behaviour and physiology of the yeast during fermentation [11–13]. In some cases [12], it was observed that the size of inoculation enhanced stress protectants like glycerol and proline production in high inoculation rate. Also in the same study, an observation of the reduction of citric acid cycle intermediate metabolites was made. On another study dealing with three inoculation concentrations 1×10^4, 1×10^5 and 1×10^6 [11] were studied. It was clear that the most favourable results, for the increase concentration desirable volatiles, like esters, and the simultaneous decrease of high concentrations of unfavourable volatiles, like higher alcohols was observed at inoculation rate of 10^5. It is obvious that through this process, of inoculation, the outcome of the fermentation can be altered but since there is not a lot of research done to this direction, is something that need to be investigated further.

4. Yeast autolysis

4.1. Yeast autolysis and contribution to wine flavour

At the final stages of winemaking, the settlement of yeast cells at the bottom of fermentation tanks is inevitable, since there is not any CO_2 production. From that stage on an autolysis of dead yeast cells is observed. During this period, the hydrolysis of cell wall is taking place, releasing various compounds that up to that moment were either part of yeast cell wall or were capture inside the cell cytoplasm. Understanding the nature of the cell wall is vital in this stage. Yeast cell wall is compromising around 15–25% of total dry cell mass [70, 71] depending on the growth conditions of yeast. Yeast cell wall consists of polysaccharides, inner layer, mannoproteins and outer layer [70]. Mannans and glucans consist of about the 94–98% of the total structural cell wall mass with a small fraction of chitin. Mannoproteins play an important role in stabilising various fractions of wine, tartaric salts and also proteins [72]. Moreover, for the red wines, this fraction can make wine to feel less astringent due to mannoproteins/tannin condensation (46, 48) and also increase the colour stability [72]. Apart from these, the autolysis can also influence the aroma character of the wine [73, 74]. First, by the absorption of volatiles like 4-ethylphenol and 4-ethylguaiacol, which they have a negative contribution to the wine aroma [74]. OAC compound profile also is affected by the yeast autolysis. The overall reduction in ester concentration is observed due to release of esterase [75]. But in some case, volatiles like diethyl succinate, vitispirane and 1,1,6-trimethyl-1,2-dihydronaphthalene (TDN) levels are increased over the period of maturation of sparkling wine with the lees [76]. Also the reduction of oak character is observed due to the absorption of this group of volatiles by the yeast [75], in some case is considered to be positive and negative in some cases.

5. Future perspectives

Deep understanding of the volatiles chemistry in the matrix of wine, by the winemaker, is one crucial step understanding of the character of wine. With the current legislation, in most of the wine producing countries, allowing major taste alterations, like acidification, deacidification and fining, wine is subject to winemaker's decision regarding the final mouth feel. Never the less any additions of adjuncts, to alternate any of the characteristics of wine aroma are forbidden. The single most important period that the winemaker has a chance to manipulate or redirect the wine flavour during the period of fermentation. During this relatively short period to the wines life, the decision-making is critical. The decisions for the fermentation temperature, addition of FAN and period of autolysis are tools for this purpose. The next step for this is monitoring the timing for the various procedures with the correlation across the various parameters. It is widely known and accepted that the quality of the wine is defined by the quality of the grapes. Although the new omics techniques employed in commercial yeast can give a lot of info not only regarding the physiology but most importantly to predict the direction of the outcome of fermentation, in terms of aroma profile, this has not yet put in industrial application. When these technologies become more available and affordable they will provide the winemaker with additional tools in order to improve the quality of the wine by addition of nutrients, adjustment of temperature, selection of commercial yeast strains in order to express at its best the character of the grapes that is handled.

In addition to the above, it is of current interest, research and development the role and use of other non-*Saccharomyces* yeasts from the native microbiota of grapes contributing to the complexity of wine aroma [59, 60, 77–79] and the geographical fingerprinting and indication of origin [57, 80, 81]. With these latest developments, it is understandable that it is impossible to proceed without good understanding of the yeasts physiological traits of many more genus, species and strains along the environment's role on them. For these reasons, high-throughput tools and instruments of molecular biology and biotechnology as well as of analytical chemistry are absolutely required to unravel the yeasts roles in aroma and flavour. These tools and instruments are in an exponential growth technologically and in downward trend their prices which are currently available in many laboratories around the globe and soon will be available for industrial application at large winery level.

Author details

Minas Mina[1,2] and Dimitrios Tsaltas[1]*

*Address all correspondence to: dimitris.tsaltas@cut.ac.cy

1 Cyprus University of Technology, Limassol, Cyprus

2 P. Photiades Group, Nicosia, Cyprus

References

[1] Bisson LF, Karpel JE. Genetics of yeast impacting wine quality. Annual Review of Food Science and Technology. 2010;**1**:139-162

[2] Sumby KM, Grbin PR, Jiranek V. Microbial modulation of aromatic esters in wine: Current knowledge and future prospects. Food Chemistry. 2010;**121**(1):1-16

[3] Fleet GH. Yeast interactions and wine flavour. International Journal of Food Microbiology. 2003;**86**(1-2):11-22

[4] Bartowsky EJ, Pretorius IS. Microbial formation and modification of flavor and off-flavor compounds in wine. In: König H, Unden G, Fröhlich J, editors. Biology of Microorganisms on Grapes, in Must and in Wine. Berlin, Heidelberg: Springer Berlin Heidelberg; 2009. p. 209-231

[5] Styger G, Prior B, Bauer FF. Wine flavor and aroma. Journal of Industrial Microbiology and Biotechnology. 2011;**38**(9):1145-1159

[6] Beltran G et al. Integration of transcriptomic and metabolic analyses for understanding the global responses of low-temperature winemaking fermentations. FEMS Yeast Research. 2006;**6**(8):1167-1183

[7] Molina AM et al. Influence of wine fermentation temperature on the synthesis of yeast-derived volatile aroma compounds. Applied Microbiology and Biotechnology. 2007;**77**(3):675-687

[8] Killian E, Ough CS. Fermentation esters—Formation and retention as affected by fermentation temperature. American Journal of Enology and Viticulture. 1979;**30**(4):301-305

[9] Fairbairn S et al. Environmental stress and aroma production during wine fermentation. South African Journal of Enology and Viticulture. 2014;**35**(2):168-177

[10] Torija MJ et al. Effects of fermentation temperature and *Saccharomyces* species on the cell fatty acid composition and presence of volatile compounds in wine. International Journal of Food Microbiology. 2003;**85**(1-2):127-136

[11] Carrau F et al. Effect of *Saccharomyces cerevisiae* inoculum size on wine fermentation aroma compounds and its relation with assimilable nitrogen content. International Journal of Food Microbiology. 2010;**143**(1-2):81-85

[12] Ding MZ et al. Inoculum size-dependent interactive regulation of metabolism and stress response of *Saccharomyces cerevisiae* revealed by comparative metabolomics. Journal of Biotechnology. 2009;**144**(4):279-286

[13] Mateo JJ et al. Yeast starter cultures affecting wine fermentation and volatiles. Food Research International. 2001;**34**(4):307-314

[14] Garde-Cerdán T, Ancín-Azpilicueta C. Effect of the addition of different quantities of amino acids to nitrogen deficient must on the formation of esters, alcohols, and acids during wine alcoholic fermentation. LWT: Food Science and Technology. 2008;**41**(3):501-510

[15] Gutiérrez A, Chiva R, Guillamón JM. Arginine addition in the stationary phase influences the fermentation rate and synthesis of aroma compounds in a synthetic must fermented by three commercial wine strains. LWT: Food Science and Technology. 2015;**60**(2):1009-1016

[16] Gutiérrez A et al. Nitrogen requirements of commercial wine yeast strains during fermentation of a synthetic grape must. Food Microbiology. 2012;**31**(1):25-32

[17] Vilanova M et al. Effect of ammonium nitrogen supplementation of grape juice on wine volatiles and non-volatiles composition of the aromatic grape variety Albariño. Food Chemistry. 2012;**133**(1):124-131

[18] Casalta E et al. Review: Characterization and role of grape solids during alcoholic fermentation under enological conditions. American Journal of Enology and Viticulture. 2016;**67**(2):133-138

[19] Groat M, Ough CS. Effects of insoluble solids added to clarified musts on fermentation rate, wine composition, and wine quality. American Journal of Enology and Viticulture. 1978;**29**(2):112-119

[20] Puig-Deu M et al. Influence of must racking and fining procedures on the composition of white wine. VITIS-GEILWEILERHOF. 1996;**35**:141-146

[21] Klingshirn LM, Liu JR, Gallander JF. Higher alcohol formation in wines as related to the particle size profiles of juice insoluble solids. American Journal of Enology and Viticulture. 1987;**38**(3):207-210

[22] Bauer F, Pretorius IS. Yeast stress response and fermentation efficiency: How to survive the making of wine—A review. South African Journal for Enology and Viticulture. 2000;**21**:27-51

[23] Hirst MB, Richter CL. Review of aroma formation through metabolic pathways of *Saccharomyces cerevisiae* in beverage fermentations. American Journal of Enology and Viticulture. 2016;**67**(4):361-370

[24] Ebeler SE, Thorngate JH. Wine chemistry and flavor: Looking into the crystal glass. Journal of Agricultural and Food Chemistry. 2009;**57**(18):8098-8108

[25] Bakker J, Clarke RJ. Wine: Flavour Chemistry. John Wiley & Sons; The Atrium, Southern Gate, Chichester, West Sussex PO19 8SQ, UK 2011

[26] Parker M et al. Aroma precursors in grapes and wine: Flavor release during wine production and consumption. Journal of Agricultural and Food Chemistry. 2017; DOI: 10.1021/acs.jafc.6b05255

[27] Ugliano M, Henschke PA. Yeasts and wine flavour. In: Moreno-Arribas MV, Polo MC, editors. Wine Chemistry and Biochemistry. New York, NY: Springer New York; 2009. p. 313-392

[28] Rojas V et al. Acetate ester formation in wine by mixed cultures in laboratory fermentations. International Journal of Food Microbiology. 2003;**86**(1-2):181-188

[29] Rodrıguez-Bencomo J et al. Determination of esters in dry and sweet white wines by headspace solid-phase microextraction and gas chromatography. Journal of Chromatography A. 2002;**963**(1):213-223

[30] Romano P et al. Function of yeast species and strains in wine flavour. International Journal of Food Microbiology. 2003;**86**(1-2):169-180

[31] Swiegers JH, Pretorius IS. Yeast modulation of wine flavor. Advances in Applied Microbiology. 2005;**57**:131-175

[32] Saerens SM et al. Production and biological function of volatile esters in *Saccharomyces cerevisiae*. Microbial Biotechnology. 2010;**3**(2):165-177

[33] Hazelwood LA et al. The Ehrlich pathway for fusel alcohol production: A century of research on *Saccharomyces cerevisiae* metabolism. Applied and Environmental Microbiology. 2008;**74**(8):2259-2266

[34] Torija MAJ et al. Effect of the nitrogen source on the fatty acid composition of *Saccharomyces cerevisiae*. Food Microbiology. 2003;**20**(2):255-258

[35] Torrea D et al. Comparison of inorganic and organic nitrogen supplementation of grape juice—Effect on volatile composition and aroma profile of a Chardonnay wine fermented with *Saccharomyces cerevisiae* yeast. Food Chemistry. 2011;**127**(3):1072-1083

[36] Fraile P, Garrido J, Ancín C. Influence of a *Saccharomyces cerevisiae* selected strain in the volatile composition of rose wines. Evolution during fermentation. Journal of Agricultural and Food Chemistry. 2000;**48**(5):1789-1798

[37] Rapp A, Versini G. Influence of nitrogen compounds in grapes on aroma compounds of wines. Developments in Food Science. 1995;**37**:1659-1694

[38] Hernández-Orte P et al. Addition of amino acids to grape juice of the Merlot variety: Effect on amino acid uptake and aroma generation during alcoholic fermentation. Food Chemistry. 2006;**98**(2):300-310

[39] Tominaga T, Darriet P, Dubourdieu D. Identification of 3-mercaptohexyl acetate in Sauvignon wine, a powerful aromatic compound exhibiting box-tree odor. VITIS-GEILWEILERHOF. 1996;**35**(4):207-210

[40] Tominaga T et al. Stereoisomeric distribution of 3-mercaptohexan-1-ol and 3-mercaptohexyl acetate in dry and sweet white wines made from *Vitis vinifera* (Var. Sauvignon Blanc and Semillon). Journal of Agricultural and Food Chemistry. 2006;**54**(19):7251-7255

[41] Tominaga T, Peyrot des Gachons C, Dubourdieu D. A new type of flavor precursors in *Vitis v inifera* L. cv. Sauvignon blanc: S-cysteine conjugates. Journal of Agricultural and Food Chemistry. 1998;**46**(12):5215-5219

[42] Peyrot d, Gachons C, Tominaga T, Dubourdieu D. Sulfur aroma precursor present in S-glutathione conjugate form: Identification of S-3-(hexan-1-ol)-glutathione in must from *Vitis vinifera* L. cv. Sauvignon blanc. Journal of Agricultural and Food Chemistry. 2002;**50**(14):4076-4079

[43] Schneider R et al. Evidence for an alternative biogenetic pathway leading to 3-mercaptohexanol and 4-mercapto-4-methylpentan-2-one in wines. Analytica Chimica Acta. 2006;**563**(1):58-64

[44] Baudouin-Cornu P et al. Glutathione degradation is a key determinant of glutathione homeostasis. Journal of Biological Chemistry. 2012;**287**(7):4552-4561

[45] Drawert F, Barton H. Biosynthesis of flavor compounds by microorganisms. 3. Production of monoterpenes by the yeast *Kluyveromyces lactis*. Journal of Agricultural and Food Chemistry. 1978;**26**(3):765-766

[46] Cordonnier R, Bayonove CL. Etude de la phase préfermentaire de la vinification: Extraction et formation de certains composés de l'arôme; cas des terpenols, des aldehydes et des alcools en C 6. OENO One. 1981;**15**(4):269-286

[47] Mateo J, Di Stefano R. Description of the β-glucosidase activity of wine yeasts. Food Microbiology. 1997;**14**(6):583-591

[48] Rosi I, Vinella M, Domizio P. Characterization of β-glucosidase activity in yeasts of oenological origin. Journal of Applied Microbiology. 1994;**77**(5):519-527

[49] Miklosy E, Pölös V. Yeasts with β-D-glucosidase activity: Properties and possible application in winemaking processes. Acta Alimentaria. 1995;**24**(2):167-179

[50] Mendes Ferreira A, Climaco MC, Mendes Faia A. The role of non-*Saccharomyces* species in releasing glycosidic bound fraction of grape aroma components—A preliminary study. Journal of Applied Microbiology. 2001;**91**(1):67-71

[51] Fernández-González M, Di Stefano R, Briones A. Hydrolysis and transformation of terpene glycosides from muscat must by different yeast species. Food Microbiology. 2003;**20**(1):35-41

[52] Fernández M, Úbeda JF, Briones AI. Typing of non-*Saccharomyces* yeasts with enzymatic activities of interest in wine-making. International Journal of Food Microbiology. 2000;**59**(1):29-36

[53] Azzolini M et al. Effects of *Torulaspora delbrueckii* and *Saccharomyces cerevisiae* mixed cultures on fermentation and aroma of Amarone wine. European Food Research and Technology. 2012;**235**(2):303-313

[54] Azzolini M et al. Contribution to the aroma of white wines by controlled *Torulaspora delbrueckii* cultures in association with *Saccharomyces cerevisiae*. World Journal of Microbiology and Biotechnology. 2015;**31**(2):277-293

[55] Sadoudi M et al. Yeast–yeast interactions revealed by aromatic profile analysis of Sauvignon Blanc wine fermented by single or co-culture of non-*Saccharomyces* and *Saccharomyces* yeasts. Food Microbiology. 2012;**32**(2):243-253

[56] Carrau FM et al. De novo synthesis of monoterpenes by *Saccharomyces cerevisiae* wine yeasts. FEMS Microbiology Letters. 2005;**243**(1):107-115

[57] Bozoudi D, microbiome TDG. Potential and opportunities as a source of starter cultures. In: Morata A, Loira I, editors. Grape and Wine Biotechnology. Rijeka: InTech; 2016 Ch. 10

[58] Sun SY et al. Selected non-*Saccharomyces* wine yeasts in controlled multistarter fermentations with *Saccharomyces cerevisiae* on alcoholic fermentation behaviour and wine aroma of cherry wines. Food Microbiology. 2014;**44**:15-23

[59] Jolly NP, Varela C, Pretorius IS. Not your ordinary yeast: Non-*Saccharomyces* yeasts in wine production uncovered. FEMS Yeast Research. 2014;**14**(2):215-237

[60] Padilla B, Gil JV, Manzanares P. Past and future of non-*Saccharomyces* yeasts: From spoilage microorganisms to biotechnological tools for improving wine aroma complexity. Frontiers in Microbiology. 2016;**7**:411

[61] Lencioni L et al. Controlled mixed fermentation at winery scale using *Zygotorulaspora florentina* and *Saccharomyces cerevisiae*. International Journal of Food Microbiology. 2016; **234**:36-44

[62] Wang XC et al. Evaluation of aroma enhancement for "Ecolly" dry white wines by mixed inoculation of selected *Rhodotorula mucilaginosa* and *Saccharomyces cerevisiae*. Food Chemistry. 2017;**228**:550-559

[63] Masneuf-Pomarède I et al. Influence of fermentation temperature on volatile thiols concentrations in Sauvignon blanc wines. International Journal of Food Microbiology. 2006;**108**(3):385-390

[64] Barbosa C et al. Nitrogen addition influences formation of aroma compounds, volatile acidity and ethanol in nitrogen deficient media fermented by *Saccharomyces cerevisiae* wine strains. Journal of Bioscience and Bioengineering. 2009;**108**(2):99-104

[65] Barbosa C, Mendes-Faia A, Mendes-Ferreira A. The nitrogen source impacts major volatile compounds released by *Saccharomyces cerevisiae* during alcoholic fermentation. International Journal of Food Microbiology. 2012;**160**(2):87-93

[66] Moreira N et al. Relationship between nitrogen content in grapes and volatiles, namely heavy sulphur compounds, in wines. Food Chemistry. 2011;**126**(4):1599-1607

[67] Procopio S et al. Significant amino acids in aroma compound profiling during yeast fermentation analyzed by PLS regression. LWT: Food Science and Technology. 2013; **51**(2):423-432

[68] Subileau M et al. Nitrogen catabolite repression modulates the production of aromatic thiols characteristic of Sauvignon Blanc at the level of precursor transport. FEMS Yeast Research. 2008;**8**(5):771-780

[69] Ancin C, Ayestaran B, Garrido J. Clarification by vacuum filtration of Grenache must. Utilization of free amino acids during fermentation and bottle-aging of wine. American Journal of Enology and Viticulture. 1996;**47**(3):313-322

[70] Walker GM. Yeast Physiology and Biotechnology. John Wiley & Sons; 1998

[71] Klis FM, Boorsma A, De Groot PW. Cell wall construction in *Saccharomyces cerevisiae*. Yeast. 2006;**23**(3):185-202

[72] Escot S et al. Release of polysaccharides by yeasts and the influence of released polysaccharides on colour stability and wine astringency. Australian Journal of Grape and Wine Research. 2001;**7**(3):153-159

[73] Comuzzo P et al. Yeast derivatives (extracts and autolysates) in winemaking: Release of volatile compounds and effects on wine aroma volatility. Food Chemistry. 2006; **99**(2):217-230

[74] Tao Y, Garcia JF, Sun DW. Advances in wine aging technologies for enhancing wine quality and accelerating wine aging process. Critical Reviews in Food Science and Nutrition. 2014;**54**(6):817-835

[75] Gambetta JM et al. Factors influencing the aroma composition of Chardonnay wines. Journal of Agricultural and Food Chemistry. 2014;**62**(28):6512-6534

[76] Kemp B et al. Effect of production phase on bottle-fermented sparkling wine quality. Journal of Agricultural and Food Chemistry. 2015;**63**(1):19-38

[77] Pretorius IS. Tailoring wine yeast for the new millennium: Novel approaches to the ancient art of winemaking. Yeast. 2000;**16**(8):675-729

[78] Ciani M et al. Controlled mixed culture fermentation: A new perspective on the use of non-*Saccharomyces* yeasts in winemaking. FEMS Yeast Research. 2010;**10**(2):123-133

[79] García M, Esteve-Zarzoso B, Arroyo T. Non-*Saccharomyces* yeasts: Biotechnological role for wine production. In: Morata A, Loira I, editors. Grape and Wine Biotechnology. Rijeka: InTech; 2016 Ch. 11

[80] Bokulich NA et al. Associations among wine grape microbiome, metabolome, and fermentation behavior suggest microbial contribution to regional wine characteristics. mBio. 2016;**7**(3):e00631-16. doi: 10.1128/mBio.00631-16

[81] Bokulich NA et al. Microbial biogeography of wine grapes is conditioned by cultivar, vintage, and climate. Proceedings of the National Academy of Sciences. 2014; **111**(1):E139-E148

Chapter 6

Yeast Ecology of Fermented Table Olives: A Tool for Biotechnological Applications

Dimitrios Anagnostopoulos, Despina Bozoudi and
Dimitrios Tsaltas

Additional information is available at the end of the chapter

http://dx.doi.org/10.5772/intechopen.70760

Abstract

The table olive is considered to be a traditional fermented vegetable in the Mediterranean countries and its production and consumption is recently spreading all around the world. The presence of yeasts is very important during olive fermentation due to their double role. On one hand, yeasts maintain desirable biochemical activities (lipase, esterase, β-glucosidase, catalase, production of killer factors, etc.) with essential technological applications in this fermented vegetable. On the other hand, spoilage activity may be shown. However, recent studies have reported that yeasts coming from table olives would be a new source of potential probiotics. Indeed, many yeast species found in table olive processing, have been reported to demonstrate such properties. Thus, starter cultures technology will play significant role, not only in olive fermentation by controlling the safety and the quality of the final product, but also in consumer's health.

Keywords: yeast, ecology, table olives, starter cultures, probiotic, spoilage, flavor, aroma

1. Introduction

Table olives are one of the most important and well known fermented vegetables, with an estimated worldwide production currently reaching 2.5 million tons per year [1]. Not only the production, but also the consumption of this food is closely related to the culture and diet of some Mediterranean countries (Greece, Spain, Italy, France, Portugal, Egypt, and Turkey), which are the main producers [2]. Other countries such as USA, Peru, Argentina and Australia, are also rising as competitive producers [3].

Yeasts play a vital role during table olives fermentation process principally on the safety, the quality and the flavor of the final product [4]. On one hand, yeasts are responsible for properly driving the fermentation, due to their constant presence throughout the fermentative process. It is generally accepted, that they can produce compounds with important organoleptic attributes determining the quality and flavor of the final product [3]. More specifically, they are able to produce desirable biochemical activities such as lipase, esterase, β-glucosidase, catalase, production of killer factors, which have several technological applications in the fermented table olives. On the other hand, yeasts may also cause spoilage of olives, during fermentation, packing and storage, due to the production of bad odors and flavors, the accumulation of CO_2 leading to swollen containers and gas pocket effects, the clouding of brines, the softening of fruits and the degradation of lactic acid. The technological implications and functional properties of yeasts have been extensively reviewed [5–9]. It has been reported that they could exert a fundamental role for the fermentation of green-treated olives and black-naturally fermented ones [6]. Moreover, many yeast species, which were found in table olive processing, such as *Wickerhamomyces anomalus, Saccharomyces cerevisiae, Pichia membranifaciens and Kluyveromyces lactis*, have been reported to exhibit some interesting properties, such as enhancement of the organoleptic characteristics of the fruits, degradation of polyphenols, enhancement of the growth of lactic acid bacteria (LAB) and antioxidant action [5, 9]. Therefore, the selection of the most appropriate yeast strains to be used as starters, in combination with LAB or not, is a promising future target which may improve the added value of the final product [5].

2. Biology and physiology of yeasts from fermented table olives

Yeasts are single-celled microbes, larger than bacteria, belonging to the kingdom of Fungi. Nowadays 1500 species are identified which corresponds approximately to 1% of all described fungal species [10]. Both their survival and growth are ensured by consuming starches and sugars. Plenty of yeasts are commercially available and can provide an inexpensive source for biochemical and biotechnological applications (classic and molecular) [11].

Generally, in the heterotrophic organisms, such as yeasts, the energy and carbon metabolism have an interconnection between them, i.e., anabolism is connected with catabolism. ATP is provided by the oxidation of organic molecules which also play an important role in the biosynthesis, because they can be used firstly as carbon sources and eventually as energetic currency for all kinds of cellular work. Although, for instance, table olive yeasts species contain a wide set of carbon sources (polyols, alcohols, organic acids and amino acids) that can help in their growth, they prefer to metabolize sugars. There are a lot of data related to the metabolism of different carbon sources, with the most widely studied sugars being hexoses (glucose, fructose, galactose or mannose) and disaccharides (maltose or sucrose), as well as compounds with two carbons (ethanol or acetate) [5, 12].

The metabolic pathway of the central carbon metabolism is mainly identical between different yeast species, suggesting that these microorganisms might constitute a metabolic homogenous group. Nevertheless, the mechanisms for nutrient uptake, the number of different isoenzymes and the regulation of fermentation and respiration differ substantially [13], making yeasts a highly heterogeneous and complex metabolic group.

Moreover, the presence of volatile compounds, such as ethanol, glycerol, higher alcohols, and, to a lesser extent, acetaldehyde in brines can be directly attributed to the metabolic activity of yeasts together with heterofermentative bacteria [12].

On the contrary, methanol, detected in brine [14], is not related with the yeast metabolism, because its activity is due to pectinolytic enzymes. The presence of these enzymes in the drupes is favored by an improper handling of raw material and by poor fermentation conditions [15].

2.1. Enzymatic activities

Yeasts ensure their survival in tissues, and are connected with the digestion of host's proteins and other organic molecules due to a large array of physiological traits and enzymatic activities [16, 17]. It is characteristic that the kind of medium used for their isolation and growth is an important adaptive factor which determines the technological activities of the strains selected [18, 19]. Although many surveys have underlined the potential of unconventional and extreme environments as a source of natural biodiversity for the isolation and selection of useful microorganisms [20], it is the secretion of extracellular prote ases which has been studied extensively in yeasts isolated from these environments [21, 22]. Esterases from yeasts are also gaining industrial interest with applications in laundry detergents and in dairy industries [23, 24], while little attention has been paid to lipases from yeasts [7, 25, 26]. Although peptic enzymes for industrial uses have been so far produced by molds and bacteria [27], the pectinolytic activity of yeasts has, however, been studied with ambiguous results [28, 29].

Yeasts can also produce plenty of other interesting enzymes like invertase, zymase, hexose phosphatase, maltase, reductase, carboxylase, melibiase, and endo-tryptase. However, all kind of yeasts do not comprise the same range of enzymes. Hence, different yeasts are functioning differently toward the degradation of sugars. It is generally accepted that one yeast can ferment one sugar [30]. Since today, it is known that three of the aldo-hexose sugars, called dextrose, d-mannose, and d-galactose, are exclusively fermentable by yeasts. It has been reported that yeasts which ferment dextrose, can also use mannose and laevulose [30]. Due to the enzyme invertase, yeasts can firstly reverse and secondly ferment cane-sugar. However, they are unable to ferment milk-sugar, because of the lack of the enzyme lactase.

2.2. Functional properties

Despite the fact that research has been focused almost exclusively on LAB, as being extremely useful in table olives microbiota [31], recent studies have been negotiating the significant contribution of yeasts to table olives fermentation [3, 5, 9], during the whole process and adding value to the final product. The selection of yeasts to be used as starter cultures is an unclear and complicated procedure, which includes three main steps. The first one is the selection of promising species, which should have some interesting properties. The second step is the validation of those microorganisms on laboratory scale and finally the third one, and the most important, is the demonstration at large scale [32].

Starter cultures must be selected for their functional traits, and for their ability to dominate at the fermentation process, as well as leading to a stable final product. Thus, the screening of those yeast strains is very important. Esterase and lipase enzymatic activities are desirable in yeasts because of their ability to improve the organoleptic characteristics of olives through the configuration of volatile compounds that can be produced by the catabolism of free fatty acids [9, 32–34].

Yeasts with β-glucosidase activity are also good candidates, because they can be used for the hydrolysis of oleuropein, removing the natural bitterness present in brined olives and thus, avoiding the placement of olives into large amounts of water [3, 35]. However, limited data are available in the literature about the contribution of these species on table olive fermentation process [7, 17, 34]. Furthermore, very few strains which produce nontoxic compounds, like biogenic amines (BA) (spermine, spermidine, agmatine) was proposed as a new criterion for the selection of yeasts as starter cultures for many fermented products [36, 37]. Many yeast species which have the ability to produce BA can be found in olives. As a coincidence, this trait should be investigated as another critical point for the selection of proper starter cultures in the near future.

It is noteworthy to mention the probiotic profile of some table olives yeasts. Generally, probiotics are microorganisms which could have beneficial impact on the human organism [38]. This positive impact would exist only if those microorganisms are safe for the host, metabolically active within the gastrointestinal tract and are being consumed, either as food components or as non-food preparations. For instance, *Saccharomyces boulardii*, a member of *S. cerevisiae* species, is the only yeast with clinical role and proven probiotic efficiency in double blind clinical studies [39]. Thus, finding other yeast strains with probiotic characteristics and especially those isolated from table olives is of great interest.

The probiotic potential of yeasts isolated from fermented table olives has been documented to some extent [8, 35]. In particular, Psani & Kotzekidou, [35] found *Torulaspora delbrueckii* and *Debaryomyces hansenii* strains to be quite tolerant in high bile salt concentrations and low pH values. Moreover, the fact that the inoculation of these strains were able to inhibit food borne pathogens such as *Listeria monocytogenes, Bacillus cereus* and *Salmonella typhimurium* is remarkable. Furthermore, data from Ref. [8] showed that *P. membranifaciens* and *Candida oleophila* strains have similar properties but with a different spectrum of inhibition zones (for *Escherichia coli, Salmonella enteritidis* and *Staphylococcus aureus*). Another significant number of yeast species, such as *K. lactis, D. hansenii, T. delbrueckii* and *S. cerevisiae,* have shown tolerance to traverse the gastrointestinal tract, inhibit the enteropathogens, adhere the intestinal $CaCO_2$ cell line and present an immunostimulatory activity [40–42]. In particular, *S. cerevisiae* has shown its ability to prevent the survival of *E. coli* O157:H7 under simulated gastrointestinal conditions by the production of ethanol [43]. Although research of olive yeast strains with the properties mentioned above is a promising task, it is important for olive yeasts to adhere on to the olive skin and survive during storage and/or packaging, in order to be ingested by consumers at elevated numbers.

Except from their probiotic properties, yeasts could positively affect human health in other ways, as well. For instance, diverse strains of *K. lactis*, *S. cerevisiae*, and *Issatchenkia orientalis* show great ability to reduce cholesterol serum levels [40]. Secondly, phytate has a strong chelating ability to form insoluble complexes with divalent minerals of nutritional importance such as zinc, calcium, magnesium and iron. Due to a lack of some required enzymes, humans cannot degrade phytate complexes in the gastrointestinal tract. It is known that dephosphorylation of phytate is catalyzed by phytases, which are widespread in yeast species such as *I. orientalis*, *W. anomalus*, *S. cerevisiae*, *T. delbrueckii*, and *K. lactis* [41, 44]. Thus, yeasts could be included in humans' diet in order to help them to do so.

Folates (vitamin B9) are considered essential co-factors in the biosynthesis of nucleotides and play an important role in cellular replication and growth. It is common that mammals cannot synthesize folates and a potential solution is for yeasts to help them to do so because they contain a folate biosynthesis pathway and can produce natural folates. Some of those species which have a high folate biosynthesis pathway are the *S. cerevisiae* and *Candida glabrata* [41].

Other species of yeasts, such as the diverse yeast strains isolated from table olives, belonging to the *P. membranifaciens* and *P. farinosa* species have the ability to produce B-complex vitamins [8, 45]. This means that yeasts can synthesize a number of bioactive compounds which can serve as natural antioxidants.

Moreover, researchers are interested in screening of yeasts for free-radical-scavenging activity. For instance, *W. anomalus* produced the highest activity in a laboratory medium [46]. The production of bioactive antioxidants may retard the oxidative degeneration of fatty substances and improve human health. In any case, probiotic yeasts could be able to adhere onto the olive epidermis, and thus, be ingested by consumers.

Once researchers are very interested in the organoleptic features of olives, yeasts have some very essential and strain-specific metabolic properties, such as esterase and lipolytic activities. The former has frequently been detected, while the isolation of strains with lipoytic activities has been reported to a lesser extent [7]. Enterase positive yeasts are covetable because the fact that they are able to meliorate the flavor of olives from the formation of esters coming from free fatty acids. Strong lipase activity, as well as, weak activity have been detected both in vitro in some yeast species. The former has been reported in *Candida boidinii*, *D. hansenii* and *T. delbrueckii* while the latter only in *P. membranifaciens* [34, 35]. Those authors have emphasized the change of the free fatty acids composition of olives in the presence of yeast populations in contrast with sterile conditions, indicating that lipases produced by these microorganisms modify the characteristics of fruit lipids and therefore its organoleptic characteristics.

3. Yeast ecology of table olives

According to Ref. [4] it has been reported, that yeasts are responsible for the fermentation process of natural black olives. However, to green olives, the fermentation is driven by LAB. Black olive fermentations have been studied in different countries, mainly in the Mediterranean (Italy, Spain, Portugal, Greece, and Morocco) and showed some yeast species biodiversity.

Despite this fact, a few species were dominant, such as *P. membranifaciens, Saccharomyces oleaginosus, Pichia anomala, C. boidinii* and *T. delbrueckii* [47–50]. However, since LAB are partially inhibited in directly brined green and natural black olives due to the presence of phenolic compounds, yeasts became highly important for the fermentation process [3].

Nowadays, several studies have focused on yeasts microbiota situated on the surface of olive fruits. Some of those studies are summarized in **Table 1**. In the past, the characterization of yeasts associated with table olives was mainly made by morphological and biochemical methods comparing the obtained results with diverse taxonomic keys [10, 61]. However, molecular methods have started to be used recently, for the identification of yeasts coming from table olives fermentation. One of the most applicable methods is the Denaturating Gradient Gel Electrophoresis (DGGE-PCR), which is more precise than any classical method and is based on (i) restriction fragment length polymorphism (RFLP) analysis obtained after cutting the amplified 5.8S rRNA gene and the associated intergenic spacers ITS with endonucleases [62], and (ii) the direct sequencing of the D1/D2 domains of the 26S rRNA gene amplified with primers NL1 and NL4 [63] or the 5.8S-ITS region amplified with primers ITS1 and ITS4 [62].

The information following below is a short description of microbial ecology studies at different cultivars and processing steps within the major table olives production countries of Mediterranean in an attempt to draw conclusions on their yeast colonization (**Figure 1**).

3.1. Spain

According to Ref. [12], *S. cerevisiae, Issatchenkia occidentalis* and *Geotrichum candidum* were identified from Spanish naturally green seasoned table olives (cv. Alorena). In the same study the researchers identified *C. boidinii* and *Hanseniaspora guilliermondii* from the preservation stage of ripe black olives. During the fermentation of Arbequina naturally green table olives in Spain [52, 53], *C. boidinii, C. sorsoba, Candida diddensiae, K. lactis, P. membranifaciens, W. anomalus, P. kluyveri,* and *Rhodotorula glutinis* were found. In a study looking at the Spanish yeast biodiversity of oleic ecosystems, the yeast biodiversity in the fresh table olive, crushed olives and olive pomace from Arbequina and Cornicabra varieties, has been found to contain *Pichia caribbica, Lachancea fermentati and Nakazawaea holstii,* as the most important isolated species [54]. According to Ref. [7], researchers carried out the molecular identification by means of a RFLP analysis and sequencing of a total of 199 yeast isolates obtained from Spanish industrial green table olive fermentation. *C. diddensiae, S. cerevisiae* and *P. membranifaciens* were the most abundant yeast species isolated from directly brined Alorena olives, but for Gordal and Manzanilla cultivars, *Candida tropicalis, P. galeiformis* and *W. anomalus* were found. Recently, other scientists [57] used a culture-independent approach based on the PCR-DGGE analysis for the identification of yeasts associated with Alorena de Malaga olive fermentation, and found that in cold fermented olives the most essential yeasts were *S. cerevisiae* and *Candida apicola*. Finally, it has been shown that the most important species throughout the storage period of table olives were *S. cerevisiae* and *Pichia galeiformis,* although *C. boidinii* was present at the last stages of the process and *P. membranifaciens* was detected at an earlier stage of the Hojiblanca cultivar storage [34].

To summarize, the hierarchically most prominent yeast on Spanish table olives are *P. membranifaciens, C. boidinii, C. diddensiae, S. cerevisiae* and *P. anomala.*

Yeasts/stage	Country/variety	Method	Ref.
Saccharomyces cerevisiae, Issatchenkia occidentalis, Geotrichum candidum (during process)	Spanish green seasoned table olives	Sequencing of 26S rRNA, D1/D2 region	[12]
Candida boidinii, Hanseniaspora guilliermondii (final product)	Ripe black olives	Sequencing of 26S rRNA, D1/D2 region	[12]
P. anomala, C. boidinii, Debaryomyces etchelsii (during process)	French black olives	RFLP, Sequencing of D1/D2 region	[51]
C. boidinii, Candida sorsoba, Candida diddensiae, P. membranifaciens, Kluyveromyces lactis, P. anomala, P. kluyveri, Rhodotorula glutinis (during process)	Spain, Arbequina table olives	RFLP	[52, 53]
Pichia caribbica, Lachancea fermentati, Nakazawaea holstii (during process)	Spain, Arbequina and Cornicabra	Sequencing of 5.8 rRNA, ITS1-ITS2 region	[54]
Metschnikowia pulcherrima, Debaryomyces hansenii, Aureobasidium pullulans (during process)	Greece, Conservolea black olives	Sequencing of 5.8 rRNA, ITS1-ITS2 region	[55]
P. membranifaciens, P. anomala (final product)	Greek black olives	Sequencing of 5.8 rRNA, ITS1-ITS2 region	[55]
Candida parapsilosis, P. guilliermondii, P. kluyveri (during process)	Sicilian green table olives.	RFLP, Sequencing of 26 s rRNA D1/D2 region	[56]
S. cerevisiae, Pichia galeiformis, C. boidinii (final product), and *P. membranifaciens* (during possess)	Hojiblanca	RFLP, Sequencing of 5.8 s rRNA, ITS1-ITS2 region	[34]
C. diddensiae, S. cerevisiae and *P. membranifaciens* (during process)	Spain/ Alorena	RFLP, Sequencing of 26 s rRNA D1/D2 region	[7]
Candida tropicalis, P. galeiformis and *P. anomala* (during process)	Spain/Gordal and Manzanilla	RFLP, Sequencing of 26 s rRNA D1/D2 region	[7]
S. cerevisiae and *Candida apicola* (during process)	Spain/ Alorena de Malaga	PCR-DGGE of 26SrRNA	[57]
S. cerevisiae, P. anomala, C. diddensiae, Issatchenkia orientalis (during process)	Sicilian green table olives	PCR-DGGE of 26 s rRNA	[58]
P. membranifaciens, Pichia fermentans, S. cerevisiae, Candida oleophila. (during process)	Portuguese brined olive	RFLP, Sequencing of 26 s rRNA D1/D2 region	[8]
Citeromyces matritensis, Zygotorulaspora mrakii, S. cerevisiae (during process)	Portugal, Manzanilla	Sequencing of 26 s rRNA D1/D2 region	[59]
P. membranifaciens (during process)	Greece, Conservolea	Sequencing of 5.8 s rRNA, ITS1-ITS2 region	[60]

Table 1. Scientific reports of microbial ecology studies reporting the isolation of yeasts from different cultivars of table olives and different fermentation processes and different Mediterranean countries.

Figure 1. (A) Cyprus naturally black olives with salt, (B) Kalamata naturally black olives in brine, (C) Cyprus green cracked olives in brine, (D) Colonized surface of Cyprus green cracked olives (arrows show the formation of microbial colonies viewed under stereomicroscope).

3.2. Greece

In relevant scientific work of Greek scientists it has been discovered that *Metschnikowia pulcherrima, D. hansenii* and *Aureobasidium pullulans* were the dominant yeast species at the beginning of the fermentation process of Greek Conservolea black olives [55]. These researchers found a new yeast species associated with this type of fermentation, named *Candida olivae*. Species heterogeneity changed during fermentation and both *P. membranifaciens* and *P. anomala* became the only dominant yeasts at the end of the fermentation. A similar work focusing on microbial heterogeneity during aerobic and modified atmosphere packaging storage of Conservolea natural black olives found that *P. membranifaciens* was dominated in all pouches regardless of gas composition with a frequency more than 80% during storage [60]. Summarizing *P. membranifaciens* is the dominant yeast on Conservolea black olives followed by *P. anomala*.

3.3. Italy

In Ref. [56], it has been revealed that in Sicilian environment in four Italian olive cultivars (Bradofina, Castriciana, Nocellara del Belice and Passalunara) and one Spanish (Manzanilla) the presence of *P. kluyveri followed by Candida parapsilosis and P. guilliermondii* during the entire fermentation period is clear. In a similar work [58] also in Sicily with green table olives of cultivar Nocellara dell' Etna, researchers found *S. cerevisiae, P. anomala, C. diddensiae,* and *I. orientalis* during the process. Finally, when researchers [6] studied the technological and spoiling characteristics of yeast microflora isolated from cultivar Bella Di Cerignola table olives, most prominent yeasts isolated were *Candida famata* and *C. guilliermondii*. Summarizing the Italian yeast biodiversity in green olives is seems that *P. guilliermondii* and *P. kluyveri* are likely to play a crucial role in their fermentation.

3.4. France

In Ref. [51] it has been identified that *P. anomala, C. boidinii* and *Debaryomyces etchelsii* were dominant species in French black olives of Nyons area (South France).

3.5. Portugal

According to Ref. [8], researchers studied the yeast population associated with Portuguese brined green olives of cultivars Galega and Cordovil, fermentation was mainly driven by *P. membranifaciens, P. fermentans, S. cerevisiae* and *C. oleophila*. In a similar work [59], it has been found that during the initial phases of cracked green Manzanilla olive fermentation a great diversity of yeasts was observed; however, as the process was evolving, the biodiversity decreased with the fermentative yeasts *Citeromyces matritensis, Zygotorulaspora mrakii* and *S. cerevisiae* becoming the dominant species. These species though are reported as high risk spoilage microorganisms contrary to *P. membranifaciens* reported by [8].

Concluding this brief overview of yeast ecological studies in fermented table olives around the Mediterranean it becomes apparent that among Candida species, *C. boidinii, C. diddensiae, C. famata* (formerly *D. hansenii*), *C. guilliermondii* (formerly *H. guilliermondii*), and *C. oleophila* are the most prominent. Among Pichia species, *P. membranifaciens, Pichia anomala* (*W. anomalus*) and *P. fermentans* are the most prominent. From other species we tend to isolate more *A. pullulans, Debaryomyces etchellsii, G. candidum, I. occidentalis, K. lactis, Rhodotorula spp., S. cerevisiae* and *Z. mrakii*. These yeasts are plentiful for detailed studies on their role during table olive fermentations and could offer great opportunities for biotechnological tools as starter cultures.

4. Biotechnological applications of table olives yeast strains

Yeast microbiota in olives is very heterogeneous and can be altered depending on the olive cultivar, region, type of fermentation process, salt concentration, pH, nutrients, oxygen and interactions with other microorganisms [5, 57, 64–69]. In table olive fermentations, yeasts are an important group of microorganisms that act as both desirable and spoilage microorganisms and it is important to evaluate their biodiversity in table olive fermentations [3, 4]. As a positive effect, some yeast species isolated from table olives, such as *Debaryomyces, Pichia* and *Candida* are known to include a considerable number of strains with a killer character and it has been found, that *W. anomalus* protects olives from unsaturated fatty acid oxidation and peroxide formation [3, 70, 71], relating also to the antioxidant activity of yeasts [7, 8]. Moreover, it has been reported that *W. anomalus* and *S. cerevisiae* isolated from diverse table olive fermentations, have phytase enzymes that are required for the degradation of phytate complexes [3, 41, 44]. On the other hand, *W. anomalus* is important yeast for olive fermentation, but it may also have a role in the deterioration of olives at the end of fermentation [72]. It has been reported that some olive-related yeast strains, such as *W. anomalus*, can produce enzymes that could cause softening of the olives as an unfavorable property [33].

According to Ref. [73], the combination of yeasts with LAB, has resulted to an improvement in growth of the LAB. As a result, the production of lactic acid was improved, thanks to the greater availability of the necessary nutrients by the yeasts activity and lysed cells. Indeed, recent studies have shown that the growth of LAB could be stimulated by the use of yeasts, as for example *L. plantarum,* improved its growth when *D. hansenii* was inoculated in olive brine [74]. Moreover, *L. pentosus's* performance was rapidly improved, with the use of *S. cerevisiae* as a starter in green table olive solutions [73]. In the same study, it has been proved that the production of lactic acid was increased, as well. Yeasts seem to be active microorganisms during the fermentation of table olives, synthesizing substances such as vitamins, amino acids and purines, or breakdown complex carbohydrates, which are essential for the growth of *Lactobacillus* species [75].

Therefore, selection of the most appropriate yeasts for starters should be firstly based on strains possessing the best enzymatic activities as mentioned earlier and secondly to their ability to predominate during fermentation. Moreover, it is needed to have a high resistance to salt and low pH values. Predictive microbiology is seemed to be a valuable tool for the discrimination and selection of the most promising strains, determining the influence of environmental variables on yeast growth [5]. A problem usually occurring in such applications is to find the appropriate methodology to manage such a large amount of data, which is necessary when researchers have to analyze several biochemical activities or growth data from a considerable number of strains.

Multivariate analysis techniques offer a viable approach in solving this setback. For instance, multivariate analysis approach to study growth and qualitative activity data of a number of yeasts isolated from Bella di Cerignola table olives [6] has been used. Principal Component Analysis (PCA) clearly differentiated and assisted the selection and discrimination of several *W. anomalus* and *C. boidinii* isolates with high global desirable activity levels [76].

However, in some cases, dominant yeasts could create products with milder taste and less self-life preservation [4, 15]. This problem was reported in [65] during fermentation of natural black olives at different NaCl levels and temperatures. Moreover, an excessive growth of fermented yeast (7 log10 CFU mL^{-1}) could produce a vigorous production of CO_2 resulting in penetrating olives and damaging the fruits [64]. The use of high levels of NaCl during fermentation (8% in the equilibrium) could privilege the growth of yeasts against LAB [4, 65]. In accordance with Ref. [33], it has been found that some strains of *Rhodotorula minuta* and *D. hansenii* in green table olive fermentations are having this ability. Finally, strains of *R. glutinis, R. minuta* and *R. rubra* could grow, form pellicles in olive brines and produce polygalacturonases causing a softening of olives kept in storage [77].

Moreover, yeasts present in packed olives can produce an excess gas (CO_2) leading to swollen containers, clouding of the brines, or produce off-flavors and off-odors [4]. Furthermore, it has been reported that yeasts identified currently as *S. cerevisiae* and *P. anomala* spoiled the olives through a combination of gas-pocket formation and softening [78]. Fortunately, yeasts from table olives are almost entirely nonpathogenic. The inhibitory effect of sorbic acid, benzoic acid and their salts on yeasts growth to stabilize table olive packing, were reported previously [5, 79, 80].

5. Future perspectives

At this moment, the scientific community has gathered significant amount of knowledge about the ecology of fermented table olives, physiology, biochemistry and genetics of yeasts isolated from table olives fermentation and more data are soon going to be added with the use of high throughput sequencing techniques. Despite the large amount of beneficial properties that these microorganisms could offer to the final product, recent findings showed that yeasts are not only in the cover brines, but also on the fruit epidermis, which indicates a promising source of probiotic strains [43, 81]. The production of an innovative and functional food having the advantages of probiotic yeasts, can contribute to a final product with added value, even higher than it already has. From the research so far, it is clear that yeasts from table olives show an interesting technological and probiotic profile. However, further research is needed as far as their physiology, ecology, biochemistry is concerned. Also the study of yeasts are needed in molecular level throughout the fermentation, in order to be explained their prevalence or not in the final product, especially when it has to do with probiotic microorganisms. Undoubtedly, the full potential of table olive related yeasts have not been fully determined and many challenges are awaiting research, dissemination and industrial exploitation.

The microbiota of olives varies somewhat from region to region, from cultivar to cultivar and from type to type of processing [5, 12, 33, 34, 76]. For this reason and due to the importance of yeasts to the final product, it is a challenge to investigate and combine the diversity of yeasts in table olives around regions as a geographical indication. This could lead to the introduction of new table olives as PDO or PGI products, especially for EU countries. Finally, starter cultures can play an essential role in olive fermentation by controlling the safety and the quality of the final product. For extensive information on selection of yeasts as starter cultures for table olives, Refs. [32, 82] are thorough reviews with step by step procedures to follow.

Next Generation Sequencing (NGS) approaches have recently started to appear in the subject with most recent the work enhancing the knowledge on fungal communities in directly brined Alorena de Malaga green olives fermentations using metabarcoding analysis [83]. Along these lines more publications are expected in the forthcoming years.

However, NGS technologies are providing massive amount of data, many times difficult to interpret at logical, applied level. Recently, a very important effort on databasing, visualizing and exploring the food bacterial communities based on network analysis has announced preliminary results [84]. FoodMicrobionet, the platform prepared by the inclusion of 17 bacterial studies on dairy products, dairy starter cultures, raw and fermented meat, doughs and sourdoughs and fermented vegetables is attempting to analyze nodes and network properties while building an interactive web-based visualization. By this the researcher can explore the relationships between Operational Taxonomic Units (OTUs) and samples in order to identify core and sample specific bacterial communities. A similar approach will be very much in use for yeast/fungal studies of various matrices. The above if combined with a high throughput screening technique for aroma formation like the one used in [85] may provide exciting results with a plethora of new interrelationships. Similarly, a recent work with Sicilian table olives (cv. Nocellara Etnea) [86] investigates the bacterial community and its dynamics during the fermentation of the olives and its effect on metabolome formation.

In the following years we expect more detailed studies in the field of table olives microbial ecology and biotechnology, as well as gaining valuable knowledge for the related microbial functions that can be applied far beyond olive fermentations.

Author details

Dimitrios Anagnostopoulos, Despina Bozoudi and Dimitrios Tsaltas*

*Address all correspondence to: dimitris.tsaltas@cut.ac.cy

University of Technology, Limassol, Cyprus

References

[1] IOC (International Olive Oil Council), 2013. Statistical of Table Olive Production. Available from: http://www.internationaloliveoil.org/estaticos/view/132-world-table-olive-figures [Accessed: Jan 24, 2013].

[2] Bonatsou S, Benítez A, Rodríguez-Gómez F, Panagou EZ, Arroyo-López FN. Selection of yeasts with multifunctional features for application as starters in natural black table olive processing. Food Microbiology. 2015;**46**:66-73

[3] Arroyo-López FN, Romero-Gil V, Bautista-Gallego J, Rodríguez-Gómez F, JiménezDíaz R, García-García P, Querol A, Garrido-Fernández A. Yeasts in table olive processing: Desirable or spoilage microorganisms? International Journal of Food Microbiology. 2012;**160**:42-49

[4] Garrido-Fernández A, Fernandez Diez MJ, Adams RM. Table Olives: Production and Processing. London: Chapman and Hall; 1997

[5] Arroyo-López FN, Querol A, Bautista-Gallego J, Garrido-Fernández A. Role of yeasts in table olive production. International Journal of Food Microbiology. 2008;**128**:189-196

[6] Bevilacqua A, Perricone M, Cannarsi M, et al. Technological and spoiling characteristics of the yeast microflora isolated from Bella di Cerignola table olives. International Journal of Food Science & Technology. 2009;**44**:2198-2207

[7] Bautista Gallego J, Rodriguez-Gomez F, Barrio E, Querol A, Garrido-Fernandez A, Arroyo-Lopez FN. Exploring the yeast biodiversity of green table olive industrial fermentations for technological applications. International Journal of Food Microbiology. 2011;**147**:89-96

[8] Silva T, Reto M, Sol M, et al. Characterization of yeasts from Portuguese brined olives, with a focus on their potentially probiotic behaviour. LWT – Food Science and Technology. 2011;**44**:1349-1354

[9] Rodriguez-Gomez F, Bautista-Gallego J, Romero-Gil V, et al. Influence of yeasts on the oil quality indexes of table olives. Journal of Food Science. 2013;**78**:1208-1217

[10] Kurtzman CP, Fell JW, Boekhout T. The Yeasts, A Taxonomic Study. 5th ed. Amsterdam: Elsevier; 2011

[11] Kurtzman CP, Fell JW. The Yeast, A Taxonomic Study. 4th ed. Amsterdam: Elsevier; 1998

[12] Arroyo-López FN, Durán-Quintana MC, Ruiz-Barba JL, Querol A, Garrido-Fernández A. Use of molecular methods for the identification of yeast associated with table olives. Food Microbiology. 2006;**23**:791-796

[13] Flores CL, Rodriguez C, Petit T, Gancedo C. Carbohydrate and energy-yielding metabolism in non-conventional yeasts. FEMS Microbiology Reviews. 2000;**24**:507-529

[14] Sanchez I, Palop L, Ballesteros C. Biochemical characterization of lactic acid bacteria isolated from spontaneous fermentation of 'Alamgro' eggplants. International Journal of Food Microbiology. 2000;**59**:9-17

[15] Panagou EZ, Tassou CC. Changes in volatile compounds and related biochemical profile during controlled fermentation of cv. Conservolea green olives. Food Microbiology. 2006;**23**:738-746

[16] Krajewska-Kulak E. Hydrolytic activity of *Candida albicans* and their susceptibility to antimycotics. Medical Science Monitor. 1998;**4**:1643-3750

[17] Krajewska-Kulak E. Enzymatic biotypes of the yeast-like fungi strains and their susceptibilities to antimycotics isolated from ontocenosis of the urogenital system. Mikologia/Medical Mycology. 2002;**9**:2083-5744

[18] Brasch J. Enzyme patterns of dermatophytes. Mycoses. 1994;**37**:1439-0507

[19] Plomer-Niezgoda E, Baran E. Evaluation of extracellular activity of hydrolytic enzymes of selected mold fungi. Mikologia Lekarska. 1997;**4**:141-145

[20] Bull AT, Goodfellow M, Slater JH. Biodiversity as a source of innovation in biotechnology. Annual Review of Microbiology. 1992;**46**:219-252

[21] Abranches J, Morrais PB, Rosa CA, Mendonca-Hagler LC, Hagler HN. The incidence of killer activity and extracecullar proteases in tropical yeast communities. Canadian Journal of Microbiology. 1997;**43**:328-336

[22] Bossi A, Bonizzato L, Zapparoli G. Acidic extracellular proteases from microrganisms of fairly acidic niche. Protein and Peptide Letters. 2006;**13**:737-741

[23] Burden DW, Eveleigh DE. Yeasts-diverse substrates and production. In:Yeast Technology, Spencer JF, Spencer DM, editors. Berlin: Springer-Verlag; 1990. p. 199-227

[24] Bordes F, Barbe S, Escalier P, Mourey L, André I, Marty A, Tranier S. Exploring the conformational states and rearrangements of Yarrowia lipolytica lipase. Biophysical Journal. 2010;**99**(7):2225-2234

[25] Basaran P, Hang YD. Purification and characterization of acetyl esterase from *Candida guilliermondii*. Letters Applied Microbiology. 2000;**30**:167-171

[26] Athenstaedt K, Daum G. Tgl4p and Tgl5p, two triacylglycerol lipases of the yeast *Saccharomyces cerevisiae* are localized to lipid particles. J BiolChem. 2005;**280**(45):37301-37309

[27] Sakai N, Mano S, Nozaki K, Takaya H. Journal of the American Chemical Society. 1993; **115**:7033

[28] Charoenchai C, Fleet GH, Henschke PA, Todd BEN. Screen of non-saccharomyces wine yeasts for the presence of extracellular hydrolytic enzymes. Australian journal of grape wine Research. 1997;**3**:2-8

[29] Strauss RS, Pollack HA. Epidemic increase in childhood overweight 1986-1998. Journal of the American Medical Association. 2001;**286**:2845-2848

[30] Kiehn E, Wong B, Edwards F, Armstrong D. Comparative recovery of bacteria and yeasts from Lysis centrifugation and a conventional blood culture system. Journal of Clinical Microbiology. 1983;**18**:300-304

[31] Hurtado A, Reguant C, Bordons A, Rozes N. Lactic acid bacteria from fermented table olives. Food Microbiology. 2012;**31**:1-8

[32] Bevilacqua A, Corbo MR, Sinigaglia M. Selection of yeasts as starter cultures for table olives: A step-by-step procedure. Frontiers in Microbiology. 2012;**3**:194

[33] Hernández A, Martín A, Aranda E, Pérez-Nevado F, Córdoba MG. Identification and characterization of yeast isolated from the elaboration of seasoned green table olives. Food Microbiology. 2007;**24**:346-351

[34] Rodriguez-Gomez F, Arroyo-Lopez FN, Lopez-Lopez A, Bautista-Gallego J, Garrido Fernandez A. Lipolytic activity of the yeast species associated with the fermentation/storage phase of ripe olive processing. Food Microbiology. 2010;**27**:604-612

[35] Psani M, Kotzekidou P. Technological characteristics of yeast strains and their potential as starter adjuncts in Greek-style black olive fermentation. Springer Science+Business Media B.V. 2006;**22**:1329-1336

[36] Caruso M, Fiore C, Contursi M, Salzano G, Paparella A, Romano P. Formation of biogenic amines as criteria for the selection of wine yeasts. World Journal of Microbiology and Biotechnology. 2002;**18**:159-163

[37] Landete JM, Ferrer S, Pardo I. Biogenic amine production by lactic acid bacteria, acetic bacteria and yeast isolated from wine. Food Control. 2007;**18**:1569-1574

[38] FAO/WHO. Health and Nutritional Properties of Probiotics in Food Including Powder Milk with Live Lactic Acid Bacteria. Córdoba, Argentina: Joint Food and Agricultural Organization of the United Nations and World Health Organization Expert Consultation Report; 2001

[39] Sazawal S, Hiremath G, Dhingra U, Malik P, Deb S, Black RE. Efficacy of probiotics in prevention of acute diarrhoea: A meta-analysis of masked, randomised, placebo-controlled trials. The Lancet Infectious Diseases. 2006;**6**:374-382

[40] Kourelis A, Kotzamanidis C, Litopoulou-Tzanetaki E, Scouras ZG, Tzanetakis N, Yiangou M. Preliminary probiotic selection of dairy and human yeast strains. Journal of Biological Research of Thessaloniki. 2010;**13**:93-104

[41] Moslehi-Jenabian S, Lindegaard Pedersen L, Jespersen L. Beneficial effects of probiotic and food borne yeasts on humanhealth. Nutrients. 2010;**2**:449-473

[42] Pennacchia C, Blaiotta G, Pepe O, Villani F. Isolation of *Saccharomyces cerevisiae* strains from different food matrices and their preliminary selection for a potential use as probiotics. Journal of Applied Microbiology. 2008;**105**:1919-1928

[43] Etienne-Mesmin L, Livrelli V, Privat M, Denis S, Cardot JM, Alric M, BlanquetDiot S. Effect of a new probiotic *Saccharomyces cerevisiae* strain on survival of *Escherichia coli* O157:H7 in a dynamic gastrointestinal model. Applied and Environmental Microbiology. 2011;**77**:1127-1131

[44] Olstorpe M, Schnüren J, Passoth V. Screening of yeast strains for phytase activity. FEMS Yeast Research. 2009;**9**:478-488

[45] Ruiz Barba JL, Jiménez Diaz R. Availability of essential B-group vitamins to *Lactobacillus plantarum* in green olive fermentation brines. Applied and Environmental Microbiology. 1995;**61**:1294-1297

[46] Gazi MR, Hoshikuma A, Kanda K, Murata A, Kato F. Detection of free radical scavenging activity in yeast culture. Bulletin of the Faculty of Agriculture of Saga University. 2001;**86**:67-74

[47] Marquina D, Peres C, Caldas FV, Marques JF, Peinado JM, Spencer Martin J. Characterization of the yeast population in olives brines. Letters in Applied Microbiology. 1992;**14**:279-283

[48] Marquina D, Toufani S, Llorente P, Santos A, Peinado JM. Killer activity in yeast isolated from olive brines. Advances in. Food Science. 1997;**19**:41-46

[49] Fernández González MJ, Gárcia P, Garrido-Fernández A, Durán Quintana MC. Microflora of the aerobic preservation of directly brined green olives from Hojiblanca cultivar. Journal of Applied Bacteriology. 1993;**75**:226-233

[50] Kotzekidou P. Identification of yeast from black olives in rapid system microtitre plates. Food Microbiology. 1997;**14**:609-616

[51] Coton E, Coton M, Levert D, Casaregola S, Sohier D. Yeast ecology in French cider and black olive natural fermentations. International Journal of Food Microbiology. 2006;**108**:130-135

[52] Hurtado A, Reguant C, Esteve-Zarzoso B, Bordons A, Rozès N. Microbial population dynamics during the processing of Aberquina table olives. Food Research. 2008; **41**:738-744

[53] Hurtado A, Reguant C, Bordons A, Rozès N. Influence of fruit ripeness and salt concentration on the microbial processing of Arbequina table olives. Food Microbiology. 2009;**26**:827-833

[54] Romo-Sánchez S, Alves-Baffi M, Arévalo-Villena M, Úbeda-Iranzo J, Briones-Pérez A. Yeast biodiversity from oleic ecosystems: Study of their biotechnological properties. Food Microbiology. 2010;**27**:487-492

[55] Nisiotou AA, Chorianopoulos N, Nychas GJE, Panagou EZ. Yeast heterogeneity during spontaneous fermentation of black Conservolea olives in different brine solutions. Journal of Applied Microbiology. 2010;**108**:396-405

[56] Aponte M, Ventorino V, Blaiotta G, Volpe G, Farina V, Avellone G, Lanza CM, Moschetti G. Study of green Sicilian table olive fermentations through microbiological, chemical and sensory analyses. Food Microbiology. 2010;**27**:162-170

[57] Abriouel H, Benomar N, Lucas R, Gálvez A. Culture-independent study of the diversity of microbial populations in brines during fermentation of naturally fermented Alorena green table olives. International Journal of Food Microbiology. 2011;**44**:487-496

[58] Muccilli S, Caggia C, Randazzo CL, Restuccia C. Yeast dy-namics during the fermentation of brined green olives treated inthe field with kaolin and Bordeaux mixture to control the olive fruit fly. International Journal of Food Microbiology. 2011;**148**:15-22

[59] Alves M, Gonçalves T, Quintas C. Microbial quality and yeast population dynamics in cracked green table olives' fermentations. Food Control. 2012;**23**:363-368

[60] Doulgeraki A, Hondrodimou O, Iliopoulus U, Panagou EZ. Lactic acid bacteria and yeast heterogeneity during aerobic and modified atmosphere packaging storage of natural black Conservolea olives in polyethylene pouches. Food Control. 2012;**26**:49-57

[61] Barnett JA, Payne RW, Yarrow D. Yeasts: Characteristics and Identification. 2nd ed. Cambridge: University Press; 1990

[62] Esteve-Zarzoso B, Belloch C, Uruburu F, Querol A. Identification of yeasts by RFLP analysis of the 5.8 rRNA gene and the two ribosomal internal transcribed spacers. International Journal of Systematic Bacteriology. 1999;**49**:329-337

[63] Kurtzman CP, Robnett C. Identification and phylogeny of ascomycetous yeast from analysis of nuclear large subunit (26S) ribosomial DNA partial sequences. Antonie Van Leeuwenhoek. 1998;**73**:331-371

[64] Duran-Quintana MC, Garcia-Garcia P, Garrido-Fernández A. Establishment of conditions for green table olive fermentation at low temperature. International Journal of Food Microbiology. 1999;**51**:133-143

[65] Tassou CC, Panagou EZ, Katsaboxakis KZ. Microbiological and physicochemical changes of naturally black olives fermented at different temperatures and NaCl levels in the brines. Food Microbiology. 2002;**19**:605-615

[66] Alvarez DME, Sánchez A, Lamarque AL. Naturally black olives: Comparison of three processes for fermenting cv. 'Farga' olives. Olivae. 2003;**97**:47-51

[67] Chorianopoulos NG, Boziaris IS, Stamatiou A, Nychas GJE. Microbial association and acidity development of unheated and pasteurised green table olives fermented using glucose or sucrose supplements at various levels. Food Microbiology. 2005;**22**:117-124

[68] Corsetti A, Perpetuini G, Schirone M, Tofalo R, Suzzi G. Application of starter cultures to table olive fermentation: An overview on the experimental studies. Frontiers in Microbiology. 2012;**3**:1-6

[69] Tofalo R, Schirone M, Perpetuini G, Angelozzi G, Suzzi G, Corsetti A. Microbiological and chemical profiles of naturally fermented table olives and brines from different Italian cultivars. Antonie Van Leeuwenhoek. 2012;**102**:121-131

[70] Llorente P, Marquina D, Santos A, Peinado JM, Spencer-Martins I. Effect of salt on the killer phenotype of yeasts from olive brines. Applied and Environmental Microbiology. 1997;**63**:1165-1167

[71] Hernández A, Martín A, Córdoba MG, Benito MJ, Aranda E, Pérez-Nevado F. Determination of killer activity in yeasts isolated from the elaboration of seasoned green table olives. International Journal of Food Microbiology. 2008;**121**:178-188

[72] Pitt JI, Hocking AD. Yeast. Fungi and Food Spoilage. New York: Springer; 2009

[73] Segovia Bravo K, Arroyo-López FN, Garcia Garcia P, Duran Quintana MC, GarridoFernández A. Treatment of green table olive solutions with ozone. Effect on the polyphenol content on *Lactobacillus pentosus* and *Saccharomyces cerevisiae* growth. International Journal of Food Microbiology. 2007;**114**:60-68

[74] Tsapatsaris S, Kotzekidou P. Application of a central composite design and response surface methodology to the fermentation of olive juice by *Lactobacillus plantarum* and *Debaryomyces hansenii*. International Journal of Food Microbiology. 2004;**95**:157-168

[75] Viljoen BC. Yeast ecological interactions. Yeast–yeast, yeast bacteria, yeast–fungi interactions and yeasts as biocontrol agents. In: Querol A, Fleet H, editors. Yeasts in Food and Beverages. Berlin: Springer–Verlag; 2006. p. 83-110

[76] Rodríguez-Gómez F, Romero-Gil V, Bautista-Gallego J, Garrido-Fernández A, Arroyo-López FN. Multivariate analysis to discriminate yeast strains with technological applications in table olive processing. World Journal of Microbiology and Biotechnology. 2012;**28**:1761-1770

[77] Vaughn RH, Jakubczyk T, MacMillan JD, Higgins TE, Dave BA, Crampton VM. Some pink yeast associated with softening of olives. Applied Microbiology. 1969;**18**:771-775

[78] Vaughn RH, Stevenson KE, Dave BA, Park HC. Fermenting yeast associated with softening and gas-pocket formation in olives. Applied Microbiology. 1972;**23**:316-320

[79] Rodriguez de la Borbolla Alcalá JM, FernándezDíez MJ, González Cancho F. El empleo de ácidosórbico y sus sales en las aceitunasaderezadas. Grasas y Aceites. 1961;**12**:10-15

[80] Marsilio V, Cichelli A. Influencia del sorbatopotásico y delbenzoatosódicosobre la estabilidad de las aceitunas de mesa ensalmuera. Grasas y Aceites. 1992;**43**:66-74

[81] Domínguez-Manzano J, Olmo-Ruiz C, Bautista-Gallego J, Arroyo-López FN, Garrido-Fernández A, Jiménez-Díaz R. Biofilm formation on abiotic and biotic surfaces during Spanish style green table olive fermentation. International Journal of Food Microbiology. 2012;**157**:230-238

[82] Bonatsu S, Tassou CC, Panagou EZ, Nychas G-JE. Table olive fermentation using starter cultures with multifunctional potential. Microorganisms. 2017;**5**:30. DOI: 10.3390/microorganisms5020030

[83] Arroyo-López FN, Medina E, Ruiz-Bellido MA, Romero-Gil V, Montes-Borrego M, Landa BB. Enhancement of the knowledge on fungal communities in directly brined Alorena de Malaga green olive fermentations by metabarcoding analysis. PloS One. 2016;**11**(9). DOI: 10.1371/journal.pone.0163135

[84] Parente E, Cocolin L, De Filippis F, Zotta T, Ferrocino I, O'Sullivan O, Neviani E, De Angelis M, Cotter PD, Ercolini D. FoodMicrobionet: A database for the visualisation and exploration of food bacterial communities based on network analysis. International Journal of Food Microbiology. 2016;**219**:28-37

[85] Gamero A, Quintilla R, Groenewald M, Alkema W, Boekhout T, Hazelwood L. High-throughput screening of a large collection of non-conventional yeasts reveals their potential for aroma formation in food fermentation. Food Microbiology. 2016;**60**:147-159

[86] Randazzo CL, Todaro A, Pino A, Pitino I, Corona O, Caggia C. Microbiota and metabolome during controlled and spontaneous fermentation of Nocellara Etnea table olives. Food Microbiology. 2017;**65**:136-148

Chapter 7

The Application of System Identification and Advanced Process Control to Improve Fermentation Process of Baker's Yeast

Zeynep Yilmazer Hitit, Baran Ozyurt and
Suna Ertunc

Additional information is available at the end of the chapter

http://dx.doi.org/10.5772/intechopen.70696

Abstract

Fermentation process of *Saccharomyces cerevisiae* has been investigated by many researchers for higher product quality and yield with lower cost. Operating parameters such as pH, dissolved oxygen (DO) concentration, temperature, substrate type and concentration, agitation speed, air flow rate should be optimized to achieve valuable products. In this point, system identification and advanced control techniques emerge to provide solutions. Dynamic analysis of pH and DO of the growth medium were performed at aerobic conditions in a batch bioreactor by applying step and square wave inputs to the base and air flow rates, respectively. Input–output data of the process and linear Auto Regressive Moving Average with eXogenous (ARMAX)-type model were used to determine the relationship between controlled and manipulated variable in baker's yeast production by system identification. The model parameters were estimated using the recursive least squares (RLS) method. The most suitable parametric model was determined by carrying out estimations with different values of initial value of the covariance matrix, forgetting factor, and order of the ARMAX model. Self-tuning generalized minimum variance (ST-GMV) control was performed with the ARMAX model for controlling pH and DO. Integrated square error (ISE) values were considered as a performance criteria for modeling and control studies.

Keywords: Baker's yeast, *S. cerevisiae*, system identification, ARMAX model, RLS, ST-GMV control, dissolved oxygen control, pH control

1. Introduction

Saccharomyces cerevisiae microorganism, also known as baker's yeast, is used in a variety of applications such as ethanol, glycerol, β-glucan, invertase enzyme, and mostly yeast production.

Using molasses or glucose as the carbon source with batch or fed-batch operation, it is possible to produce ethanol and the yeast itself on commercial scale under anaerobic or aerobic conditions, respectively [1]. Due to the process economy, cells with high volumetric efficiency should be obtained at the growth phase of microorganisms [2]. *S. cerevisiae* research is continuing in biotechnology and genetics fields as well as R&D works in the food and pharmaceutical industry [3]. Many studies on β-glucan production emphasize the importance of *S. cerevisiae* production for use in the pharmaceutical field [4]. Production data must be examined well in order to better understand the process [3]. Microorganisms used as biocatalysts can be produced at high concentrations with high enzyme activity at suitable values of bioreactor operating conditions such as pH, temperature, dissolved oxygen (DO) concentration, air flow rate, agitation speed, and substrate concentration [2].

If the facts of bioprocessing are mentioned, the operating parameters of the systems change with time, and these changes are not linear. In addition to that, they lack the mathematical models that define the complex reactions, which take place during cell growth and product formation [5, 6]. Furthermore, there are limited online sensors that can detect state variables such as cell, substrate, and product concentrations. Also, inhibitory effects of the substrate, oxygen or formed product, on the activity of the biocatalyst are present. Alpbaz et al. emphasized that *S. cerevisiae* is very sensitive to changes in the growth environment [7]. Because of these constraints in bioprocesses, it is necessary and important to determine the optimal operating conditions and to control the operating parameters at the determined optimum values to ensure economic gain, high-quality product, and safe operation [8]. Despite installation, operation and modeling studies are present in the literature for batch and fed-batch operation, and it is very difficult to control them due to both their biological process nature and the dynamics throughout that process. [9].

The control performance can be affected by the controller-tuning parameters and the process model parameters besides, the choice of control algorithm and the structure of the process model. Parameters of parametric and nonparametric models are calculated using dynamic analyses of different disturbance effects and system identification algorithms. Parametric models are generally constructed using a discrete time polynomial model [10]. Step, square wave, pseudorandom binary sequence (PRBS), impulse, pulse, and random inputs are generally applied as input variables (such as flow rates of acid/base for pH, cooling/heating fluid flow rate for temperature, air flow rate for DO, and substrate feed for substrate concentration) to perform dynamic analyses and obtain input–output data from process [11]. Various system identification algorithms such as Biermann, Levenberg–Marquardt, genetic, least squares (LS), and recursive least squares (RLS) have been studied for the calculation of the model parameters [10, 11]. The parameters of nonparametric models are calculated using the corresponding curves, such as reaction curve and Bode diagrams [12, 13]. The actual part of the system can be modeled with approximated structure and estimated parameters of it [3]. During control, the closed-loop performance of the system is largely depend on incompatibility of the actual process with the model. Therefore, a particular model structure should include all the known information about operating conditions and approximate the system to a chosen degree. Also, it should be flexible and lead to fast parameter estimation procedures [14]. Auto Regressive Moving Average with eXogenous (ARMAX)-type input polynomial model has been widely used in the literature due to the basic structure to describe the process dynamics [2, 3, 10–12, 15–18].

The need for self-regulating controllers stems from the desire to control processes whose parameters are unknown or slowly changing over time [11]. The fundamentals of this method were based on the self-tuning regulator (STR) developed by Åström and Wittenmark [19]. The control objective of this method is to reduce the variance of the output variable (such as pH, temperature, DO, and substrate concentration) to a minimum. The STR predicts the future output variance and then tries to implement a control action that forces the estimated variance to be zero [11]. However, in the applications of the STR technique, some difficulties have been experienced such as lack of online tuning parameters, weakness for control of nonminimum phase systems and poor control on changing or unknown time-delayed systems [2]. Later, Clarke, and Gawthrop modified the STR to the self-tuning generalized minimum variance (ST-GMV) to overcome these difficulties [20]. ST-GMV is an adaptive algorithm based on GMV cost function and a predictive form of the process model. This formulation leads to an easier tuning [21]. ST-GMV method has become very popular nowadays and is widely used in the industrial applications [22, 23].

In this chapter, dynamic analysis and system identification of the most important operating parameters for the baker's yeast production process which are pH and DO of the growth medium were investigated. Dynamic analysis were conducted at an optimal temperature for baker's yeast production determined from previous studies as 32°C were done in order to explain the process behavior and obtain the data to be used in the system identification step. It is clear that this study must be performed as process specific due to the data depends on mostly the physical structure of the process. This chapter especially focused on the success of the system identification step on the control applications. For examining the controller performance dependence on the process model structure and the model parameters, two contrary models which were called as suitable and unsuitable models were used in the ST-GMV simulation studies. Controller performances were evaluated according to the constant set point trajectory with various noisy conditions for both controlled variables. Similarly, as most of the studies given in the literature, simulation study results conducted at the MATLAB environment and interpreted in the base of intergrated squared errors.

2. Effects of operating parameters on baker's yeast production

S. cerevisiae yeast is undoubtedly one of the most important microorganisms that have been consumed safely throughout human history. Yeast cells need both nutrients and energy for growth and product formation. They use sugars such as glucose, maltose, and sucrose, vitamins such as biotin, pantothenate, inositol, and minerals such as Cu, Zn, Fe, Mo, and Mn to provide the necessary nutrients and energy [22]. High substrate concentrations may cause inhibition on microorganism production due to that substrate is binding to a second, nonactive site on a form of enzyme [22]. In the same case, ethanol production also occurs due to oxygen deficiency. This is known as the Crabtree effect, which is undesirable because it causes low yield in the fermentative growth of the cells [22]. Substrate level that inhibits the production of yeast is essentially dependent on the cell and substrate type. Glucose concentration over 200 g/L inhibits the microorganism growth in yeast production [2].

Another essential requirement is oxygen. Yeast cells use oxygen together with sugar to grow without ethanol production. As in the case of beer and wine production, if there is not enough oxygen in the environment, yeast will continue to grow by producing ethanol. In order to produce the yeast in the desired way, the oxygen and sugar transfer to the growth medium should be good and the ethanol formation should be low. Oxygen is a limiting substrate due to its lower solubility in water. The solubility of oxygen in water is 8 mg/L at 30°C [24]. It is known that the lower the DO concentration, the lower the substrate consumption and the rate of carbon dioxide formation, and this is called as Pasteur effect [25]. When working at aerobic conditions, it is not enough to feed only the oxygen source to the system. Providing a homogeneous distribution of oxygen in the liquid medium is also an important parameter [24]. The transfer of oxygen from the gas phase to the microorganism in the feed medium is of great importance in determining bioreactor design and operating conditions. Depending on whether the medium condition is aerobic or anaerobic, the following reactions occur during yeast production.

$$C_6H_{12}O_6 + 6O_2 \xrightarrow{\text{Aerobic conditions}} 6CO_2 + 6H_2O + 688\text{kcal} \quad (1)$$

$$C_6H_{12}O_6 \xrightarrow{\text{Anaerobic conditions}} 2C_2H_5OH + 6CO_2 + 56\text{kcal} \quad (2)$$

Carbon dioxide and water form in the medium when the yeast can grow in a suitable environment under aerobic conditions. If the goal is to produce yeast, aerobic conditions are required [12]. Otherwise, ethanol production occurs under anaerobic conditions [12]. Since ethanol itself is a carbon source, yeast cells can also use the ethanol to grow primarily during the diauxic phase [25]. Since ethanol is toxic, yeast prefers sugars to ethanol in order to grow. During growth, the yeast cell produces carbon dioxide. Another reliable variable that can be used to describe the state of the yeast since it does not use oxygen while producing ethanol is the ratio of the carbon dioxide production rate (CPR) to oxygen uptake rate (OUR). This ratio is defined as respiratory quotient (RQ) [2].

The pH of the growth medium is an important operating parameter because it affects the activity of enzymes in the microorganisms. Yeast is resistant to acidic environments, as well as being very sensitive to alkali environments. It is preferable for the growth medium to be between pH 3 and 6 [24]. It is difficult for cations and anions to pass through the cells when not working at the appropriate pH for the microorganism. Disruption of cell permeability affects enzyme activity and causes protein synthesis to stop. Besides, cells become more susceptible to toxic substances. For these reasons, the pH of the growth medium is influential on the substrate consumption and production yield of the yeast.

Like all microorganisms, yeasts have minimum (5–25°C), maximum (40–50°C), and optimum (30–40°C) growth temperatures [26]. The activity of enzymes involved in microorganisms' structures is greatly affected by temperature. In general, the temperature of the growth medium has a great influence on the growth of microorganisms, respiration, and product formation. According to the chemical kinetics, rise of temperature increases the reaction rate; however, enzymatic reactions adhere to this rule until a certain temperature. The temperature increase caused by the heat generated during the production of the baker's yeast under aerobic conditions is undesirable in terms of product efficiency.

3. System identification

Before working on a system, it is necessary to know the system's upper and lower limits in order to be aware of the possible situations that can be faced. To control the system, firstly the relations between the input variables and the output variables must be obtained and the model of the system must be developed. While modeling the systems, it is possible to use input–output data obtained by experimental studies or mass-energy balances. However, it is difficult to obtain a model by using mass-energy balances in complex systems, and in some cases these balances may be insufficient to accurately identify the system. In such cases, it is more useful to create models using system identification methods from the experimental input–output data [27]. System identification might be described as a method based on giving a disturbance to the input variable and consequently obtaining the output variable data of the system and determining the model. The unknown parameters in the parametric model are found by an appropriate method using input–output data. Then, the model is compared with the experimental data and the fitness is tested.

3.1. Signals used in system identification

The first step of system identification is the selection of input signals that will affect the system. Step, square wave, sinusoidal wave, PRBS, impulse, pulse, and random signals are generally applied as inputs [11]. Often, discrete time models are used to describe the system [8–11]. For this purpose, input and output variable signals are sampled and recorded at a suitable time interval. This chapter was focused on step effect as input signal.

3.2. System models

Linear model of an open-loop discrete time system can be written in terms of u(t) as input variable and x(t) as output variable as shown in Eq. (3).

$$x(t) + a_1 x(t-1) + \ldots + a_{n_a} x(t-n_a) = b_0 u(t-1) + \ldots + b_{n_b} u(t-n_b) \tag{3}$$

Backward shift operator is defined as Eq. (4).

$$z^{-i} x(t) = x(t-i) \tag{4}$$

Here, x(t) represents the x value at time t, x(t − 1) represents the x value at time (t − Δt) for Δt = 1, and x(t − i) represents the x value at time (t − iΔt). Eq. (5) is defined as a discrete time transfer function.

$$x(t) = \frac{B}{A} u(t) \tag{5}$$

Here, polynomials of A and B can be written as Eqs. (6) and (7).

$$A = 1 + a_1 z^{-1} + \ldots + a_{n_a} z^{-n_a} \tag{6}$$

$$B = b_0 + b_1 z^{-1} + \ldots + b_{n_b} z^{-n_b} \tag{7}$$

Roots of polynomials A and B are poles and zeros of the system, respectively. If one of the poles or zeros of the system is placed outside the unit circle of the z-plane, system is defined as unstable or nonminimum phase, respectively [28]. In a self-tuning system, disturbance effects can occur in a wide variety of forms. The disturbance signal s(t) can be a part of the control system and is often treated as an additional disturbance factor at the output of the controlled process. In this case, the self-tuning controller will attempt to eliminate this disturbance effect. Such signals can be summed up in two groups as defined signals and random signals. In general, model of a constant random signal source is shown as follows.

$$s(t) = \frac{C}{A} e(t) \tag{8}$$

Here, C polynomial is defined as Eq. (9).

$$C = 1 + c_1 z^{-1} + c_2 z^{-2} + \ldots + c_{n_c} z^{-n_c} \tag{9}$$

Eq. (8) is also described as auto regressive moving average (ARMA) model. Whole system output can be written in various forms as Eqs. (10–12).

$$y(t) = x(t) + s(t) \tag{10}$$

$$y(t) = \frac{B}{A} u(t-1) + \frac{C}{A} e(t) \tag{11}$$

$$\begin{aligned} y(t) = a_1 y(t-1) + \ldots + a_{n_a} y(t-n_a) = b_0 u(t-1) + \ldots + b_{n_b} u(t-n_b) + e(t) \\ + c_1 e(t-1) + \ldots + c_{n_c} e(t-n_c) \end{aligned} \tag{12}$$

This model is obtained by adding a control input to the ARMA-type signal, and the system is called ARMAX model.

3.3. Estimation of model parameters

3.3.1. LS method

Eq. (11) is written in matrix form, and transpose of data and parameter vectors are defined as (φ^T) and (θ^T), respectively.

$$\varphi^T = \left[-y(t-1), \ldots, -y(t-n_a), u(t-1), \ldots, u(t-n_b), e(t-1), \ldots, e(t-n_c)\right] \tag{13}$$

$$\theta^T = \left[a_1, \ldots, a_{n_a}, b_0, \ldots, b_{n_b}), c_1, \ldots, c_{n_c}\right] \tag{14}$$

Using Eqs. (13) and (14), output variable can be written as follows.

$$y(t) = \varphi^T\theta + e(t) \tag{15}$$

Eq. (15) is known as linear in the parameter model. There are N measurement values, where N is the number of data samples. Eq. (15) can be redefined as follows.

$$\begin{aligned} y(1) &= \varphi^T(1)\theta + e(1) \\ y(2) &= \varphi^T(2)\theta + e(2) \\ &. \\ &. \\ y(N) &= \varphi^T(N)\theta + e(N) \end{aligned} \tag{16}$$

Eq. (16) can be written in vector form as follows.

$$Y = \begin{bmatrix} y(1) \\ y(2) \\ . \\ . \\ y(N) \end{bmatrix}, \phi = \begin{bmatrix} \varphi^T(1) \\ \varphi^T(2) \\ . \\ . \\ \varphi^T(N) \end{bmatrix}, E = \begin{bmatrix} e(1) \\ e(2) \\ . \\ . \\ e(N) \end{bmatrix} \rightarrow Y = \phi\theta + E \tag{17}$$

The estimated parameter matrix ($\widehat{\theta}$) is written as given in Eq. (18), with the output variable ($\widehat{\theta}$) obtained from the model.

$$\widehat{y}(t) = \varphi^T(t)\widehat{\theta} \tag{18}$$

The difference between the measured output variable and the output variable calculated from the model is defined as the estimation error ($\varepsilon(t)$) and given in Eq. (19).

$$\varepsilon(t) = y(t) - \widehat{y}(t) = \varphi^T(t)\widehat{\theta} \tag{19}$$

Eq. (19) is rearranged for N measurements, and Eq. (20) is obtained.

$$\begin{aligned} \varepsilon(1) &= y(1) - \widehat{y}(1) = \varphi^T(1)\widehat{\theta} \\ \varepsilon(2) &= y(2) - \widehat{y}(2) = \varphi^T(2)\widehat{\theta} \\ &. \\ &. \\ \varepsilon(N) &= y(N) - \widehat{y}(N) = \varphi^T(N)\widehat{\theta} \end{aligned} \tag{20}$$

Eq. (20) can be written in vector form as follows.

$$\varepsilon(t) = Y - \phi\widehat{\theta} \tag{21}$$

In order to calculate $\hat{\theta}$ using LS method, cost function (J) given below in Eq. (22) must be minimized. ε value of Eq. (21) is written in Eq. (22) and must be rearranged as follows.

$$J = \sum_{t=1}^{N} \varepsilon^2(t) = \varepsilon^T \varepsilon = \|\varepsilon\|^2 \tag{22}$$

$$J = \left(Y - \phi\hat{\theta}\right)^T \left(Y - \phi\hat{\theta}\right) = Y^T Y - \phi^T \hat{\theta}^T Y - \left(Y^T \phi\hat{\theta} + \phi^T \phi\hat{\theta}^T \hat{\theta}\right) \tag{23}$$

To find the value that makes the Eq. (23) minimum, the derivative is taken, then equalized to zero, and rearranged and the predicted parameter vector calculated by the LS method is found as follows.

$$\frac{\partial J}{\partial \hat{\theta}} = -2\phi^T \left(Y - \phi\hat{\theta}\right) = 0 \tag{24}$$

$$\hat{\theta} = \left(\phi^T \phi\right)^{-1} \phi^T Y \hat{\theta} = \left[\sum_{t=1}^{N} \varphi(t)\varphi^T(t)\right]^{-1} \sum_{t=1}^{N} \varphi(t) y(t) \tag{25}$$

Although the LS is a widely used method, it is not suitable for self-tuning and predictive control methods because the parameter calculation is not made in real time. In such control methods, the data and parameters must be solved at every t instant and updated.

3.3.2. RLS method

Real-time parameter estimation is possible with RLS method and can be easily applied to self-tuning and predictive control algorithms for calculating time-varying model parameters. In the RLS method, the new value of the output variable is calculated by using the model parameters based on past data and the new input–output variable values. The actual value (y(t)) is compared with this estimated value, and the error (ε(t)) is found. The model parameters calculated in the previous step are updated with the newly calculated model parameters [27]. If Eq. (25) is written for any t instant, the parameter calculation equation will be as follows.

$$\hat{\theta}(t) = \left(\phi^T(t)\phi(t)\right)^{-1} \phi^T(t) Y(t) \tag{26}$$

Description of terms in Eq. (26) is in terms of vector forms as follows.

$$Y(t) = \begin{bmatrix} y(1) \\ y(2) \\ . \\ . \\ y(t) \end{bmatrix}, \phi(t) = \begin{bmatrix} \varphi^T(1) \\ \varphi^T(2) \\ . \\ . \\ \varphi^T(t) \end{bmatrix} \tag{27}$$

If Eq. (26) is written for the next sampling time (t + 1), parameter calculation equation will be as follows.

$$\widehat{\theta}(t+1) = \left(\phi^T(t+1)\phi(t+1)\right)^{-1}\phi^T(t+1)Y(t+1) \tag{28}$$

Terms $\phi(t+1)$ and $Y(t+1)$ of Eq. (28) are written in vector form as follows.

$$\widehat{\theta}(t+1) = \begin{bmatrix} \phi(t) \\ \phi^T(t+1) \end{bmatrix}, Y(t+1) = \begin{bmatrix} Y(t) \\ y(t+1) \end{bmatrix} \tag{29}$$

Using these equations, the terms in Eq. (28) are updated.

$$\phi^T(t+1)\phi(t+1) = \left[\phi^T(t)\ \ \varphi(t+1)\right]\begin{bmatrix} \phi(t) \\ \varphi^T(t+1) \end{bmatrix} = \phi^T(t)\phi(t) + \varphi(t+1)\varphi^T(t+1) \tag{30}$$

$$\phi^T(t+1)Y(t+1) = \left[\phi^T(t)\quad \varphi(t+1)\right]\begin{bmatrix} Y(t) \\ y(t+1) \end{bmatrix} = \phi^T(t)Y(t) + \varphi(t+1)y(t+1) \tag{31}$$

Covariance matrix (P(t)) is defined and written in Eq. (30) as follows.

$$P(t) = \left[\phi^T(t)\phi(t)\right]^{-1} \tag{32}$$

$$P(t+1)^{-1} = P(t)^{-1} + \varphi(t+1)\varphi^T(t+1) \tag{33}$$

Using covariance matrix definition, Eqs. (26) and (28) can be rewritten as follows.

$$\widehat{\theta}(t) = P(t)\phi^T(t)Y(t) \tag{34}$$

$$\widehat{\theta}(t+1) = P(t+1)\phi^T(t+1)Y(t+1) \tag{35}$$

The term of $\phi^T(t+1)Y(t+1)$ in Eq. (31) is written in Eq. (35) and rearranged as follows.

$$\widehat{\theta}(t+1) = P(t+1)\left[\phi^T(t)Y(t)\varphi(t+1)y(t+1)\right] = P(t+1)\phi^T(t)Y(t) + P(t+1)\varphi(t+1)y(t+1) \tag{36}$$

Eq. (37) is obtained using Eq. (34) as follows.

$$\phi^T(t)Y(t) = P^{-1}(t)\widehat{\theta}(t) \tag{37}$$

Eq. (33) is combined with Eq. (37) and rearranged as follows.

$$\phi^T(t)Y(t) = P(t+1)\widehat{\theta}(t) - \varphi(t+1)\varphi^T(t+1)\widehat{\theta}(t) \tag{38}$$

Eq. (38) is written in place at $\phi^T(t)Y(t)$ term of Eq. (36).

$$\begin{aligned} \widehat{\theta}(t+1) &= P(t+1)\left[P(t+1)^{-1}\widehat{\theta}(t) - \varphi(t+1)\varphi^T(t+1)\widehat{\theta}(t)\right] + P(t+1)\varphi(t+1)y(t+1) \\ &= \widehat{\theta}(t) + P(t+1)\varphi(t+1)\left[y(t+1) - \varphi^T(t+1)\widehat{\theta}(t)\right] \end{aligned} \tag{39}$$

Estimation error at time (t + 1) is defined as Eq. (40), and model parameter vector at time (t + 1) is found as Eq. (41).

$$\varepsilon(t+1) = y(t+1) - \varphi^T(t+1)\widehat{\theta}(t) \tag{40}$$

$$\widehat{\theta}(t+1) = \widehat{\theta}(t) + P(t+1)\varphi(t+1)\varepsilon(t+1) \tag{41}$$

Matrix inversion is applied to Eq. (33), and future value of covariance matrix is obtained as Eq. (42).

$$P(t+1) = P(t) - \frac{P(t)\varphi(t+1)\varphi^T(t+1)P(t)}{1+\varphi^T(t+1)P(t)\varphi(t+1)} \tag{42}$$

RLS method consists of Eqs. (40–42), and the algorithm used is given below.

At time t + 1,

1. Form $\varphi(t+1)$ vector using new data.

2. Calculate $\varepsilon(t+1)$ using Eq. (40).

3. Calculate $P(t+1)$ using Eq. (42).

4. Update model parameter vector $\widehat{\theta}(t+1)$ using Eq. (41).

5. Return to step 1.

4. ST-GMV control

Cost function of STR, defined as the difference between the set point and the measured value for an input–output model, is as follows.

$$J(u,t) = \Xi\left\{(y(t+k) - r(t+k))^2\right\} \tag{43}$$

where y is the output variable, r is the set point, u is the manipulated variable (input), Ξ is the expectation, and k is the default time delay [16]. It is possible to minimize this cost function by choosing u(t) which can be defined as an appropriate control output at time t. At the next sampling time step (t + Δt), a new situation occurs between y and r, and u will need to get a new value. If the default time delay is smaller than the time delay to be encountered in the real system, then the control output will try to remove the noise components before being transmitted to the system with the time delay in the real system. This would result in large feedback gains, resulting in an unrealizable controller that would make the system unstable. On the other hand, if the default time delay is greater than the time delay of the real system, then the lowest possible noise value will not be obtained since the highest rate for manipulation is not provided [16]. Clarke and Gawthrop set out the ST-GMV method using the control cost of the STR of Aström and Wittenmark to remove the difficulties in the STR altogether [29]. ST-GMV control is a one-step ahead optimal control strategy. The cost function of this technique is expressed by the following equation.

$$J(u,t) = \Xi\left\{(y(t+k) - r(t+k))^2 + \lambda(u(t))^2\right\} \tag{44}$$

This type of controller design can internally stabilize the system, and the stability depends on the selected λ values. ST-GMV algorithm has a good set point tracking characteristic and has the ability to control nonminimum phase systems. If the default time delay is implemented within the generalized system, then the control signal compensates the pseudo output $\phi(t)$ accordingly and directs the feed-forward path.

Using Eq. (44), ST-GMV method relies on maintaining closed-loop stability by taking λ as small as possible while maintaining a minimum output change to stay reasonably close to the expectation. Cost function can be generally expressed as follows.

$$J(u,t) = \Xi\{\phi^2(t+k)\} \tag{45}$$

ST-GMV method uses a system pseudo output $\phi(t + k)$ given by the following equation to minimize the cost function expressed in general by Eq. (45).

$$\phi(t+k) = Py(t+k) + Qu(t) - Rr(t) \tag{46}$$

Here, r(t) is the set point, P, Q, and R are the transfer functions with backward shift operator (z^{-k}). Pseudo output of the system includes a feed-forward feed term (Q) and filters (P, R) of output and the set point. ST-GMV algorithm uses the feed-forward polynomial Q to prevent output noise removal problem before signal transmission. $\phi(t + k)$ term of Eq. (46) can be expressed using Eq. (11) with the implementation of default time delay as follows.

$$\phi(t+k) = \frac{PB + QA}{A}u(t) + \frac{PC}{A}e(t+k) - Rr(t) \tag{47}$$

According to this equation, cost function to be minimized given by Eq. (45) will be the pseudo output variation. ST-GMV control algorithm divides the system into parts. For this, firstly, the error term is fragmented to include past, current, and future data.

$$\frac{\mathrm{PC}}{\mathrm{A}}\mathrm{e}(\mathrm{t}+\mathrm{k}) = \mathrm{Ee}(\mathrm{t}+\mathrm{k}) + \mathrm{z}^{-\mathrm{k}}\frac{\mathrm{G}}{\mathrm{A}}\mathrm{e}(\mathrm{t}) \tag{48}$$

Both sides of Eq. (48) are multiplied by A and rearranged as follows.

$$\mathrm{PC} = \mathrm{AE} + \mathrm{z}^{-\mathrm{k}}\mathrm{G} \tag{49}$$

Polynomials are written as follows.

$$A = 1 + a_1 z^{-1} + \dots + a_{n_a} z^{-n_a} \tag{50}$$

$$B = b_0 + b_1 z^{-1} + \dots + b_{n_b} z^{-n_b} \tag{51}$$

$$C = 1 + c_1 z^{-1} + \dots + c_{n_c} z^{-n_c} \tag{52}$$

$$E = 1 + e_1 z^{-1} + \dots + e_{k-1} z^{-(k-1)} \tag{53}$$

$$G = g_0 + g_1 z^{-1} + \dots + g_{n_g} z^{-n_g} \tag{54}$$

$$P = 1 + p_1 z^{-1} + \dots + p_{n_p} z^{-n_p} \tag{55}$$

AE term of Eq. (49) is expressed in terms of ARMAX model including offset as follows.

$$Ay(t+k) = Bu(t) + Ce(t+k) + d \tag{56}$$

If both sides of Eq. (54) is multiplied by E and written in Eq. (49), Eq. (55) is obtained.

$$PCy(t+k) = BEu(t) + CEe(t+k) + Ed + Gy(t) \tag{57}$$

Both sides of Eq. (55) are divided to C, and Eq. (56) is obtained.

$$Py(t+k) = \frac{BE}{C}u(t) + Ee(t+k) + \frac{Ed}{C} + \frac{G}{C}y(t) \tag{58}$$

Eq. (56) is combined with Eq. (46) and rearranged as follows.

$$\phi(t+k) = \frac{1}{C}[(BE+QC)u(t) + Gy(t) - CRr(t) + Ed] + Ee(t+k) \tag{59}$$

Eq. (57) is the sum of current and future terms. Current terms can be expressed as follows and represents the best $\phi(t + k)$ estimation made by using the data until time t.

$$\phi(t+k|_t) = \frac{1}{C}[(BE+QC)u(t) + Gy(t) - CRr(t) + Ed] \tag{60}$$

The second term is the estimation error caused by the noise source, e (t + 1), e (t + 2), …, e (t + k). The second term cannot be removed using control signal u(t) as mentioned before.

$$Ee(t+k) = \phi(t+k) + \phi(t+k|_t) \tag{61}$$

So, J is minimized by equalizing Eq. (58) to zero.

$$\phi(t+k|_t) = 0 \rightarrow \frac{[(BE+QC)u(t) + Gy(t) - CRr(t) + Ed]}{C} = 0 \tag{62}$$

Eq. (60) is rearranged using following definitions.

$$F = BE + QC \tag{63}$$

$$H = CR \tag{64}$$

ST-GMV control law can be expressed as follows.

$$Fu(t) + Gy(t) - Hr(t) + Ed = 0 \tag{65}$$

Calculation of input variable using ST-GMV control law is made using following equation.

$$u(t) = \frac{Hr(t) - Gy(t) - Ed}{F} \tag{66}$$

Application of ST-GMV algorithm consists of following steps [16, 29]:

1) Apply a PRBS to the system as a forcing function and obtain the plant output.

2) Estimate F, G, H from Eq. (63), implementing the RLS algorithm.

3) Employ Eq. (64) to evaluate the control signal.

4) Apply the control signal.

5) The system output is obtained.

6) Return to step 1.

5. Material and methods

5.1. Microorganism, inoculum preculture, and growth medium

S. cerevisiae NRRL Y-567 was obtained from NRRL-Agricultural Research Service Culture Collection. Preculture and growth media consist of 2% glucose, 0.6% yeast extract, 0.3% K_2HPO_4, 0.335% $(NH_4)_2SO_4$, 0.376% NaH_2PO_4, 0.052% $MgSO_4{\cdot}7H_2O$, and 0.0017% $CaCl_2{\cdot}4H_2O$ which were sterilized by autoclaving under 1.2 atm at 121°C for 20 min. Microorganisms were incubated for 8 hours at 32°C at 120 rpm, and inoculum ratio of 1:10 was used for scale enlargement.

5.2. Experimental system

In order to observe the change of DO and pH over time during baker's yeast production using a 2-L working volume of laboratory-scale bioreactor which was operated continuously, the input–output data were recorded and the ARMAX model parameters were determined by RLS method written in MATLAB. Experimental system is given in **Figure 1**. Most suitable parametric model was estimated using different values of α (covariance matrix), λ (forgetting factor), and order of parametric model. During experiments, DO and pH were measured with a WTW Oxi 340 with polarographic DO sensor and WTW pH340i pH meter, respectively. The DO and pH probes immersed in the bioreactor measure the online DO and pH values of the

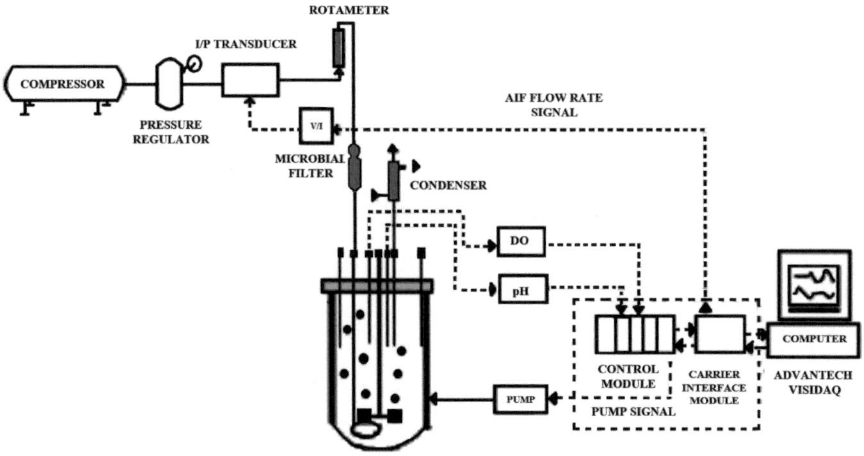

Figure 1. Experimental system.

Temperature (°C)	Air flow rate (vvm)	Cooling water flow rate (mL/min)	Cooling water temperature (°C)	Agitation speed (rpm/min)
32	1	55	21	600

Table 1. Operating conditions of the bioreactor.

growth medium, and these values were converted to the electrical signal with DO, and pH meters reach the I/O card in the computer via the carrier interface modules. The signals arriving to card are interpreted by the algorithm written in Visual Basic in the ADVANTECH VISIDAQ package program and was sent to the system online. Operating conditions of the bioreactor was given in **Table 1**.

6. Results and discussion

6.1. Dynamic analysis

6.1.1. Step input given to air flow rate

In baker's yeast production, the oxygen concentration must not fall below the critical value (0.7 mg/L); therefore, firstly manipulated variable must be selected to control the DO [12]. For that purpose, the air flow rate was chosen as the manipulated variable for the control of the DO. However, the effect of air flow rate on the DO in the bioreactor was investigated in order to observe effective control of this variable. For this purpose, while the system was in steady state at 1 mg/L DO for 0.5 L/min air flow rate, the positive step input was given to air flow rate as 3.4 L/min, and the change in DO over time was observed. In this case, DO was increased to 3 mg/L as can be seen from **Figure 2**.

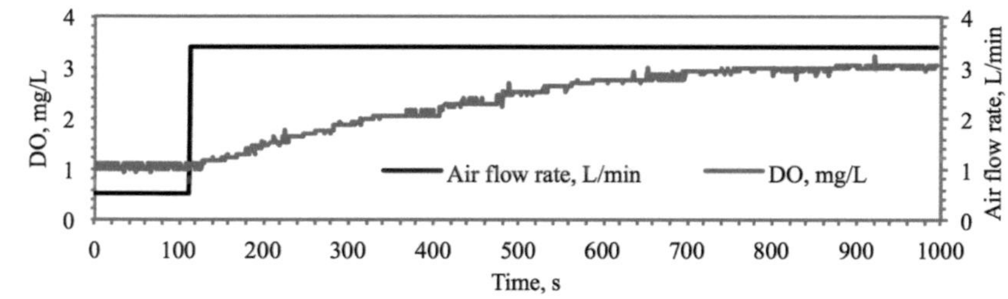

Figure 2. Positive step input given to air flow rate from 0.5 to 3.4 L/min.

6.1.2. Step input given to base flow rate

During the yeast growth, due to the degradation process of glucose in aerobic conditions to save the chemical energy in ATP molecules cause an increase in the concentration of H^+ ions resulted with pH decrease. Decreasing the pH of the medium affects not only the cell division, but also the cleavage rate and the production of many products from yeast and the activity of enzymes. For this purpose, it is necessary to determine the manipulated variable in order to achieve pH control. Therefore, the base flow rate was selected as the manipulated variable for pH control. However, in order to observe effective control of this variable, the effect of the base flow rate on the pH in the bioreactor was investigated. The bioreactor was settled at pH 3.90 under the specified operating conditions, and then, microorganism was fed to the bioreactor. Positive step input was given to base flow rate from 0.26 to 1.41 mL/min with 0.05 M NaOH solution. The acid (H_2SO_4) flow rate was kept constant at 0.22 mL/min. The change in pH value over time under such an effect is shown in **Figure 3**.

6.2. System identification results

6.2.1. Determination of model parameters for controlled variable of DO

In order to find the most appropriate ARMAX model using data of the manipulated variable air flow rate and the controlled variable DO obtained from dynamic analysis, the various forgetting factors (0.96–1), the initial value of the covariance matrix (1100,1000,10,000), and the order of the model (na = 2 nb = 1, na = 2 nb = 2, na = 3 nb = 1, na = 3 nb = 2) were run with the RLS algorithm, and integrated square error (ISE) was used for comparison. The models with the lowest and highest ISE values were used in the ST-GMV control algorithm to demonstrate the effect of the model structure on the control performance which will be explained in the next section. The compared values in terms of estimation performance are given in **Table 2**.

The estimation performance criterion λ value shows a decrease in ISE values between 0.96 and 0.97, but an increase in ISE values is observed after 0.97 of λ. The initial values of the covariance matrix between 1 and 10,000 resulted in a decrease in ISE values (**Table 2**).

Consequently, the most suitable ARMAX model was obtained with the order na = 3 nb = 2, the forgetting factor of 0.97, and the initial value of the covariance matrix of 1000. In the least successful ARMAX model case, the order was na = 2, nb = 1, the forgetting factor was 1, and

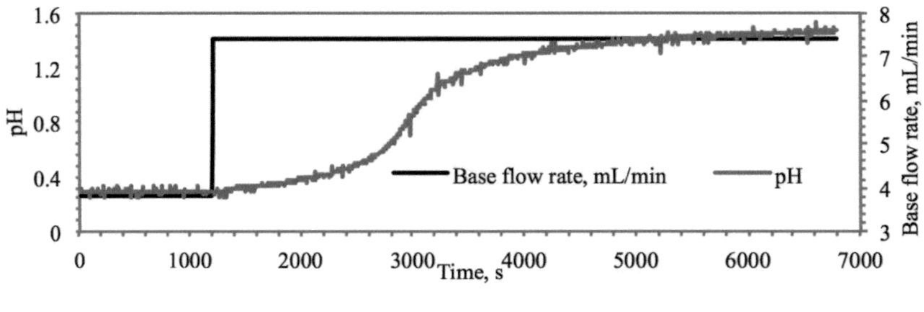

Figure 3. Positive step input given to base flow rate from 0.26 to 1.41 mL/min.

Estimation performance criteria		Order of the model			
		na = 2, nb = 1	na = 2, nb = 2	na = 3, nb1	na = 3, nb = 2
λ	P	ISE	ISE	ISE	ISE
0.96	1	1.8720	1.2037e + 36	1.5937	1.9718e + 61
	100	1.6297	3.8648e + 05	1.4280	2.2419e + 34
	1000	1.6110	9.9163e + 23	1.4124	8.5341e + 04
	10,000	1.6080	7.4019e + 25	1.4089	1.7541e + 25
0.97	1	2.0045	1.9599	1.7180	1.6972
	100	1.7544	1.7441	1.5485	1.5402
	1000	1.7337	1.7304	1.5306	1.5278
	10,000	1.7308	1.7295	1.5267	1.5254
0.98	1	2.1602	2.1154	1.8570	1.8365
	100	1.9039	1.8920	1.6852	1.6752
	1000	1.8803	1.8772	1.6641	1.6610
	10,000	1.8776	1.8765	1.6597	1.6585
0.99	1	2.3344	2.2891	2.0055	1.9853
	100	2.0702	2.0559	1.8307	1.8186
	1000	2.0427	2.0395	1.8052	1.8018
	10,000	2.0404	2.0392	1.8002	1.7991
1	1	2.4483	2.4060	2.1171	2.0993
	100	2.1663	2.1490	1.9384	1.9243
	1000	2.1336	2.1304	1.9077	1.9040
	10,000	2.1317	2.1305	1.9020	1.9010

Table 2. Estimated performance criteria as a result of positive step from 0.5 to 3.4 L/min for air flow rate and ISE values.

the initial value of the covariance matrix was 1. At the end of this approach, it was decided that the type of ARMAX model to be developed for GMV control was given in Eq. (67).

$$y(t) - 0.1928y(t-1) + 0.0338y(t-2) - 0.0168y(t-3) = 0.1853u(t-1) + 0.5794u(t-2) + e(t-1) \quad (67)$$

As a conclusion, the most suitable ARMAX model was obtained with the model order as na = 3, nb = 2, forgetting factor of 0.97, and initial value of the covariance matrix as 1000, and the RLS estimation of DO in the growth media by alterations with the air flow rate is given in **Figure 4**.

6.2.2. Determination of model parameters for controlled variable of pH

In order to find the most appropriate ARMAX model using data of the manipulated variable base flow rate and the controlled variable pH obtained from dynamic analysis, the various forgetting factors (0.96–1), the initial value of the covariance matrix (1, 100, 1000, 10,000), and the order of the model (na = 2, nb = 1; na = 2, nb = 2; na = 3, nb = 1, na = 3, nb = 2) were run with the RLS algorithm and ISE values of the prediction were used for comparison. The models with the lowest and highest ISE values were used in the ST-GMV control algorithm, which will be explained in the next section. The compared values in terms of estimation performance are given in **Table 3**.

ISE values were raised with the increase of the estimation performance criterion λ. For the same forgetting factor values, the ISE values decrease with the increase of covariance matrix initial value. The lowest ISE was obtained when the initial value of the covariance matrix was

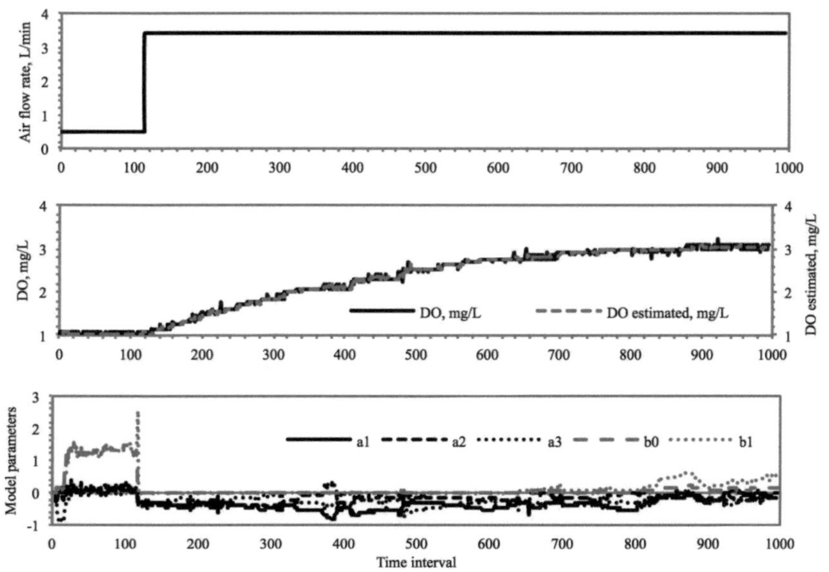

Figure 4. Experimental system identification estimation of DO (na = 3 nb = 2, λ = 0.97, P = 1000).

Estimation performance criteria		Order of the model			
		na = 2, nb = 1	na = 2, nb = 2	na = 3, nb = 1	na = 3, nb = 2
λ	P	ISE	ISE	ISE	ISE
0.96	1	1.2954	1.2948	1.1136	1.2699
	100	1.2394	1.2278	1.0715	1.0658
	1000	1.2096	1.2029	1.0521	1.0479
	10,000	1.1952	1.1943	1.0410	1.0383
0.97	1	1.3644	1.3631	1.1858	1.1851
	100	1.3160	1.3062	1.1474	1.1423
	1000	1.2815	1.2731	1.1260	1.1204
	10,000	1.2620	1.2595	1.1117	1.1091
0.98	1	1.4439	1.4428	1.2706	1.2699
	100	1.4020	1.3940	1.2351	1.2312
	1000	1.3670	1.3558	1.2149	1.2077
	10,000	1.3405	1.3373	1.1961	1.1929
0.99	1	1.5302	1.5292	1.3676	1.3669
	100	1.4910	1.4852	1.3326	1.3298
	1000	1.4602	1.4468	1.3159	1.3079
	10,000	1.4257	1.4213	1.2930	1.2887
1	1	1.6207	1.6202	1.4828	1.4820
	100	1.5827	1.5785	1.4484	1.4460
	1000	1.5582	1.5441	1.4355	1.4276
	10,000	1.5171	1.5108	1.4104	1.4044

Table 3. Estimated performance criteria as a result of positive step from 0.26 to 1.412 mL/min for base flow rate and ISE values.

10,000. As the order of polynomial A increases, the ISE values decrease, and as the order of polynomial B increases, a significant change in ISE values cannot be observed.

$$y(t) - 0.2235y(t-1) + 0.1077y(t-2) - 0.0737y(t-3) = 1.2723u(t-1) + 1.921u(t-2) + e(t-1) \quad (68)$$

As a result, the most suitable ARMAX model was obtained with the order na = 3 nb = 2, the forgetting factor of 0.96, and the initial value of the covariance matrix of 10,000. In the least successful ARMAX model case, the order was na = 2, nb = 1, the forgetting factor was 1, and the initial value of covariance matrix was 1. At the end of this screening, it was decided that the suitable ARMAX model structure in order to develop the GMV control algorithm for pH control with by manipulating the base flow rate is given in Eq. (68).

As a conclusion, the most suitable ARMAX model was obtained with the model order of na = 3, nb = 2, forgetting factor as 0.96, and initial value of the covariance matrix of 10,000, and the RLS estimation is given in **Figure 5**.

6.3. ST-GMV control applications of baker's yeast production

The suitable and unsuitable ARMAX models of the yeast production process expressed the relationship between the controlled variables of DO and pH, with the manipulated variables of air flow rate and base flow rate in system identification results section. After this step, ST-GMV control performances were evaluated with the suitable and unsuitable ARMAX models determined for each controlled variable in the case of constant set point trajectory for various noise levels. The control performance criterion was selected as ISE and values were evaluated for the ST-GMV control simulations of both DO and pH control cases. By this way, how much the system identification step, including the determination of model structure and model parameter settings, has affected the success of process control is demonstrated by using control simulations.

6.3.1. DO control

In the baker's yeast production process, in which DO was controlled variable and the air flow rate was selected as the manipulated variable, the most suitable and unsuitable ARMAX models obtained from system identification have been used in ST-GMV control algorithm. In the case of positive step input from 0.5 to 3.4 L/min for the air flow rate, the order of the most suitable model was na = 3 nb = 2, $\lambda = 0.97$, and P = 1000. By the same way, the model that does not identify the system (largest ISE value) was found as na = 2 nb = 1, $\lambda = 1$, and P = 1. When the both suitable and unsuitable obtained models were used in the ST-GMV control algorithm

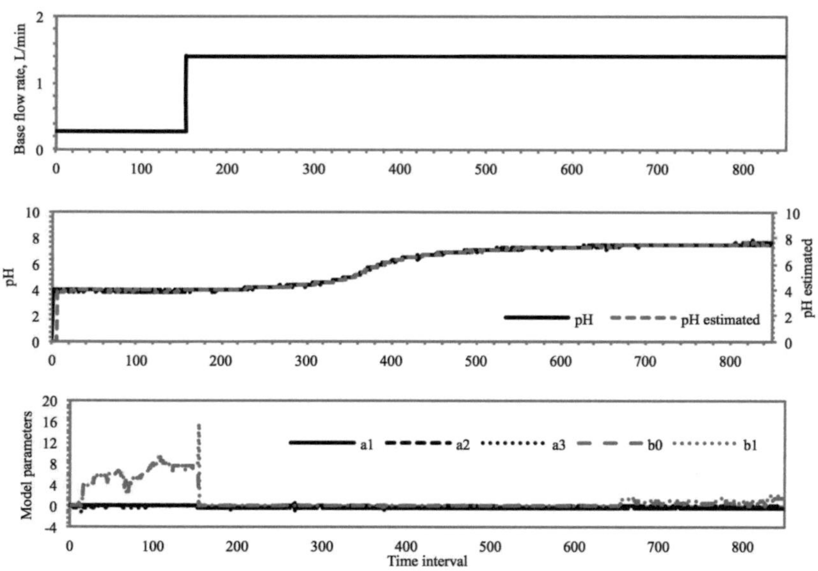

Figure 5. Experimental system identification estimation of pH (na = 3 nb = 2, $\lambda = 0.96$, P = 10,000).

with the controller parameters of P = 1, Q = 0.9975, R = 2.0885 in the presence of two different noises. It was observed that the suitable model with calculated ISE values of 50.14 and 52.37 was definitely able to identify the system as expected and provided a good control (**Figure 6**) in contrast to unsuitable model with ISE values of 502 and 609.56, respectively (**Figure 7**).

6.3.2. pH Control

In the baker's yeast production process, in which the pH value was controlled variable and the base flow rate was selected as the manipulated variable, the most suitable and unsuitable ARMAX models obtained from system identification have been used in the ST-GMV control algorithm. In the case of positive step input from 0.26 to 1.41 mL/min for the base flow rate, the order of the most suitable model was na = 3, nb = 2, λ = 0.96, and P = 10,000. By the same way, the model that does not identify the system (largest ISE value) was found as na = 2 nb = 1, λ = 1, and P = 1. When the both suitable and unsuitable obtained models were used in the ST-GMV control algorithm with the controller parameters of P = 1, Q = 0.9375, R = 1.1885 in the presence of two different noises. It was observed that the suitable model with calculated ISE values of 110.2 and 113.51 was definitely able to identify the system as expected and provided a good control (**Figure 8**) in contrast to unsuitable model with ISE values of 245.69 and 213.69, respectively (**Figure 9**). ST-GMV control simulation results are summarized in **Table 4**.

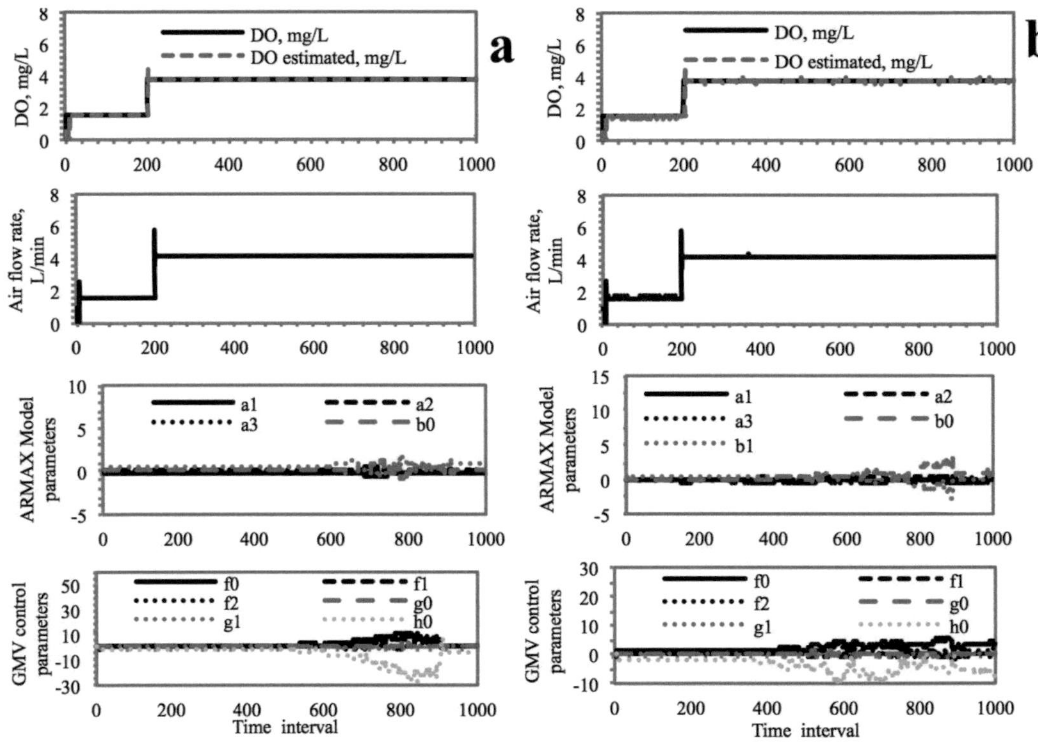

Figure 6. ST-GMV control simulation results with the most appropriate ARMAX model at the desired constant DO set point (a) having 0.005 noises, (b) having 0.05 noises.

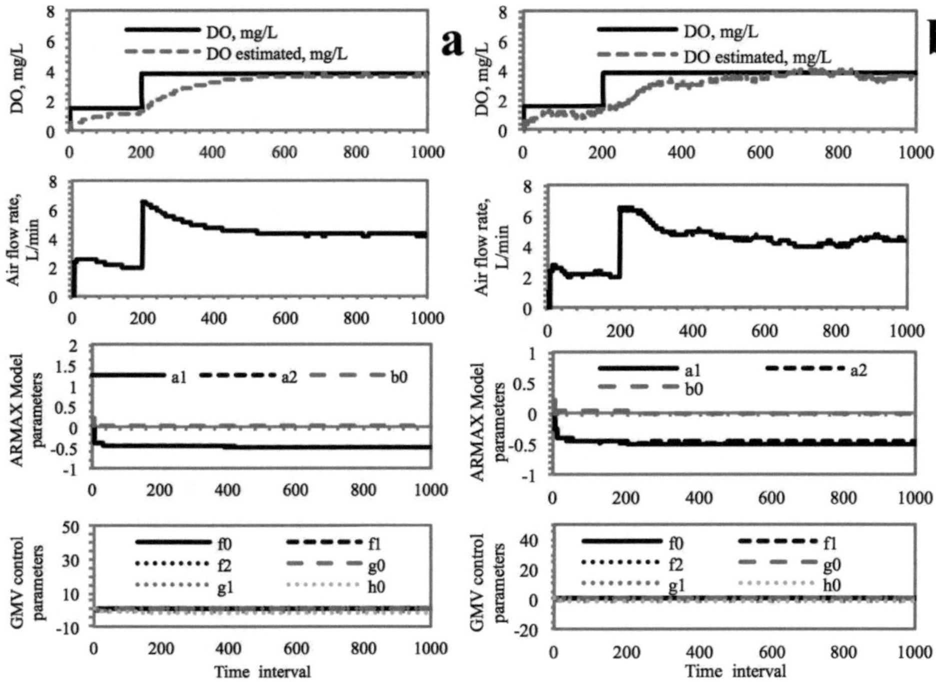

Figure 7. ST-GMV control simulation results with unsuitable ARMAX model at the desired DO set point (a) having 0.005 noises, (b) having 0.05 noises.

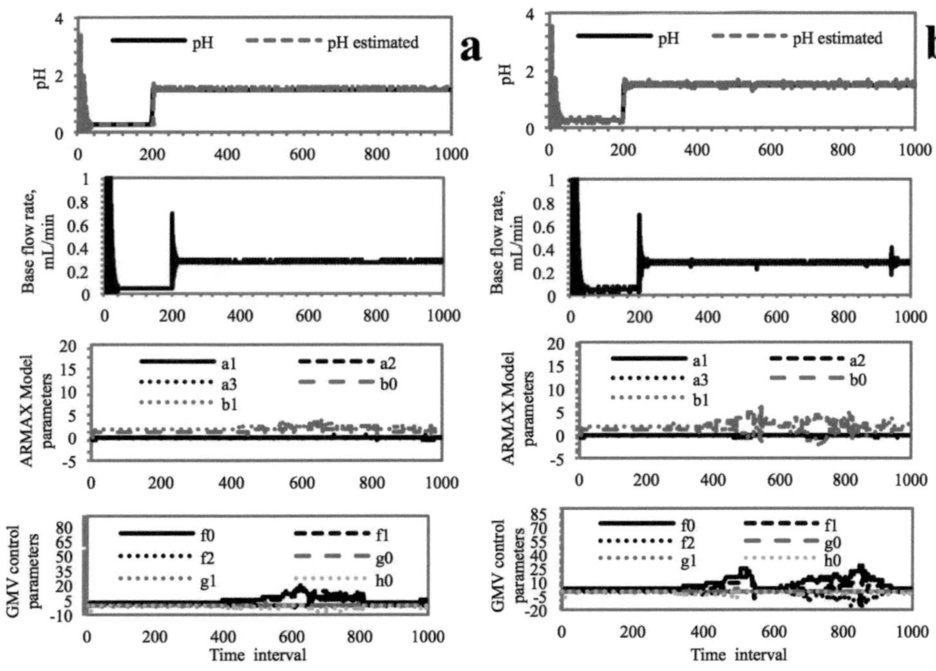

Figure 8. ST-GMV control simulation results with the most appropriate ARMAX model at the desired constant pH set point (a) having 0.005 noises, (b) having 0.05 noises.

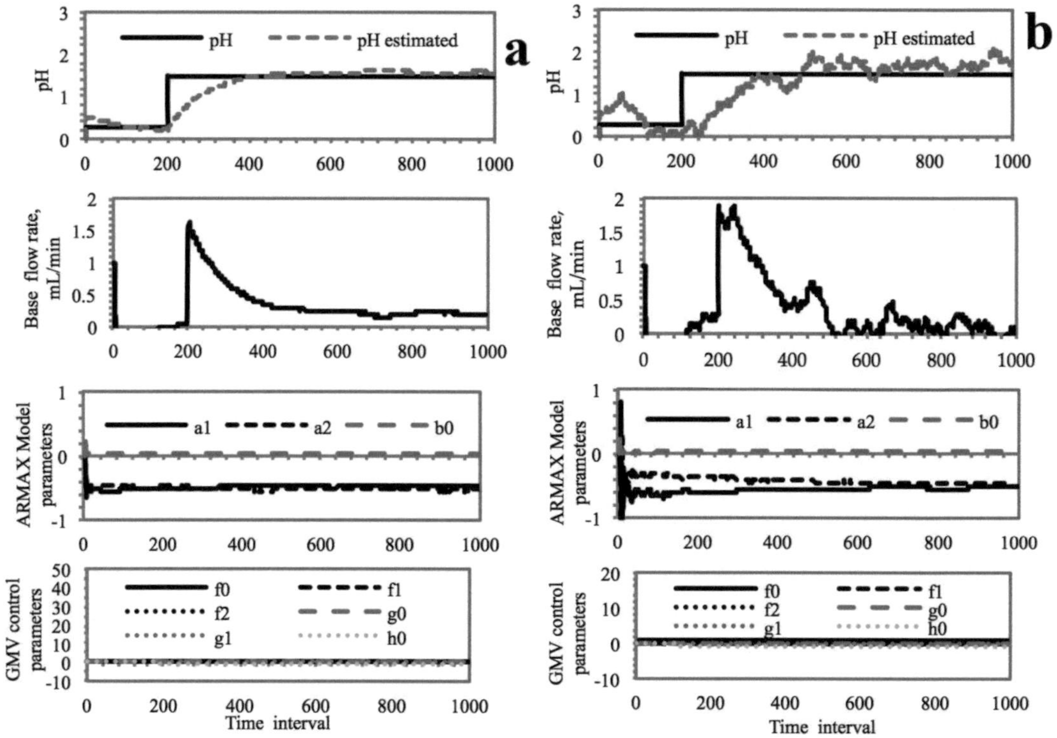

Figure 9. ST-GMV control simulation results with unsuitable ARMAX model at the desired pH set point (a) having 0.005 noises, (b) having 0.05 noises.

Controlled variable	Noises	Control with suitable model ISE	Control with unsuitable model ISE
DO	e = 0.005	50.1361	502.0056
	e = 0.05	52.3745	609.5629
pH	e = 0.005	110.2026	245.6958
	e = 0.05	113.5143	213.6950

Table 4. Theoretical ST-GMV control performance summary with suitable and unsuitable ARMAX models.

7. Conclusion

Understanding the dynamic behavior of biotechnological processes, in which living cells are used as biocatalysts, is one of the most challenging issues nowadays due to the fact that thousands of biochemical reactions are taking place simultaneously. It is clear that process operation in the batch mode will be difficult due to time-varying parameters. In this case, estimation procedure is gaining the main importance to express the real process behavior by the mathematical models. For this purpose, various methods and the approaches exist. Selecting the most appropriate method for the system identification is the next critical step.

It also affects the success of the process dynamic behavior estimation. In the production process of baker's yeast in a batch operational mode with aerobic conditions using *S. cerevisiae* microorganism, system identification studies carried out easily by RLS algorithm and were found successful for identifying DO and pH variations with the air flow rate and the base flow rate manipulations. Prediction error defined as the ISE demonstrates that the estimation performance was good.

Selection of the model structure is crucial in expressing process behavior accurately. In this study, order of the model was found as na = 3, nb = 2 for both polynomial-type ARMAX model structure by examining the different order of the models. As the order of the polynomial A increases, the difference between the actual value and the predicted value decreases, which is desirable. However, the increase in B polynomial does not show any significant change. The forgetting factor was found as 0.96 and 0.97, while the initial value of covariance matrix was not as effective as the value of the forgetting factor, and the 1000 value was observed as appropriate for all experiments.

The theoretical ST-GMV control of DO and pH was successfully performed with the most suitable ARMAX models obtained from system identification. When the noise level is increased in the theoretical ST-GMV control, it is possible to achieve successful control under the constant set point condition with obtained models. In addition, the performance of a controller that uses unsuitable models decreases with the increase of noise levels. So as a conclusion, successful control can only be accomplished with a good system identification.

Author details

Zeynep Yilmazer Hitit*, Baran Ozyurt and Suna Ertunc

*Address all correspondence to: zyilmazer@ankara.edu.tr

Department of Chemical Engineering, Engineering Faculty, Ankara University, Ankara, Turkey

References

[1] Akay B, Ertunç S, Kahvecioğlu A, Hapoğlu H, Alpbaz M. Adaptive control of *S. cerevisiae* production. Food and Bioproducts Processing. 2002;**80**(1):28-38. DOI: 10.1205/09603080 2753479089

[2] Hitit ZY, Boyacioglu H, Ozyurt B, Ertunc S, Hapoglu H, Akay B. Self-tuning GMV control of glucose concentration in fed-batch baker's yeast production. Applied Biochemistry and Biotechnology. 2014;**172**(8):3761-3775. DOI: 10.1007/s12010-014-0794-5

[3] Boyacıoğlu H, Ertunç S, Hapoğlu H. Modelling of baker's yeast production. International Journal of Secondary Metabolite (IJSM). 2017;**4**(1):10-17. DOI: 10.21448/ijsm.252053

[4] Huang B, Guo J, Yi B, Yu X, Sun L, Chen W. Heterologous production of secondary metabolites as pharmaceuticals in *Saccharomyces cerevisiae*. Biotechnology Letters. 2008; **30**(7):1121-1137. DOI: 10.1007/s10529-008-9663-z

[5] Cinar A, Parulekarb SJ, Undey C, Birol G. Batch Fermentation: Modelling: Monitoring, and Control. New York: CRC press; 2003. DOI: 10.1201/9780203911358.ch7

[6] Bailey JE, Ollis DF. Biochem. Eng. Fund: New York: Mc Graw Hill; 1986

[7] Alpbaz M, Bursali N, Ertunç S, Akay B. Application of a statistical technique to the production of *Saccharomyces cerevisiae* (baker's yeast). Biotechnology and Applied Biochemistry. 1997;**26**(2):91-96. DOI: 10.1111/j.1470-8744.1997.tb00452.x

[8] Bursali N, Akay B, Ertunc S, Hapoglu H, Alpbaz M. New tuning method for generalized predictive control of the production of *S. cerevisiae*. Food and Bioproducts Processing. 2001;**79**(1):27-34. DOI: 10.1205/09603080151123335

[9] Ertunc S, Akay B, Hapoglu H, Alpbaz M. A new computer algorithm approach to identification of continuous-time batch bioreactor model parameters. Computational Science–ICCS 2005. 2005;154-147. DOI: 10.1007/11428831_19

[10] Bursali N, Ertunc S, Akay B, Pamuk V, Hapoğlu H, Alpbaz M. The effect of the tuning parameters on the performance of the parametric and nonparametric model based control methods for growth medium temperature of baker's yeast production. Food and Bioproducts Processing. 2001;**79**(4):242-249. DOI: 10.1205/096030801753252315

[11] Akay B, Ertunc S, Boyacioglu H, Hapoglu H, Alpbaz M. Discrete system identification and self-tuning control of dissolved oxygen concentration in a stirred reactor. Korean Journal of Chemical Engineering. 2011;**28**(3):837-847. DOI: 10.1007/s11814-010-0450-x

[12] Ertunc S, Akay B, Boyacioglu H, Hapoglu H. Self-tuning control of dissolved oxygen concentration in a batch bioreactor. Food and Bioproducts Processing. 2009;**87**(1):46-55. DOI: 10.1016/j.fbp.2008.04.003

[13] Dorf RC, Bishop RH. Modern Control Systems. New Jersey: Pearson; 2011

[14] Warwick K, Rees D. Industrial digital control systems (no. 37). In: IET; 1988

[15] Soderstrom T, Stoica P. System Identification. Hertfordshire: Prentice hall international; 1989

[16] Ertunc S, Akay B, Bursali N, Hapoğlu H, Alpbaz M. Generalized minimum variance control of growth medium temperature of baker's yeast production. Food and Bioproducts Processing. 2003;**81**(4):327-335. DOI: 10.1205/096030803322756411

[17] Camcıoğlu Ş, Özyurt B, Zeybek Z, Hapoğlu H. Experimental application of one step ahead advanced pH control to water-based paint wastewater treatment. Journal of the Faculty of Engineering and Architecture of Gazi University. 2016;**31**(3):655-664. DOI: 10.17341/gummfd.52934

[18] Camcıoğlu Ş, Ozyurt B, Hapoglu H. pH control of paper wastewater treated with electrocoagulation method. Anadolu University Journal of Science and Technology–A Applied Sciences and Engineering. 2015;**16**(2):107-115

[19] Åström KJ, Wittenmark B. Adaptive Control:Courier Corporation; 2013.

[20] Clarke DW, Gawtthrop PJ. Self-tuning control. IEEE Proc. 1979;**126**(6):633-640

[21] Olmstead DR. Computer Control of Fermentation Processes. Florida: CRC Press; 1989

[22] Shuler ML, Kargi F. Bioprocess Engineering: Basic Concepts. New Jersey: Prentice Hall; 1992

[23] Sales KR, Billings SA. Self-tuning control of non-linear ARMAX models. International Journal of Control. 1990;**51**(4):753-769. DOI: 10.1080/00207179008934096

[24] Perry J-G. Introduction to Food Biotechnology. Florida: CRC Press; 2002

[25] Furukawa K, Heinzle E, Dunn IJ. Influence of oxygen on the growth of *Saccharomyces cerevisiae* in continuous culture. Biotechnology and Bioengineering. 1983;**25**(10):2293-2317. DOI: 10.1002/bit.260251003

[26] Walker G. Yeast Physiology and Biotechnology. Chichester: Wiley; 1998

[27] Wellstead PEE, Zarrop MB. Self-Tuning Systems: Control and Signal Processing. New York: John Wiley & Sons Inc; 1991

[28] Hapoğlu H, Balas CE, Wardle AP. Self-tuning and conventional control of an industrial scale packed distillation column. Turkish Journal of Engineering and Environmental Sciences. 1998;**22**(1):1-8

[29] Karagöz AR, Hapoğlu H, Alpbaz M. Generalized minimum variance control of optimal temperature profiles in a polystyrene polymerization reactor. Chemical Engineering and Processing: Process Intensification. 2000;**39**(3):253-262. DOI: 10.1016/S0255-2701(99)00104-X

Chapter 8

Probiotic Yeast: Mode of Action and Its Effects on Ruminant Nutrition

Shakira Ghazanfar, Nauman Khalid,
Iftikhar Ahmed and Muhammad Imran

Additional information is available at the end of the chapter

http://dx.doi.org/10.5772/intechopen.70778

Abstract

The main purpose of yeast supplementation is to treat rumen microbial dysbiosis which may enhance the nutrient utilization leading to enhanced animal growth and productivity. Yeast improves rumen ecosystem by two ways: by direct production of digestive enzymes and growth stimulator and by promoting the growth and function of beneficial microbiota. Yeasts have potential to produce metabolites, which stimulate the growth, like rumen acetogens and antimicrobial compounds which inhibit potential pathogens. The yeast probiotic impact on animals depend on different interacting factors including animal breed, supplemented dose, type, diet, strain, physiological stage and feeding system. In the situation of a high feed cost all over the world, probiotic yeast gives a useful nutritional strategy which allows increasing diet digestibility and consequently enhances the performance in ruminants in cost-effective manner. Many yeast culture-based products are commercially available worldwide, but their effectiveness as probiotic dietary supplement in a particular breed is mostly questionable. Therefore, exploration of the new indigenous probiotic strain is of great interest in this context. The probiotic strains of same ecological origin may be more compatible with rumen microbiome giving maximum outputs. Moreover, the breed specific probiotic yeast is an economical and viable option for farmers to overcome the effects of malnutrition.

Keywords: indigenous probiotics, viability, stability, gastrointestinal tract, rumen microbiota

1. Introduction

Ruminants can eat different types of feed that are digested by microbial biomass resulting in better metabolism, which ultimately impacts the dairy animal productivity. The microbial flora in the gastrointestinal tract (GIT) has a major impact on the productive efficiency, health status, and well-being of the dairy animals [1–3]. The diversity and function of ruminal GIT microbes are very important in feed digestion. The way the nutrients are digested in GIT in ruminants have a crucial impact on growth, health, and productivity [4]. The GIT inhabits multifarious microbial diversity that helps in generating impassive response regarding nutritious health, physiology, and productivity of animals [1]. The existing gut microbiota regulates food safety through the shedding of pathogens, interaction with organisms and resource competition in the GIT [5]. Gastrointestinal tract microflora aids in stimulation of the immune system that acts as a barrier against infectious pathogens. It also restrains the injurious and pathogenic bacteria in gut colonization [6]. Different strategies have been used to enhance the microbiota of gastrointestinal tract, which ultimately affect the production potential and growth efficiency of dairy animals. Nowadays, the improvement of microbiota of the gastrointestinal tract by using probiotic has become a useful and economical method to enhance the health and productive performance of animals. A live microorganism which beneficially influences the host by improving microbial flora of its intestine is called probiotic [7]. Numerous microorganisms have been sanctioned as probiotics that are used in diet of ruminants to upgrade nutrient utilization and animal performance [8]. Bacterial probiotics give better results in young calves, chickens, and pigs, whereas yeast/fungal probiotics are effective in adult ruminants [9]. During the last decades, *Saccharomyces cerevisiae* have been used as preventer supplement against diarrhea and other digestive system problems in livestock [10]. They also give production benefits, reduced digestive problems, and better health of animals in cost-effective manners [11]. Dietary supplementation of yeast culture has a positive effect on feed intake, which ultimately enhances ruminant growth [12] and production efficiency [13]. It also has positive effect on milk fat content [13–15] and milk urea nitrogen [16]. Moreover, it decreases lactate production [17], increases desirable bacterial population [13], prevents the rumen acidosis [18], increases the hemicelluloses degradability, and some important nutrient (NDF, ADF) digestibilities [19–21]. Another advantage of use of the *S. cerevisiae* is that the benefit to cost ratio of *S. cerevisiae* supplementation in dairy cattle is 4:1 [22]. This chapter explains the yeast effects on the ruminants and proposes guidelines to develop the breed specific probiotic yeast for animal use.

2. Probiotic yeast

Yeasts are eukaryotic microorganisms and are different from bacteria from the structure and functional aspects [23]. Yeasts are facultative anaerobes and differ in terms of their location, shape, reproducing activities, subtracts they utilize and are highly resistant to different antibiotics, such as sulfamides and other antibacterial substrates [24]. The resistance capability of the yeast cells is natural and genetically encoded. This resistance cannot be changed or transmitted to other microbial species. The size of the yeast cell (5 × 10 μm) is also higher than bacteria (0.5 × 5 μm). The study of antagonistic yeast to block bacterial pathogenicity in the early stage of development is mainly due to following steps; (1) competition for nutrients, (2) pH changes in the medium, (3) high concentrations of ethanol production, (4) secretion of antibacterial compounds and release of antimicrobial compounds (toxins or "mycocins"). However, the effectiveness of probiotic organism is viewed as population-specific due to variation in gastrointestinal microbial flora, feeding habits, and precise interaction between host animal and microbes. As most of the probiotic yeast strains accessible in the market are of Western or European origin, hence, there is a dire need to explore new indigenous probiotic strains. Yeast cells produce many important fermentation metabolites and different types of important minerals and enzymes that make it useful and highly nutritive feed supplement for ruminants [25–27]. It also provides improved production, reduced digestive problem, and better health in cost-effective manners.

3. Understating the ruminal gut microbiology for development of new target-specific probiotic strain

Uses of molecular techniques have changed the study of the rumen ecosystem [10]. A better understanding of the rumen microbiology is an important step to select and prepare a new yeast strain affecting on functional-specific microbes. Ruminants' stomach consists of reticulum, rumen, omasum, and abomasums [28]. The rumen is an anaerobic chamber that harbors an immense diversity of microbial community including bacteria, archaea, fungi, and single-celled ciliated protozoa (**Figure 1**) [29].

This microbial ecosystem has been used for better feed digestion. Bacteria are numerous microbes in rumen [30]. Mostly bacteria are associated with feed; some are free living, attached with mucous membrane and associated with fungi and protozoa. The shape of rumen bacteria are mostly cocci, rod spirochete budding, and filamentous. Rumen bacteria are 1–2 μm in size. The majority of the rumen bacterial species are Gram-negative. The structure of this microbial community is influenced by many factors, including host species, age, seasons [31], type of feed, geographical location, and whether the animal has received any treatment [32]. The balance in rumen microbial flora plays a crucial role in feed utilization and could result

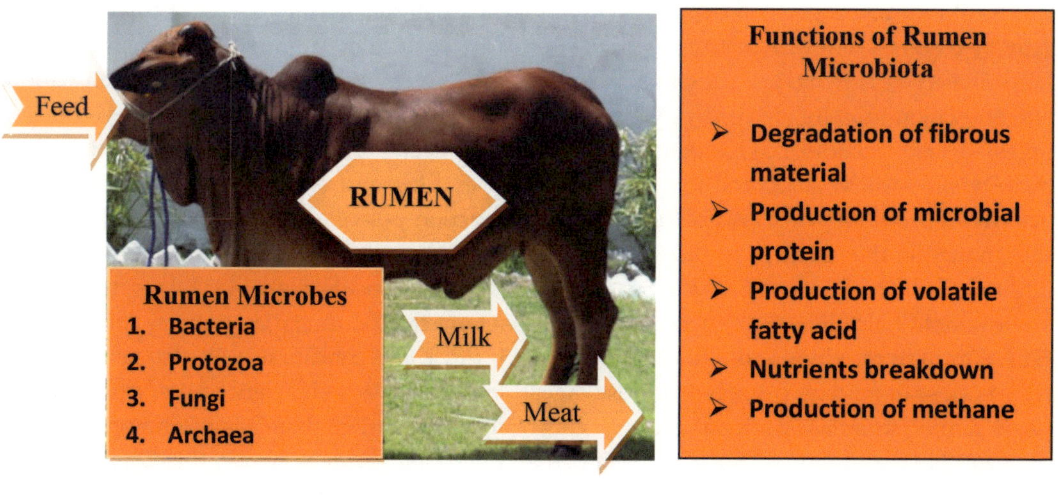

Figure 1. Estimated rumen microbial ecosystems.

in better productivity [33]. The rumen microbial profile directly depends on the type of feed [29]. Ruminants can eat from different types of feed sources that are digested by microbial biomass, which helps in better metabolism [34]. This ultimately impacts the productivity of dairy animals. The feed microbial flora could be managed by using beneficial microbial supplementation. The management and modification of ruminal fermentation to improve animal performance have been the main objective of several studies on ruminant species. From this line of research, we will use method to manipulate the rumen fermentation for improving nutrient utilization and productivity of animals. The banning of the use of antibiotics as animal growth promoters in the European Union in 2006 has increased the demand from producers for alternative feed additives that can be used to manipulate the ruminal fermentation and improve animal production [31, 35, 36]. The modulation in the rumen population for better nutrients metabolism can be achieved made by manipulating the feed, antibiotic and some microbial inoculants. The diet-shift effect such as high-forage diet, increases the rumen pH and consequently improves the stability and viability of cellulolytic and hemicellulolytic bacteria and protozoa. On the other hand, high-concentrate diet that decreases the rumen pH, resultantly decreases the cellulolytic and hemicellulolytic bacteria and increases the amylolytic bacteria and lowers the rumen protozoa. Microbial inoculants may alter the stability and viability of microbial population of the rumen and hindgut in a better way.

4. Role of probiotic yeast in ruminant nutrition

The balance in rumen microbial flora plays a crucial role in feed utilization and could result in better animal productivity [37]. Several hypotheses concerning the mode of action of probiotic yeast in animal nutrition have been proposed; however, a majority of them emphasize positive effects by modifying rumen microbial population. The first and most widely supported mode of action is that the yeast stimulates the growth of bacteria (cellulolytic, amylolytic, and proteolytic) and protozoa [38, 39]. The rumen dissolved oxygen (O_2) can be measured in situ [40]. Loesche [41] found that a majority of rumen microbial flora are highly sensitive to O_2. Probiotic yeasts remove oxygen from rumen and provide a better anaerobic environment for bacterial growth [42]. Sixteen liters of oxygen can enter inside rumen daily, by the mean of feeding, rumination, and salivation [43]. Inside rumen, yeast cells use oxygen for their metabolic process. Freshly ingested feed particles have sugars and small oligosaccharides. Probiotic yeast metabolizes these small particles and produces peptides, polypeptides, and amino acids. This respiratory activity of probiotic yeast lowers the oxidation-reduction potential inside rumen [44]. A negative change in the redox potential (−20 mV) has been observed in rumen with probiotic yeast addition [36]. This change gives a better anaerobic condition inside rumen [33]. Aforementioned environment helps in the protection of rumen bacteria from damage by oxygen and stimulation of growth of cellulose degrading bacteria [45, 46]. These conditions will also be helpful in the cellulose degrading process (cellulose digestion). Respiratory-deficient mutants of probiotic yeast cannot stimulate bacterial growth. As we mention earlier that O_2 scavenging property of yeasts is very important for growth of rumen microbial biomass, hence, this O_2 scavenging property should be considered when probiotic yeast is selected for ruminants (**Figure 2**).

http://dx.doi.org/10.5772/intechopen.70778

Probiotic yeasts have beneficial effects on the lactate-metabolizing bacterial species. *S. cerevisiae* provides different growth factors essential for the growth of lactic acid fermenting bacterial species, such as *Megasphaera elsdenii* or *Selenomonas ruminantium*. In dairy animals, a reduction of lactic acid concentration was seen inside the rumen with live yeast addition (**Figure 3**).

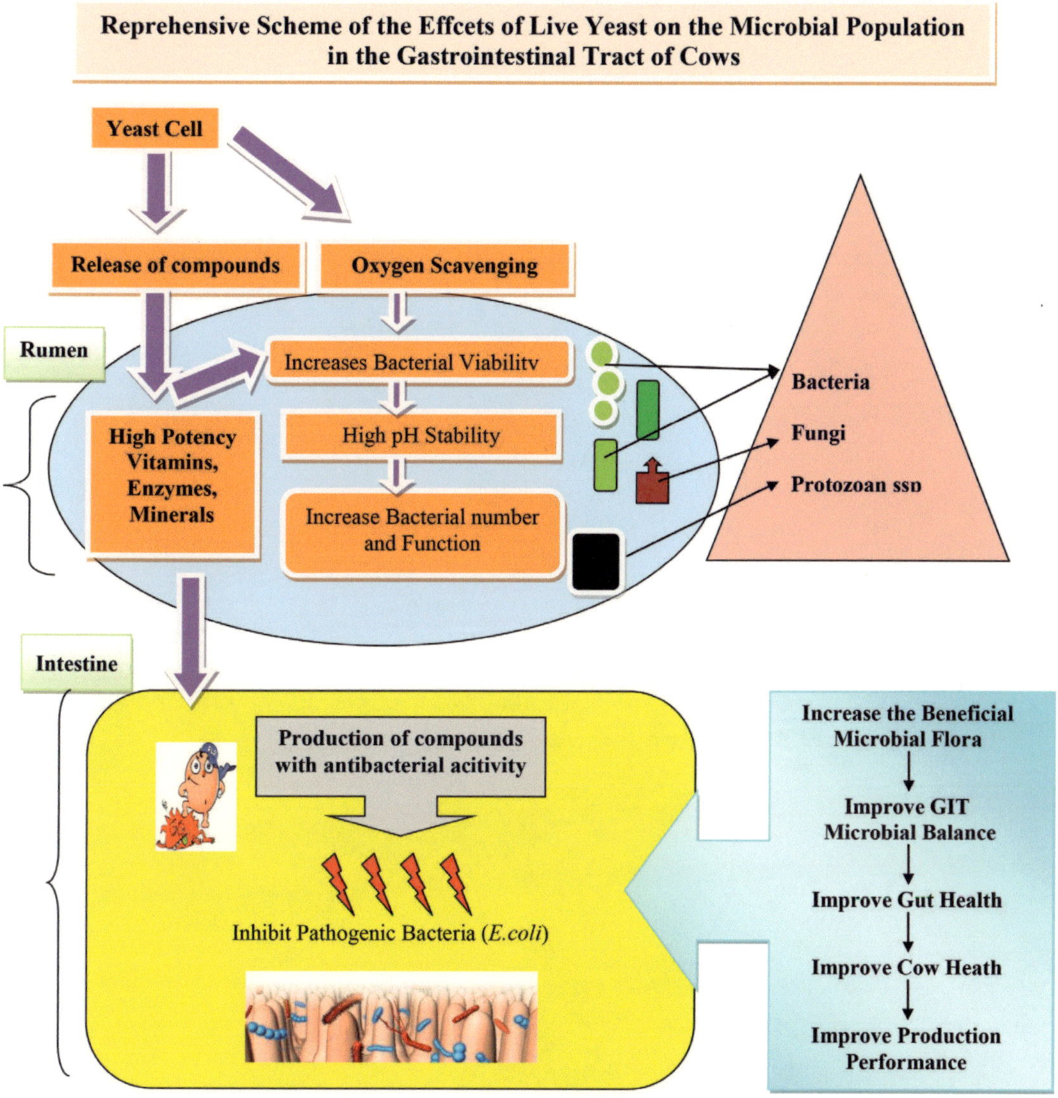

Figure 2. Representative scheme of effect of live yeast on the microbial flora of the gastrointestinal tract in ruminants: the probiotics yeast can improve the composition of the intestinal microbiota through the production of antimicrobial substances which inhibit the pathogenic bacteria ultimately improving the gut health. Although there is a difference between the probiotic colonization microbiota and the target rumen microbiota, many researchers suggested that there is a relationship between the GIT microbiota and other tissues of the host body.

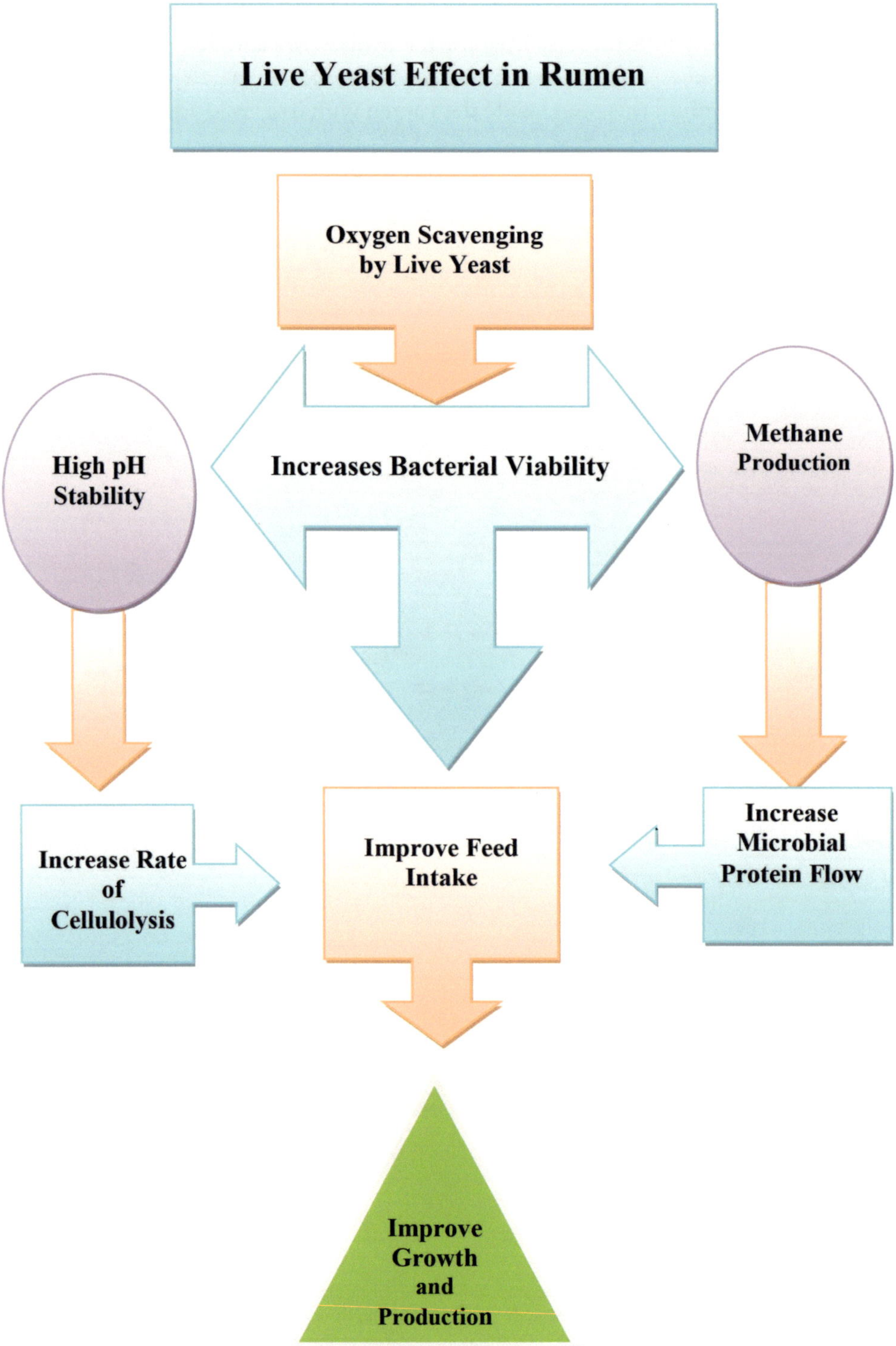

Figure 3. A scheme describing the mode of action of yeast culture.

4.1. Role of probiotics in the establishment of rumen and hindgut micro-flora establishment

The key of the rumen development is to provide supporting conditions to the microbiota to ensure its optimal establishment. It has been well studied that live yeast can help in establishing different types of micro-flora in neonate by positively modulating rumen colonization, by important functional microbial population. A newborn ruminant digestive system is sterile but with passage of time when they contact with their mother and other animals, they get microbes from their saliva and feces [10]. In contrary, the mother and her young connection are more common in small-scale farming systems. On the other hand, in intensive dairy farming systems, the neonate is alienated from the mother and is fed on solid feeding that provides a negative situation in the development of rumen microflora [47]. The early maternal separation has a negative effect on the rumen colonization by important microbial species. This negative situation leads to poor rumen microbial development making the neonate to suffer from different digestive diseases like diarrhea. Different diseases of digestive system are most important factors of low income heifers rearing. It has been well studied that live yeast culture can help the establishment of key microbial communities (*Bacteroides-Prevotella* and *Clostridium coccoides-Eubacterium rectale* group) in neonate by removing the oxygen from rumen. The rate of cellulose degrading microflora population was greater in lambs fed on *S. cerevisiae* addition as compared to the control [10]. Live yeast could be used as a nutrition tool for maturation of the rumen microbial ecosystem, which can result in a positive effect on animal performance, and health both before and after weaning, with an increase in grain intake and reduced frequency of diarrhea. *S. cerevisiae* had the ability to provide different types of organic acids or vitamins, those stimulating ruminal populations of cellulolytic bacteria and lactic acid utilizing bacteria (LUB) [48]. The cellulose degrading microbial population was also much stable in the animals fed on yeast addition because protozoa comes in rumen only once the bacterial species are present inside the rumen. It has been also noted that protozoa appeared earlier in those animals who fed on *S. cerevisiae* addition [49]. Amylolytic bacterial population is also affected by yeast in the rumen [38]. It is because the protozoan concentrations are proliferated and are able to store starch and postpone bacterial fermentation [50]. Proteolytic bacterial activity was highest in the yeast supplemented animals. Proteins in the feed are quickly broken down into peptides, amino acids and ammonia (NH3) by different protozoa and fungi inside rumen [51]. Some NH_3 is converted into microbial protein (MP) and some ammonia is used by the animal in the form of urea. An important portion of rumen ammonia is excreted and represents an indication of nitrogen (N) loss of the dietary nitrogen intake (20–25%) [30]. Amino acids and peptides issued from dietary proteins cannot be directly slipped in the animal intestine, if the diet has highly nutritious value. The same effect on ammonia concentration was observed with daily yeast culture supplementation in adult ruminants [52]. In vitro findings tell that probiotic yeast could alter the growth and activities of protein-degrading bacteria, which ultimately enhanced the protein digestion inside rumen [53]. The mode of action of yeast can be explained by a fight between live *S. cerevisiae* cells and different bacterial species for energy utilization [54]. A study on 14 dairy cows field trials addition of yeast strain in the diet revealed that the soluble nitrogen of the diet was a key factor to drive the production parameters to the probiotics yeast [55]. However, with other yeast strains, no significant effect was observed on the concentration and fraction of microbial nitrogen in dairy cattle [56]. Further study is needed to investigate the effect of probiotic yeast on the nitrogen microbial metabolism [10]. Many studies have shown that increased feed intakes are driven by increased flow of absorption nitrogen [43, 57]. This step stems simultaneously from the proliferation and stimulation of viable cell counts of anaerobic bacteria population. A higher ammonia nitrogen concentration measured for vessel in which live yeast was added compared to autoclaved yeast suggest that, the live yeast stimulated the proteolytic activity of the rumen bacterial species that ultimately influenced rumen fermentation [58]. It was noted that digestibility of crude protein was significantly higher in animals fed on the mixed fungal (yeast and *Aspergillus*) supplementation and it is suggesting that fungal supplementation might promote proteolytic activities by supplying some types of stimulatory factors [59]. Many studies have shown that animals fed on the yeast supplementation have been associated with higher concentration of ammonia nitrogen, which might suggest that proteolytic bacterial activity has been stimulated by yeast culture [60, 61]. The second proposed mechanism is that yeast cell provides the soluble growth factors such as, organic acids, branched-chain volatile fatty acids, vitamins, and amino acids, which have a positive effect in stimulating cellulolytic, proteolytic and lactic acid utilizing bacteria [59, 62].

4.2. Effect of yeast and yeast cultures on rumen fiber digestion

Fiber is non-digestible polysaccharides (a complex form of carbohydrate) [63]. In nutrition, the term fiber defines as a component of plant that is not digestible by mammalian enzyme [64]. Cellulose, hemicellulose, and lignin are the primary components of fiber. Cellulose and hemicelluloses constitute 15–70% of most ruminant diet [65]. Cellulose is the most abundant carbohydrate in plant cell wall. Chemically, cellulose is made up of linear chains of sugar molecules. In cellulose, glucose molecules are linked together in a β-1,4 links, and this linkage can only be digested by microbial cellulolytic enzymes (**Table 1**).

Cellulose makes up about 40% of plant cell walls. Hemicellulose also can only be digested by microbial enzymes because it also has β-1,4 linkages. Hemicellulose has a strong negative effect on fiber degradation because of close association with lignin [66]. The rumen is an important part of the ruminant's stomach because cellulose is broken down into simple sugar that can be used by the animal body inside rumen. The rumen represents a mobile, self-sustaining fermentation system for plant material [67, 68]. It is a complex microbial ecosystem that contain many types of microorganisms such as, bacteria (1010–1011 cells per ml), protozoa (104–106 per ml) and fungi species (103–105 zoospores per ml) [69, 70].

4.2.1. Rumen fibrolytic bacteria

Rumen bacteria (10^{11} viable cells per ml) dominate the fermentation, both in terms of numbers and metabolic processes. The rumen bacteria are 99.5% obligatory anaerobic. In rumen, 200 species with many subspecies of bacteria are present. There are different kinds of bacteria in the rumen, which aid in fermentation process [71]. *Fibrobacter* and *Ruminococcus* are the main rumen fibers degrading bacteria in cattle [72, 73]. *Fibrobacter succinogenes* is a Gram-negative and rod-shaped anaerobe first isolated from the cattle [74]. Despite their important role, cellulose degrading bacteria are argued to only comprise of 0.3% of the total bacteria population inside rumen [75]. Rumen bacteria are classified into fibrolytic, amylolytic, pectinolytic, proteolytic, lipolytica,

Strain	Diet type and dose	Animal type	Effect	References
Saccharomyces cerevisiae, QAU03, (locally isolated yeast from *Sahiwal* cow dung sample)	3 kg concentrate feed, 8 kg silage and 20 kg fodder 3.13 × 10^{07} (CFU/g) yeast	Lactating cows	1. Increase fiber digestibility and improve milk and its fat contents 2. Improve gastrointestinal tract microbial balance	[21]
Saccharomyces cerevisiae Yea-sacc [(1026)] (Alltechinc., Nicholasville, KY)	1 g/kg as fed High concentration or low concentration diets 2.5 × 10^9 (CFU/g)	Dairy Holstein heifers	1. Improved feed efficiency of HC-fed heifers. 2. Yeast culture increased dry matter digestibility in HC- and LC-fed heifers	[21]
Saccharomyces cerevisiae CNCM l-1077, Pasteur institute), (Levucell Sc),	1 × 10^{10} CFU/head/day Total mixed ration (TMR)	Non-lactating dairy Holstein cows	1. Enhanced rumen fermentation 2. Lower CH_4 emissions	[48]
Saccharomyces cerevisiae, Yea-sacc [(1026)] (Alltechinc., Nicholasville, KY)	4.5 × 10^9 CFU High starch low starch diet	Holstein heifers	1. Positive effect on DM, NDF, ADF, and hemicellulose digestibility	[21]
Dry yeast (CNCM-1077, Levucell Sc 20. Sc, Lallemand, animal nutrition)	0.5 g/hd/day basal diet consisting	Holstein dairy cows in late lactation	1. Decreased time spent in subacute rumen acidosis	[49]
Levucell SC 10 ME	Maize silage, 1 × 10^{10} CFU/g yeast	Holstein dairy cows in Non-lactation	1. Lower the risk of rumen acidosis 2. Increased fiber degradation of low quality maize silages	[18]
Saccharomyces cerevisiae, Yea-sacc [(1026)] (Alltechinc., Nicholasville, KY)	Balanced TMR or pasture 2.5 × 10^9 CFU/g yeast	Multiparous dairy cows	1. Improve the metabolic status	[38]

Table 1. The effects of various probiotic yeast strains on ruminant performance.

lactate using bacteria and hydrogen-using bacteria. Amylolytic bacteria ferment starch while fibrolytic bacteria involve in the fermentation of fiber. Different bacterial populations dominate the rumen fermentation depending on the type of feed. Cattle that are fed on high-fiber diet will have a ruminal bacterial population that is high in fibrolytic bacteria especially *Ruminococcus* ssp. Rumen bacteria are mainly involved in the fermentation of fiber, starch, and sugar in the feed.

4.2.2. *Rumen fibrolytic fungi*

Ruminal anaerobic fungi, an emerging group of animal probiotics, account for approximately 8% of the total rumen microbial biomass in ruminants [76]. Rumen fungi have a crucial role in the degradation of fiber material [77–80]. The fungi have an important role in fiber digestion because of the vegetative thallus rhizoids. The rhizoids have a more penetrating capability to plant cell wall as compared to the bacteria and protozoa. Degradation of lignin of the plant cell wall is a main characteristic of rumen fungi [81, 82]. Fungi degraded 37–50% of barley straw. The fungi fibrolytic activity enhanced by hydrogen-utilizing methanogens decreases the cruel effect of hydrogen [76, 83]. Fungi play an active and significant role in fiber digestion of low quality roughages by breaking the beta-1,4 linkages between lignin and hemicelluloses inside the plant cell [84]. Fungi have a positive role in fiber degradation as evidenced by producing a wide array of potential hydrolytic enzymes [79, 85, 86].

4.2.3. Rumen protozoa

In vitro studies have suggested that 19–28% of the total cellulosic activity in fiber digestion can be attributed to protozoa [87]. However, digestion seems to be limited to very susceptible tissue, for instance, mesophyll cells [88]. Studies have demonstrated that defaunation (removal of protozoa) reduces the rate of fiber/cell wall degradation digestion [89, 90]. However, in the absence of protozoa, there is an increased requirement for non-protein nitrogen (NPN) because of an increased bacterial population. A reduction of N may therefore result in the reduction in fiber digestion [91].

4.3. Mode of action of probiotic yeast in the post-ruminal GIT

The GIT inhabits multifarious microbial diversity that helps in generating impactive response regarding nutritious, health, physiology, and productivity of animals [1]. The existing gut microbiota regulates food safety through shedding of pathogens, interaction with organisms, and resource competition in the GIT [5]. The physiological, anatomical, and immunological status of the host is strongly dependent upon microbiota of GIT which facilitates essential functions to host. GIT microflora aids in the stimulation of immune system that acts as a barrier against infectious pathogens. It also restrains the injurious and pathogenic bacteria gut colonization [6]. The microflora that resides in GIT mostly belongs to *Bacteroides, Bifidobacterium, Clostridium, Eubacterium, Fusobacterium,* and *Lactobacillus* families. Among all of the intestinal microbiota, *Enterococcus* and *Escherichia coli* represent the least contribution (upto 1%) whereas, anaerobes show dominancy over microaerophiles and facultative anaerobes by 1000:1 [92]. *Lactobacillus* and *Bifidobacteria* are marked as predominant flora which counts for 90% of the total population in GIT. The fluctuating flora represents their existence in trace, i.e., less than (0.01%) that is usually considered as more diversified and pathogenic ones [93]. The GIT microbiota protects the host from pathogen, which produces digestive diseases like diarrhea. The performance of the calf is directly related to the efficient growth along with the improved health status [94]. Gut microbial flora play an important role in the growth and health of the animal. Probiotics put beneficial effects on the health of gut by improving its microbial balance. They have antidiarrheal capability and enhance the growth performance of animals [94, 95]. The intestinal microbiota of cattle performs its vital role in the fermentation process. They help in methane emission by means of fermentation both from rumen and large intestine [96]. The microbial diversity in the GIT of the dairy cattle has lot of impact on the productivity and well-being of the cattle [1–3, 97]. There is no direct evidence that yeast or fungal extracts affect digestion or metabolism in the lower gut. However, the potential for such effects should not be ignored [98]. This improvement can be due to either the effect of mannan-oligosaccharides (MOS, a component of yeast cell wall) on the immune modulation or direct effect of yeast on the reduction of pathogenic bacteria and toxic metabolites. According to the findings of Heinrichs et al. [99], MOS has an ability to bind selected pathogen by blocking the microbial lecithin and preventing the pathogens from colonization in host GIT. As noted, previous inquiries regarding feeding direct fed microbial products (DFM) to ruminant animals focused on its potential beneficial effects on the post-ruminal GIT (**Figure 4**).

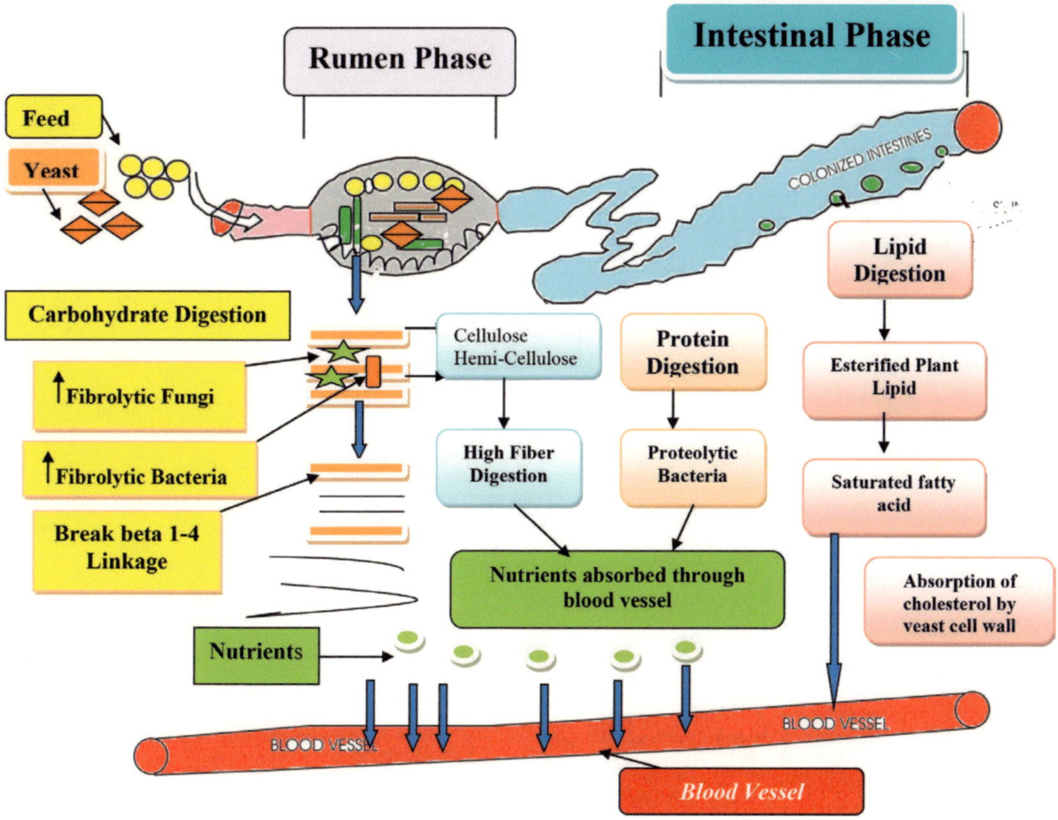

Figure 4. Simple scheme proposed to explain mode of action of probiotic yeast in rumen and post-ruminal GIT: live yeast improves carbohydrates, protein, and lipid digestion rate by improving the production of cellulolytic, hemicellulolytic and proteolytic, and lipolytic bacteria and fungi.

5. Experimental proofs

5.1. Experiment 1: effect of probiotic yeast on the growth performance and fecal biomarkers of dairy heifers

Poor growth performance in growing animals is associated with imbalanced nutrition. The use of probiotic yeast would minimize the expenditure of replacement heifers with optimum growth rate. In our experiment, young animals fed on the diet supplemented with yeast culture gain more weight than non-supplemented animals. In this experimental study, eight dairy heifers (87 ± 5 kg and 6–7 months) were divided into two equal groups of four animals each (control and probiotic) following completely randomized design [100]. During the trial, heifers in control group were offered control diet (NRC recommended diet), while in the probiotic group fed with control diet plus commercial available probiotic yeast (Yea-Sac[1026]; 5 g/animal) for a period of 120 days. Results reveal that dairy heifers fed on the probiotic feed gained significantly ($P < 0.05$) higher average weights than dairy heifers fed on control feed (**Figures 5** and **6**) [34].

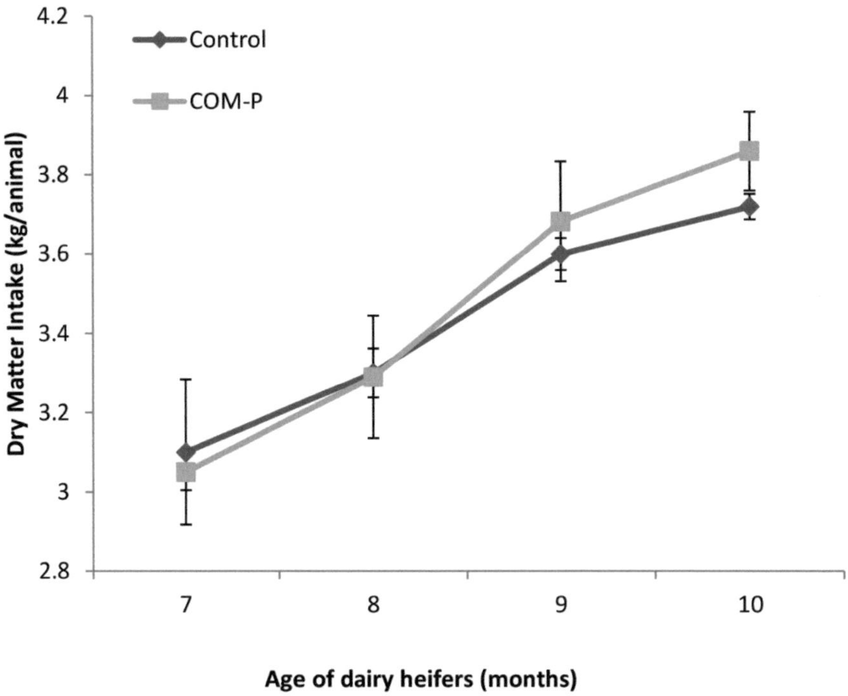

Figure 5. Average monthly dry matter intake pattern (kg) of dairy heifers fed on control feed (control, ◆; no yeast) or commercial probiotic feed (COM-P, ■; control feed plus commercial yeast).

Probiotic yeast decreases the pathogenic bacteria and increase the beneficial bacteria in present study (**Figure 7**) [34].

5.2. Experiment 2: development of indigenous probiotic yeast for local breed

From the aforementioned discussion, we found that an important step to establish a breed probiotic strain is that the origin of the isolated strain should be animal based for their better adhesion and colonization in the animal GIT. We hypothesize that, breed-specific probiotic yeast gives better results in terms of milk production. From this line of research, we conduct an experimental study to develop the indigenous probiotic yeast for local breed and to evaluate its effect on the lactating animals. A *S. cerevisiae* strain was isolated from dung samples of the dairy animals. After careful assessment of its probiotic test, that yeast strain (animal probiotic) was further used in same dairy cattle feed (**Figure 8**) [100].

5.3. Experiment 3: impact of indigenously isolated *Saccharomyces cerevisiae* probiotics on milk production and gut microbial species of lactating cows

Nine lactating dairy cattle of mix breed (*Sahiwal* and *Sahiwal* × *Jersey*) at their first and second lactation (producing 4–5 l/day) were used for the experiment. Cows were fed a concentrate feed, maize silage, and oat fodder. The neutral detergent fiber digestibility was improved in the presence of 3.13 × 10^{07} CFU/g of our laboratory produced live yeast and milk production was improved by 0.7 kg/d in the laboratory produced probiotic fed dairy cows (**Figure 9**).

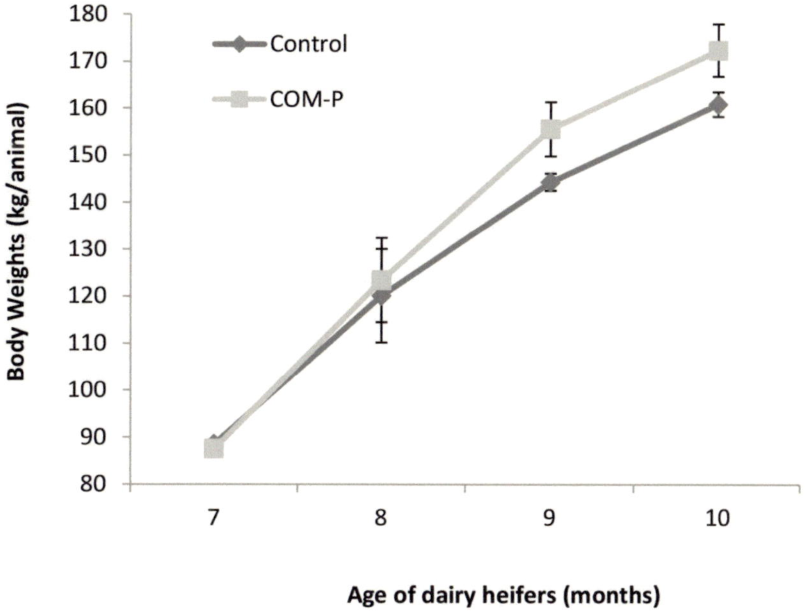

Figure 6. Average monthly growth pattern (kg) of dairy heifers fed on control feed (control, ♦; no yeast) or commercial probiotic feed (COM-P, ■; control feed plus commercial yeast).

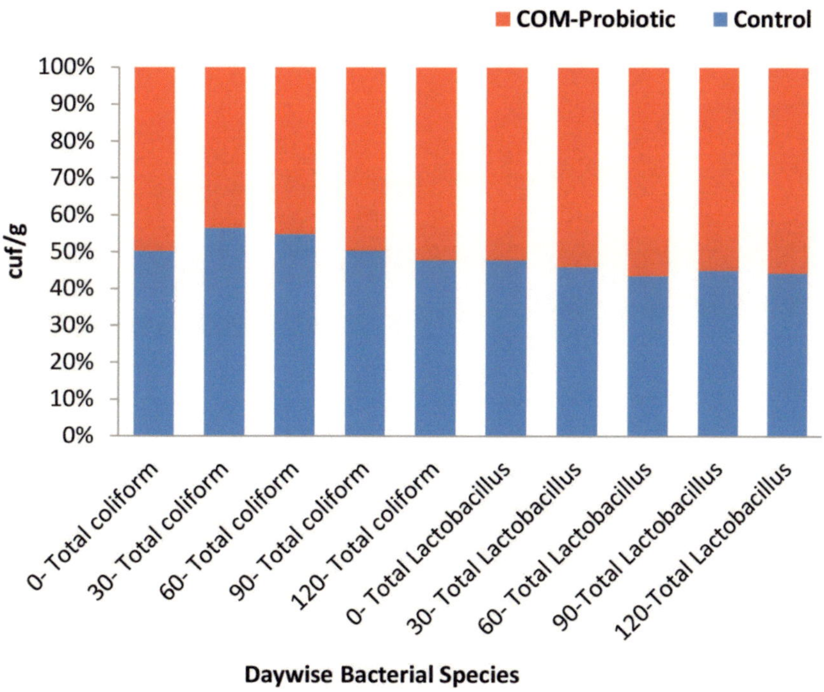

Figure 7. Total *E. coli* and *Lactobacillus* count (CFU/g) in the ruminal gut of dairy heifers fed on control feed (control, ♦; no yeast) or commercial probiotic feed (COM-P, ■; control feed plus commercial yeast).

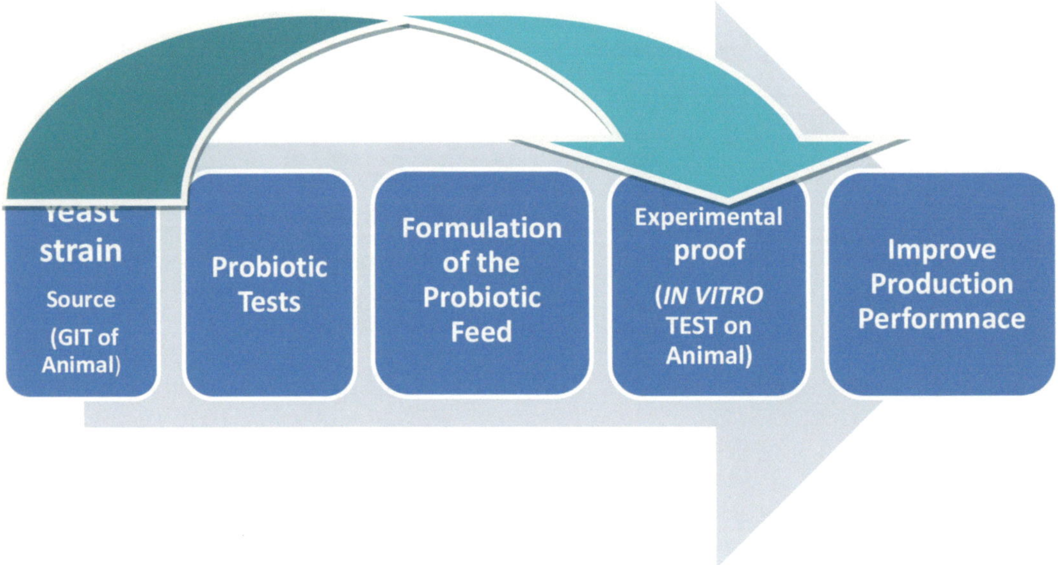

Figure 8. Flow sheet of development of indigenous probiotic yeast for local breed.

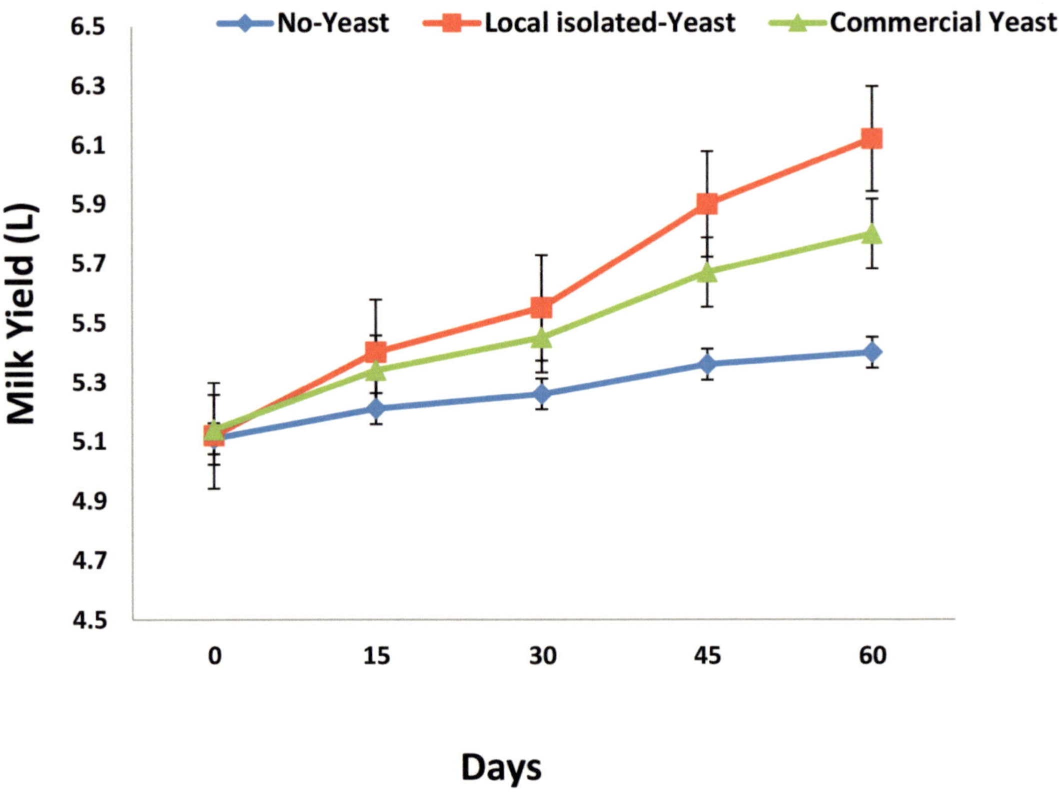

Figure 9. Effect of yeast on milk yield (Means ± SEM) in lactating dairy cattle fed on the diet supplemented with no yeast (control, ♦), laboratory yeast (LAB-Y, ■) or commercial yeast (COM-Y, ▲).

We assumed that improved performance is might be due to cellulolytic activity of the *S. cerevisiae* which was isolated from dairy animal dung sample [101]. This activity enhanced the cellulose digestion rate and helped in the milk synthesis [14].Yeast culture significantly ($P < 0.05$) increased the fiber digestibility, resulting in an increased supply of absorbed nutrients for milk synthesis in our experiment [101]. Results of the ruminal gut microflora showed that the average beneficial *Lactococcus* species (CFU/g) counts were increased while pathogenic *Enterococcus* species (CFU/g) counts were lower in laboratory produced yeast (LAB-Y) fed lactating cows than other groups which lead to improve GIT microbial balance. The economic efficiency of LAB-Y fed group that was 4.7% better than the control group, fed no yeast culture. It can be concluded that indigenous isolated probiotic yeast strain improves the production performance, gut health, and well-being of lactating dairy cattle in cost-effective manner. Locally isolated yeast strain may be adopted well in the cattle gut than exotic probiotics [100].

6. Challenges in preparation of suitable probiotics yeast

Traditionally, as ruminates live in different parts of the world, hence, different yeast strains may exhibit different effects upon the ruminal fermentation depending on their location. Therefore, we should identify new yeast strains for getting best results for the rumen fermentation in their

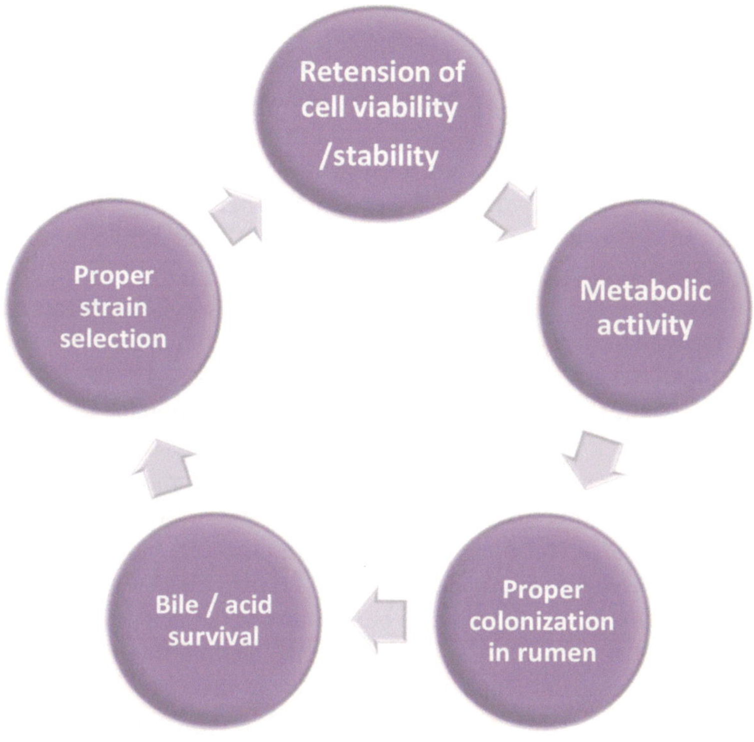

Figure 10. Challenges in preparation of suitable probiotics yeast.

own living condition. For getting positive results in the ruminants, probiotic strains should be breed specific. Latest knowledge related to modes of action of probiotic yeast and its beneficial effects on rumen fermentation, may aid in selection of new breed-specific strains which act on specific key target microorganisms (Cellulolytic, hemicellulolytic bacteria and fungi) and areas of rumen fermentation. Inside the rumen fluid, certain probiotic yeast candidate cannot remain active for longer periods of time. On the other hand, probiotic strains viability and stability are the more advanced technological challenges faced by the livestock industry holders. There is a strong interaction between the host animal and microbial population. To overcome these challenges, further empirical studies are needed on the study of probiotic candidates as well as the ruminal gut microbiota activities to enhance the information of the host-specific interactions. Then the goal is to apply the knowledge of ruminal gut health normal microbial species composition in comparison with microbiota present during disease to select the right breed-specific probiotic strain (**Figure 10**).

7. Conclusions and future research

Probiotic supplemented animal feed has promising effects on the remains to a bright livestock industry future. However, formulating the cost-effective bioactive feed for the dairy animals is remaining as the main challenge faced by the rumen microbiologist. In this regards, search for novel probiotic strains will be the key research and development spot for future livestock markets all over the world. The target oriented applications of specific strains may have huge potential application in treating many chronic disorders in animals. This would lead to have more economical and biological farming. The probiotic strains of same ecological origin may be more compatible with rumen micro-biome fielding maximum outputs. Therefore, the future feed supplementation may be of breed specific. Recently, consumer's demand about safe and healthy food products has been increased worldwide. Hence, the advantage of using probiotics is not only to enhance the productive performance but also to (contribute to) lower the risk of ruminant GIT carriage of human pathogen and to reduce excretion of polluting outputs such as nitrogen-based compounds and methane. The *S. cerevisiae* received the Generally Recognized as Safe (GRAS) status from Food and Drug Administration (FDA) and thus, is appropriate for use in animal feeds. Some factors, such as, expected response, net profit, ongoing research, and field responses should be considered to determine when a feed additive is used for experiment. Fermented yeast culture has emerged as a cost-effective product that has many benefits to ruminants. One of the major benefits of the probiotic yeast is that yeast has no antibiotic resistance gene. It also has ability to colonize in the GIT, to neutralize enterotoxin, and to tolerate bile salt and gastric acid, resultantly improving the health status and production efficiency of the dairy animals. Normally, the feed cost is high, however, probiotic yeast gives a useful nutritional strategy which provides increasing diet digestibility and consequently enhances the performance parameters of the dairy animals in cost-effective manner. Future experimental studies are needed to investigate the impact of the yeast cells in the GIT of the dairy animals. The future research will also need to address the behavior of the yeast cells in the digestive environment. We look forward to the development of the new breed-specific probiotic strains, which will hopefully mean that the rumen microbiologist instead of following the nutritious in an exploratory mood as has been the role for so long, will instead lead advances in ruminant nutrition in the year to come.

8. Recommendations

The recommendations are outlined as follows:

- Isolation of new indigenous bacterial and yeast strains.
- Study the probiotic characterization and genetic potential of the probiotic strains.
- Complete nutritional profile of the probiotic strains for preparation of probiotic feed.
- Application of probiotic strains for more milk and meat production of local breed animals.
- Amino acid profile of the milk of dairy animals fed on the probiotic feed.

Author details

Shakira Ghazanfar[1]*, Nauman Khalid[2], Iftikhar Ahmed[1] and Muhammad Imran[3]

*Address all correspondence to: shakira_akmal@yahoo.com

1 Institute of Microbial Culture Collection of Pakistan, Islamabad, Pakistan

2 Department of Global Agricultural Sciences, The University of Tokyo, Tokyo, Japan

3 Department of Microbiology, Faculty of Biological Sciences, Quaid-e-Azam University, Islamabad, Pakistan

References

[1] Guarner F, Malagelada JR. Gut flora in health and disease. The Lancet. 2003;**361**:512-519

[2] Dowd SE, Callaway TR, Wolcott RD, Sun Y, McKeehan T, Hagevoort RG, Edrington TS. Evaluation of the bacterial diversity in the feces of cattle using 16S rDNA bacterial tag-encoded FLX amplicon pyrosequencing (bTEFAP). BMC Microbiology. 2008;**8**:125

[3] Eckburg PB, Bik EM, Bernstein CN, Purdom E, Dethlefsen L, Sargent M, Gill SR, Nelson KE, Relman DA. Diversity of the human intestinal microbial flora. Science. 2005;**308**:1635-1638

[4] Fuller R. Probiotics in man and animals. The Journal of Applied Bacteriology. 1989;**66**: 365-378

[5] Nurmi E, Rantala M. New aspects of *Salmonella* infection in broiler production. Nature. 1973;**241**:210-211

[6] Cebra JJ. Influences of microbiota on intestinal immune system development. The American journal of clinical nutrition. 1999;**69**(5):1046-1051

[7] Chaucheyras-Durand F, Chevaux E, Martin C, Forano E. Use of yeast probiotics in ruminants: Effects and mechanisms of action on rumen pH, fibre degradation, and microbiota according to the diet. In: Probiotic in Animals. InTech Open Book Publisher; 2012

[8] Grochowska S, Nowak W, Mikula R, Kasprowicz-Potocka M. The effect of *Saccharomyces cerevisiae* on ruminal fermentation in sheep fed high-or low-NDF rations. Journal of Animal and Feed Sciences. 2012;**21**:276-284

[9] Musa H, Wu S, Zhu C, Seri H, Zhu G. The potential benefits of probiotics in animal production and health. Journal of Animal and Veterinary Advances. 2009;**8**:313-321

[10] Chaucheyras-Durand F, Walker ND, Bach A. Effects of active dry yeasts on the rumen microbial ecosystem: Past, present and future. Animal Feed Science and Technology. 2008;**145**(1):5-26

[11] Huber JT. Probiotics in Cattle. London: Chapman and Hall; 1997. p. 162-186

[12] Robinson PH, Erasmus LJ. Effects of analyzable diet components on responses of lactating dairy cows to *Saccharomyces cerevisiae* based yeast products: A systematic review of the literature. Animal Feed Science and Technology. 2009;**149**:185-198

[13] Poppy GD, Rabiee AR, Lean IJ, Sanchez WK, Dorton KL. A meta-analysis of the effects of feeding yeast culture produced by anaerobic fermentation of *Saccharomyces cerevisiae* on milk production of lactating dairy cows. Journal of Dairy Science. 2012;**95**:6027-6041

[14] Dolezal P, Dolezal J, Szwedziak K, Dvoracek J, Zeman L, Tukiendorf M, Havlicek Z. Use of yeast culture in the TMR of dairy Holstein cows. Iranian Journal of Applied Animal Science. 2012;**2**:51-56

[15] Ferraretto LF, Shaver RD, Bertics SJ. Effect of dietary supplementation with live-cell yeast at two dosages on lactation performance, ruminal fermentation, and total-tract nutrient digestibility in dairy cows. Journal of Dairy Science. 2012;**95**:4017-4028

[16] Sabbia JA, Kalscheur KF, Garcia AD, Gehman AM, Tricarico JM. Soybean meal substitution with a yeast-derived microbial protein source in dairy cow diets. Journal of Dairy Science. 2012;**95**:5888-5900

[17] Hristov AN, Varga G, Cassidy T, Long M, Heyler K, Karnati SK, Corl B, Hovde CJ, Yoon I. Effect of *Saccharomyces cerevisiae* fermentation product on ruminal fermentation and nutrient utilization in dairy cows. Journal of Dairy Science. 2010;**93**:682-692

[18] Guedes CM, Gonçalves D, Rodrigues MAM, Dias-da-Silva A. Effects of a *Saccharomyces cerevisiae* yeast on ruminal fermentation and fibre degradation of maize silages in cows. Animal Feed Science and Technology. 2008;**145**:27-40

[19] Lesmeister KE, Heinrichs AJ, Gabler MT. Effects of supplemental yeast (*Saccharomyces cerevisiae*) culture on rumen development, growth characteristics, and blood parameters in neonatal dairy calves. Journal of Dairy Science. 2004;**87**:1832-1839

[20] Lehloenya KV, Krehbiel CR, Mertz KJ, Rehberger TG, Spicer LJ. Effects of propionibacteria and yeast culture fed to steers on nutrient intake and site and extent of digestion. Journal of Dairy Science. 2008;**91**:653-662

[21] Lascano GJ, Heinrichs AJ, Tricarico JM. Substitution of starch by soluble fiber and *Saccharomyces cerevisiae* dose response on nutrient digestion and blood metabolites for precision-fed dairy heifers. Journal of Dairy Science. 2012;**95**:3298-3309

[22] Hutjens MF. Feed additives. Veterinary Clinics of North America: Food Animal Practice. 1991;**7**:525

[23] Faria-Oliveira F, Ferreira C, Puga S. Yeast: World's Finest Chef. Intech Open Access Publisher; 2013

[24] Stone CW. Yeast Products in the Feed Industry: A Practical Guide for Feed Professionals. Cedar Rapids, IA, USA: Diamond V Mills Inc; 1998

[25] Yalcin S, Yalcin S, Can P, Gurdal AO, Bagci C, Eltan O. The nutritive value of live yeast culture *(Saccharomyces cerevisiae*) and its effect on milk yield, milk composition and some blood parameters of dairy cows. Asian Australas Journal of Animal Sciences. 2011;**24**:1377-1385

[26] Sontakke U. Benefits of *Saccharomyces cerevisiae* as a Feed Additive in Ruminants. Vol. 1714. Karnal: National Dairy Research Institute; 2012

[27] Szucs JP, Suli A, Halasz T, Arany A, Bodor Z. Effect of live yeast culture *Saccharomyces cerevisiae* on milk production and some blood parameters. Scientific Papers Animal Science and Biotechnologies. 2013;**46**:40-44

[28] Jouany J-P, Morgavi D. Use of 'natural'products as alternatives to antibiotic feed additives in ruminant production. Animal. 2007;**1**(10):1443-1466

[29] Carberry CA, Kenny DA, Han S, McCabe MS, Waters SM. Effect of phenotypic residual feed intake and dietary forage content on the rumen microbial community of beef cattle. Applied and Environmental Microbiology. 2012;**78**:4949-4958

[30] Fonty G, Chaucheyras-Durand F. Effects and modes of action of live yeasts in the rumen. Biologia. 2006;**61**:741-750

[31] Weinberg ZG. Effect of lactic acid bacteria on animal performance. Indian Journal of Biotechnology. 2003;**2**:378-381

[32] Wcimer P, Waghorn GC. Effect of her on populations of three species of ruminal cellulolytic bacteria in lactating cows. Journal of Daily Science. 1999;**82**:122-134

[33] Dehority BA, Orpin CG. Development of, and natural fluctuations in, rumen microbial populations. In: Hobson PN, Stewart CS, editors. The Rumen Microbial Ecosystem. London, United Kingdom: Blackie Academic and Professional; 1997. p. 196-245

[34] Shakira G, Atiya A, Ahmed I. Effects of dietary supplementation of yeast (*Saccharomyces cerevisiae*) culture on growth performance, blood parameters, nutrient digestibility and fecal flora of dairy heifers. The Journal of Animal and Plant Science. 2015;**25**:53-59

[35] Dehority B, Orpin C. Development of, and natural fluctuations in, rumen microbial populations. In: The Rumen Microbial Ecosystem. Springer; 1997. p. 196-245

[36] Weimer P, Waghorn G, Odt C, Mertens D. Effect of diet on populations of three species of ruminal cellulolytic bacteria in lactating dairy cows. Journal of Dairy Science. 1999;**82**:122-134

[37] Santra A, Karim S. Rumen manipulation to improve animal productivity. Asian Australasian Journal of Animal Sciences. 2003;**16**:748-763

[38] Arakaki L, Stahringer R, Garrett J, Dehority B. The effects of feeding monensin and yeast culture, alone or in combination, on the concentration and generic composition of rumen protozoa in steers fed on low-quality pasture supplemented with increasing levels of concentrate. Animal Feed Science and Technology. 2000;**84**:121-127

[39] El-Ghani AA. Influence of diet supplementation with yeast culture (*Saccharomyces cerevisiae*) on performance of Zaraibi goats. Small Ruminant Research. 2004;**52**:223-229

[40] Hillman K, Lloyd D, Williams A. Use of a portable quadrupole mass spectrometer for the measurement of dissolved gas concentrations in ovine rumen liquor in situ. Current Microbiology. 1985;**12**:335-339

[41] Loesche WJ. Oxygen sensitivity of various anaerobic bacteria. Applied Microbiology. 1969;**18**:723-727

[42] Rose AH. Live Yeast, A Microorganism for all Species: A Theoretical Look at its Mode of Action. Nicholasville, Kentucky: Alltech Technical Publications; 1987. p. 113-118

[43] Newbold C, Wallace R, Chen X, McIntosh F. Different strains of *Saccharomyces cerevisiae* differ in their effects on ruminal bacterial numbers in vitro and in sheep. Journal of Animal Science. 1995;**73**:1811-1818

[44] Wallace R. Ruminal microbiology, biotechnology, and ruminant nutrition: Progress and problems. Journal of Animal Science. 1994;**72**:2992-3003

[45] Dawson K, Newman K, Boling J. Effects of microbial supplements containing yeast and *lactobacilli* on roughage-fed ruminal microbial activities. Journal of Animal Science. 1990;**68**:3392-3398

[46] Jouany JP, Mathieu F, Sénaud J, Bohatier J, Bertin G. Influence of protozoa and fungal additives on ruminal pH and redox potential. South African Journal of Animal Science. 1999;**29**:65-66

[47] Fonty G, Gouet P, Jouany J-P, Senaud J. Establishment of the microflora and anaerobic fungi in the rumen of lambs. Journal of General Microbiology. 1987;**133**:1835-1843

[48] Chaucheyras F, Fonty G, Bertin G, Gouet P. Effects of live *Saccharomyces cerevisiae* cells on zoospore germination, growth, and cellulolytic activity of the rumen anaerobic fungus, *Neocallimastix frontalis* MCH3. Current Microbiology. 1995;**31**:201-205

[49] Chaucheyras-Durand F, Fonty G. Yeasts in ruminant nutrition. Experiences with a live yeast product. Kraftfutter. 2002;**85**:146-150

[50] Enjalbert F, Garrett J, Moncoulon R, Bayourthe C, Chicoteau P. Effects of yeast culture (*Saccharomyces cerevisiae*) on ruminal digestion in non-lactating dairy cows. Animal Feed Science and Technology. 1999;**76**:195-206

[51] Wallace RJ, Onodera R, Cotta MA, editors. The Rumen Microbial Ecosystem. London: Chapman and Hall; 1997. p. 283-328

[52] Kumar U, Sareen VK, Singh S. Effect of *Saccharomyces cerevisiae* yeast culture supplement on ruminal metabolism in buffalo calves given a high concentrate diet. Animal Science. 1994;**59**:209-215

[53] Beev G, Todorova P, Tchobanova S. Yeast cultures in ruminant nutrition. Bulgarian Journal of Agricultural Science. 2007;**13**:357-374

[54] Chaucheyras-Durand F, Masseglia S, Fonty G. Effect of the microbial feed additive *Saccharomyces cerevisiae* CNCM I-1077 on protein and peptide degrading activities of rumen bacteria grown in vitro. Current Microbiology. 2005;**50**:96-101

[55] Sniffen C, Chaucheyras-Durand F, De Ondarza M, Donaldson G. Predicting the impact of a live yeast strain on rumen kinetics and ration formulation. 2004:53-59.

[56] Putnam D, Schwab C, Socha M, Whitehouse N, Kierstead N. Effect of yeast culture in the diets of early lactation dairy cows on Ruminal fermentation and passage of nitrogen fractions and amino acids to the small intestine 1, 2. Journal of Dairy Science. 1997;**80**:374-384

[57] Kamel H, Sekine J, El-Waziry A, Yacout M. Effect of *Saccharomyces cerevisiae* on the synchronization of organic matter and nitrogen degradation kinetics and microbial nitrogen synthesis in sheep fed Berseem hay (*Trifolium alexandrinum*). Small Ruminant Research. 2004;**52**:211-216

[58] Oeztuerk H. Effects of live and autoclaved yeast cultures on ruminal fermentation in vitro. Journal of Animal and Feed Sciences. 2009;**18**:142-150

[59] Wiedmeier R, Arambel M, Walters J. Effect of yeast culture and *Aspergillus oryzae* fermentation extract on ruminal characteristics and nutrient digestibility. Journal of Dairy Science. 1987;**70**:2063-2068

[60] Kung L, Kreck E, Tung R, Hession A, Sheperd A. Effects of a live yeast culture and enzymes on in vitro ruminal fermentation and milk production of dairy cows. Journal of Dairy Science. 1997;**80**:2045-2051

[61] Moallem U, Lehrer H, Livshitz L, Zachut M, Yakoby S. The effects of live yeast supplementation to dairy cows during the hot season on production, feed efficiency, and digestibility. Journal of Dairy Science. 2009;**92**:343-351

[62] Newbold C, Wallace R, McIntosh F. Mode of action of the yeast *Saccharomyces cerevisiae* as a feed additive for ruminants. British Journal of Nutrition. 1996;**76**:249-261

[63] Callaway E, Martin S. Effects of a *Saccharomyces cerevisiae* culture on ruminal bacteria that utilize lactate and digest cellulose. Journal of Dairy Science. 1997;**80**:2035-2044

[64] Tungland B, Meyer D. Nondigestible oligo and polysaccharides (dietary fiber): Their physiology and role in human health and food. Comprehensive Reviews in Food Science and Food Safety. 2002;**1**:90-109

[65] Moore KJ, Hatfield RD. Carbohydrates and Forage Quality. Madison: ASA-CSSA-SSSA; 1999. p. 229-280

[66] Hobson PN, Stewart CS. The Rumen Microbial Ecosystem. Springer Science and Business Media; 1997

[67] Fernando SC, Purvis HT, Najar FZ, Sukharnikov LO, Krehbiel CR, Nagaraja TG, Roe BA, DeSilva U. Rumen microbial population dynamics during adaptation to a high-grain diet. Applied Environmental Microbiology. 2010;**76**:7482-7490

[68] Flint HJ. The rumen microbial ecosystem-some recent developments. Trends in Microbiology. 1997;**5**:483-488

[69] Shakira G, Mirza I, Latif A. Scope of common DNA based methods for the study of rumen bacterial population. Bangladesh Journal of Animal Science. 2013;**41**:141-146

[70] Kamra D. Rumen microbial ecosystem. Current Science. 2005;**89**:124-135

[71] Cho SJ, Cho KM, Shin EC, Lim WJ, Hong SY, Byoung RC, Jung MK, Sun ML, Yong HK, Kim H, Yun HD. 16S rDNA analysis of bacterial diversity in three fractions of cow rumen. Journal of Microbiology and Biotechnology. 2006;**16**:92-101

[72] Kobayashi Y, Shinkai T, Koike S. Ecological and physiological characterization shows that *Fibrobacter succinogenes* is important in rumen fiber digestion – review. Folia Microbiologia (Praha). 2008;**53**:195-200

[73] Kim M, Morrison M, Yu Z. Evaluation of different partial 16S rRNA gene sequence regions for phylogenetic analysis of microbiomes. Journal of Microbiological Methods. 2011;**84**:81-87

[74] Hungate RE. The anaerobic mesophilic cellulolytic bacteria. Bacteriological Reviews. 1950;**14**:1-49

[75] Brulc JM, Yeoman CJ, Wilson MK, Miller ME, Jeraldo P, Jindou S, Goldenfeld N, Flint HJ, Lamed R, Borovok I, Vodovnik M. Cellulosomics, a gene-centric approach to investigating the intraspecific diversity and adaptation of *Ruminococcus flavefaciens* within the rumen. PLoS One. 2011;**6**(10):25329

[76] Orpin CG, Joblin KN. The rumen anaerobic fungi. In: The Rumen Microbial Ecosystem. London, UK: Elsevier Applied Science Publishers; 1988. p. 129-150

[77] Theodorou MK, Longland AC, Dhanoa MS, Lowe SE, Trinci AP. Growth of *Neocallimastix* sp. strain R1 on Italian ryegrass hay: Removal of neutral sugars from plant cell walls. Applied and Environmental Microbiology. 1989;**55**:1363-1367

[78] Paul SS, Kamra DN, Sastry VRB, Sahu NP, Agarwal N. Effect of administration of an anaerobic gut fungus isolated from wild blue bull (*Boselaphus tragocamelus*) to buffaloes (*Bubalus bubalis*) on in vivo ruminal fermentation and digestion of nutrients. Animal Feed Science and Technology. 2004;**115**:143-157

[79] Lee SS, Choi CK, Ahn BH, Moon YH, Kim CH, Ha JK. In vitro stimulation of rumen microbial fermentation by a rumen anaerobic fungal culture. Animal Feed Science and Technology. 2004;**115**:215-226

[80] Thareja A, Puniya AK, Goel G, Nagpal R, Sehgal JP, Singh PK, Singh K. In vitro degradation of wheat straw by anaerobic fungi from small ruminants. Archiver Animal Nutrition. 2006:412-417

[81] Mountfort DO, Asher RA, Bauchop T. Fermentation of cellulose to methane and carbon dioxide by a rumen anaerobic fungus in a triculture with *Methanobrevibacter* sp. strain RA1 and *Methanosarcina barkeri*. Applied and Environmental Microbiology. 1982; **44**:128-134

[82] Akin DE, Benner R. Degradation of polysaccharides and lignin by ruminal bacteria and fungi. Applied and Environmental Microbiology. 1988;**54**:1117-1125

[83] Joblin K. Physical disruption of plant fibre by rumen fungi of the Sphaeromonas group. In: The Roles of Protozoa and Fungi in Ruminant Digestion. Armidale, NSW: Penambul; 1989. p. 259-260

[84] Tripathi VK, Sehgal JP, Puniya AK, Singh K. Hydrolytic activities of anaerobic fungi from wild blue bull (*Boselaphus tragocamelus*). Anaerobe. 2007;**13**:36-39

[85] Williams AG, Orpin CG. Polysaccharide-degrading enzymes formed by three species of anaerobic rumen fungi grown on a range of carbohydrate substrates. Canadian Journal of Microbiology. 1987;**33**:418-426

[86] Paul SS, Kamra DN, Sastry VR, Sahu NP, Kumar A. Effect of phenolic monomers on biomass and hydrolytic enzyme activities of an anaerobic fungus isolated from wild nil gai (*Baselophus tragocamelus*). Letters in Applied Microbiology. 2003;**36**:377-381

[87] Gijzen HJ, Lubberding HJ, Gerhardus MJT, Vogels GD. Contribution of rumen protozoa to fibre degradation and cellulase activity in vitro. FEMS Microbiology Letters. 1988;**53**:35-43

[88] Akin DE. Histological and physical factors affecting digestibility of forages. Agronomy Journal. 1989;**81**:17-25

[89] Bonhomme A. Rumen ciliates: Their metabolism and relationships with bacteria and their hosts. Animal Feed Science and Technology. 1990;**30**:203-266

[90] Yang CM, Varga GA. The effects of continuous ruminal dosing with dioctyl sodium sulphosuccinate on ruminal and metabolic characteristics of lactating Holstein cows. British Journal of Nutrition. 1993;**69**:397-408

[91] Ushida K, Jouany JP. Effect of defaunation on fibre digestion in sheep given two isonitrogenous diets. Animal Feed Science and Technology. 1990;**29**:153-158

[92] Mestecky J, Russell M. Passive and active protection against disorders of the gut. Veterinary Quarterly. 1998;**20**:83-87

[93] Tournut J. The digestive flora of the pig and its variations. Recueil De Medecine Veterinaire. 1993;**169**:645-652

[94] Soberon F, Raffrenato E, Everett R, Van Amburgh M. Preweaning milk replacer intake and effects on long-term productivity of dairy calves. Journal of Dairy Science. 2012;**95**:783-793

[95] Donovan D, Franklin S, Chase C, Hippen A. Growth and health of holstein calves fed milk replacers supplemented with antibiotics or enteroguard 1, 2. Journal of Dairy Science. 2002;**85**:947-950

[96] Johnson KA, Johnson DE. Methane emissions from cattle. Journal of Animal Science. 1995;**73**:2483-2492

[97] Engelbrektson A, Kunin V, Wrighton KC, Zvenigorodsky N, Chen F, Ochman H, Hugenholtz P. Experimental factors affecting PCR-based estimates of microbial species richness and evenness. The ISME Journal. 2010;**4**:642-647

[98] Elghandour MM, Salem AZ, Castañeda JS, Camacho LM, Kholif AE, Chagoyán JC. Direct-fed microbes: A tool for improving the utilization of low quality roughages in ruminants. Journal of Integrative Agriculture. 2015;**14**:526-533

[99] Heinrichs A, Jones C, Heinrichs B. Effects of mannan oligosaccharide or antibiotics in neonatal diets on health and growth of dairy calves. Journal of Dairy Science. 2003;**86**:4064-4069

[100] Ghazanfar S. Study on the effects of dietry supplmentation of *Saccharomyvces cerevisiae* on performance of dairy cattle and heifers. [PhD thesis]. Islamabad. Pakistan: Quaid-e-Azam University; 2016

[101] Ghazanfar S, Qubtia M, Hassan F, Afzal M, Ahmed I, et al. Effect of indigenously isolated *Saccharomyces cerevisiae* probiotics on milk production, nutrient digestibility, blood chemistry and fecal microbiota in lactating dairy cows. The Journal of Animal and Plant Sciences. 2017 (Accepted)

Section 3

Energy Production

Chapter 9

Phenotypic Characterization of Yeasts Aiming at Bioethanol Production

Natália Manuela Strohmayer Lourencetti,
Flávia Danieli Úbere, Maria Priscila Franco Lacerda,
Maria José Soares Mendes-Giannini,
Cleslei Fernando Zanelli,
Ana Marisa Fusco-Almeida and
Edwil Aparecida de Lucca Gattás

Additional information is available at the end of the chapter

http://dx.doi.org/10.5772/intechopen.70042

Abstract

Worldwide, the production of bioethanol is derived through first-generation technology, where plants, vegetables, and cereals, that have high levels of sucrose, are fermented by yeast. Brazil, for the production of bioethanol from sugarcane, is among the world's leading producers. The process for bioethanol production is a complex that involves a variety of environmental factors, resulting in different phenotypic profiles of strain used. It has been evidenced that the interaction between environmental factors and microorganism can influence in the identification of different characteristics of *Saccharomyces cerevisiae*. Also, the bioethanol is developed by the second and third generations, and new yeast strains may also contribute to the feasibility of production. Successful performance of fermentation depends on the ability of the yeast to deal with a number of factors that occur during the fermentation, such as concentration of sugar, ethanol, nitrogen, pH, resistance to contaminants, stress protein, temperature change, and osmotic pressure.

Keywords: *Saccharomyces cerevisiae*, bioethanol, phenotypic, characterization, resistance

1. Introduction

In recent decades, recurrent crises in world oil have resulted in serious economic crises, leading to the search for alternative fuels [1]. In 1930, Brazil presented the first National Congress on Industrial Applications of Alcohol that was aimed at establishing the infrastructure for the production and use of bioethanol. This led Brazil to start production in the early twentieth century, while other countries started the production of fuel from grains, using its potential for bioenergy production. The remainder of residues with high protein content is a source of nutrition in agriculture, as well as being a rich source of sustainability [2].

The three major world powers producing bioethanol are Brazil (sugarcane), the USA (corn), and China (wheat and corn), where Brazil is the largest producer through a direct source of sugar, as production by grains requires an additional step with the liquefaction and hydrolysis of the starch. Estimates indicate that around 85% of all bioethanol worldwide is the responsibility of Brazilian and North American production, as well as inferior productions in all parts of the continents, as shown in **Figure 1** [3].

In Brazil, the bioethanol had low volume of consumption compared to the use of conventional fuels, maintaining the Brazilian dependence on imported oil [4]. This made the national government launch the National Alcohol Program (ProÀlcool) in the 1970s, which established a new behavior for air quality and the development of technologies in the area of alternative sources of energy [5]. ProÀlcool represented the largest increase in bioethanol production, from 500 million liters at the beginning of the program to about 13 billion liters per year [6]. Since then, Brazil has been characterized as a potential producer of bioethanol, with a well-developed domestic consumption policy [6, 7].

Currently, Brazil has an estimated bioethanol production with the 2016–2017 crops of 33.2 billion liters [8] and has kept that figure since 1986. All the production comes from sugarcane, representing a large-scale technology characterized by the development of new cane varieties,

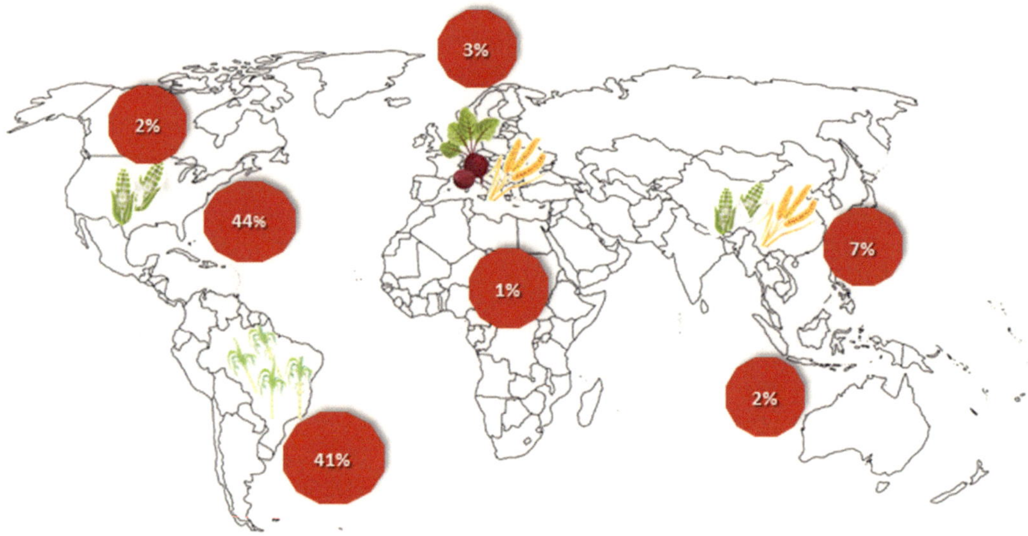

Figure 1. Global distribution of production and percentage of production per continent: Americas: South America—Brazil 43% (sugarcane); North America: USA 44% (corn) and Canada 2%; Europe 3% (vegetables and wheat); Africa 1% and Oceania 1%.

favorable climate, fertile soil, and advanced agricultural technologies [4, 6, 9]. The production of Brazilian bioethanol is derived from first-generation technologies, where a natural source of sugar from the sugarcane extraction, sucrose, is fermented by yeast with the primary product ethanol [10].

The sugarcane plant used for the Brazilian bioethanol production is derived the crossing from 637 species of the genus *Saccharum*, family *Poaceae*, *Andropogoneae* tribe, and native of hot temperate climate and with morphology characterized by stem and straw [11, 12]. The stem is the material from which the sugarcane juice is derived and is later used for the production of sugar and bioethanol. The bagasse is composed of all post-grind materials and the trash, characterized by the dry, green leaves of the plants, which serve as products of fermentation in second-generation processes for the formation of bioethanol [13].

The fermentation has been known since antiquity, being characterized as a biochemical and biological complex process, which has the objective of transforming sugar into ethanol (anhydrous and hydrated), carbonic gas, succinic acid, and volatile acids and esters [14].

The Brazilian fermentation process is differentiated and unique due to the fact that it is fed-batch in most states, being these short fermentation cycles and cell treatments with sulfuric acid [10]. This process uses cane juice as raw material, with a final product of 9–12% (v/v) and an efficiency of 90–92% [15]. The ratio of bioethanol produced to the amount of raw material used varies according to the amount of sugar present in the must, which consists of a mixture of molasses (sugar manufacturing residue), water, and sugarcane juice. The process starts with an action of invertase exoenzyme, in the process of breaking the sugar (sucrose, a disaccharide) into glucose and fructose (structural monosaccharaides), which are absorbed by facultative aerobic microorganisms, which under anaerobic conditions form the pyruvic acid cycle, the enzymes pyruvate with the help of decarboxylase and alcohol dehydrogenase, producing the bioethanol and its subproducts at the end of the fermentation [9].

The main key of the national fermentation process is that, at each end of the fermentation cycle, the yeasts are subjected to a centrifugation and sulfuric acid wash in order to minimize the risk of contamination [10]. At the end of this treatment, the cells are returned to the fermenters as a new inoculum for the subsequent cycle, this stage being repeated twice daily throughout the crop for 6–9 months, during the year, as shown in **Figure 2** [16].

The fermentation with grains (the USA and China) is rich in carbohydrates so it is essential to the stage of liquefaction and hydrolysis of this raw material, where the molecules of starches are broken down into fermentable sugars, and thus fermentation can occur, as shown in **Figure 3** [17]. One of the main characteristics of the grain fermentations, besides the additional stage of liquefaction and hydrolysis of the starch, is that the mills do not use recycled yeast cells, like the Brazilian mills, which is due to the fact that the whole concentration of residues and fermentable products is retained for distillation, decreasing the fermentation process when compared to the cell recycle process [18].

Significant changes are also observed when comparing the Brazilian and North American fermentation processes. In the fermentation of sugarcane, we have a lower concentration of solid residues, a concentration of larger yeast cells, and a much shorter time for bioethanol

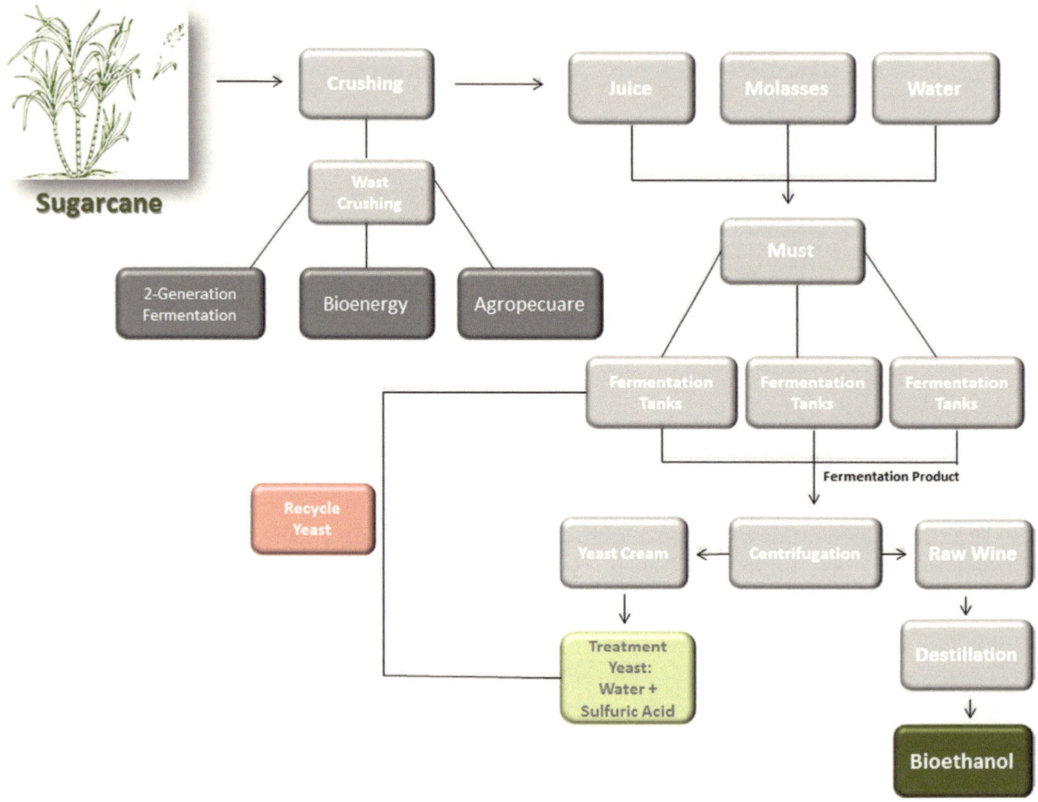

Figure 2. Simplified scheme of a fed-batch fermentation process with the recycling of yeast cells in Brazilian distillers by fermenting of sugarcane.

production, 6–12 h. The US process, that is derived for approximately 45–60 h, presents the advantages of a final concentration of bioethanol of 12–18%, against the 7–12% of the Brazilian process, and the raw material, that comes from corn plantations, lasts approximately one year, as opposed to the sugarcane harvest and its losses with rains that last around 200–240 day per year [19].

The main microorganisms used for the fermentation process are yeasts, such as *Saccharomyces* sp., *Schizosaccharomyces* sp., *Kluyveromyces* sp., among others [13]. Currently, the most used yeast in the sugar and alcohol sector, for fermentation processes in the production of bioethanol is the specie *Saccharomyces cerevisiae* [20]. The methodologies used for the identification of yeasts based on morphology, biochemical characteristics, and sexual reproduction require the evaluation of 70–90 tests to obtain the identification of species. Macroscopic and microscopy features may be the first method of identification of *S. cerevisiae* yeasts, as presented in **Figure 4** [21, 22].

The molecular techniques have been developed as alternatives to traditional techniques for the identification and characterization of yeasts, with the advantage of building an independent expression of the genes that allows quick and accurate identification of yeast species [23].

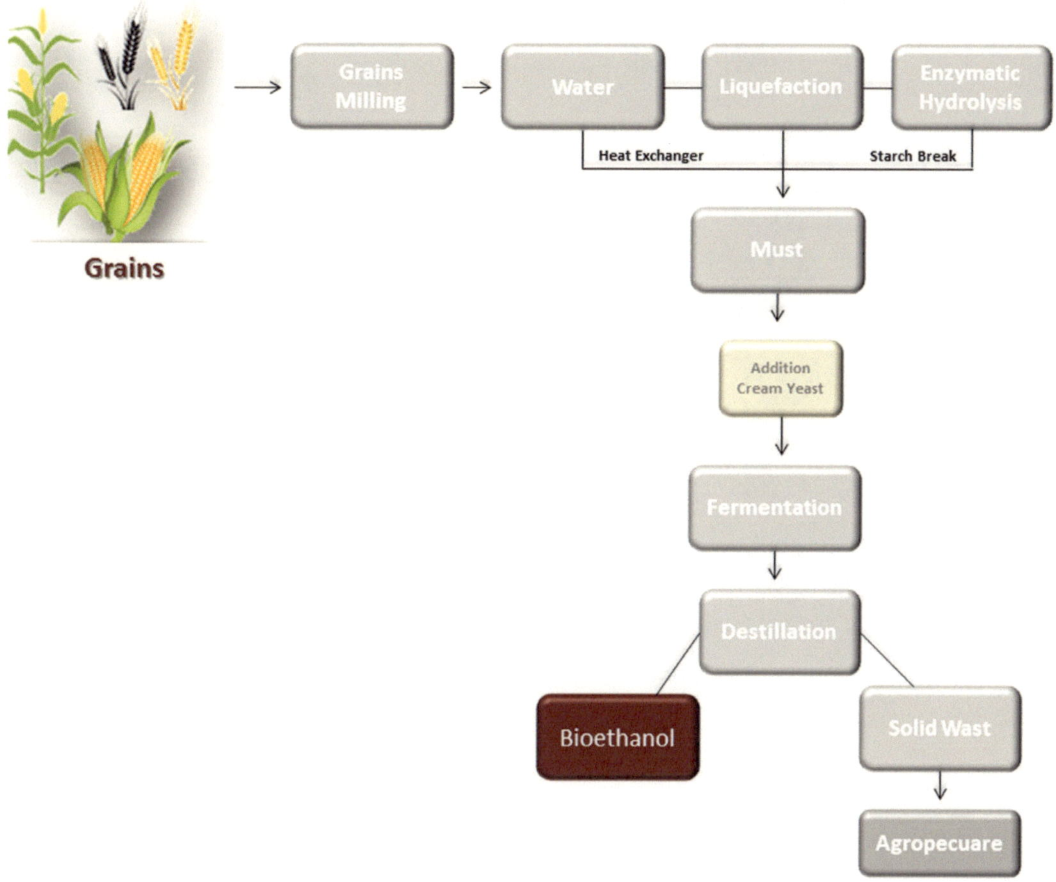

Figure 3. Simplified scheme of grains fermentation process with the liquefaction and hydrolysis in North American's and Chinese's distillers by fermenting of corns and wheat.

Due to the high mutation capacity of wild yeasts, molecular techniques for characterization and analysis of polymorphisms are being developed [24]. Genetic analyses of DNA, electrophoretic karyotyping, rRNA sequencing, rDNA restriction analysis, and polymerase chain reaction (PCR) and restriction fragment length polymorphism (RFLP) have been used as different tools to distinguish *Saccharomyces* sp. strains from strict sense group [25].

Studies of Melo Pereira [26] developed two new pairs of specific primers of the species, homologous to the HO gene of the species *Saccharomyces bayanus, S. cerevisiae*, and *Saccharomyces pastorianus*, offering a rapid method of PCR amplification, resulting in the correct identification of these species in less than 3 h. Guillamón [27] and Oliveira [28], by ribosomal DNA RFLP of ITS1, ITS2, and 5.8S identified different yeast species isolated from wine fermentation, and could also analyze the diversity of yeast species during spontaneous fermentation.

S. cerevisiae is characterized by being yeast with growth in media containing simple sugars and disaccharides, high genetic transformations, and qualities of high resistance to adverse

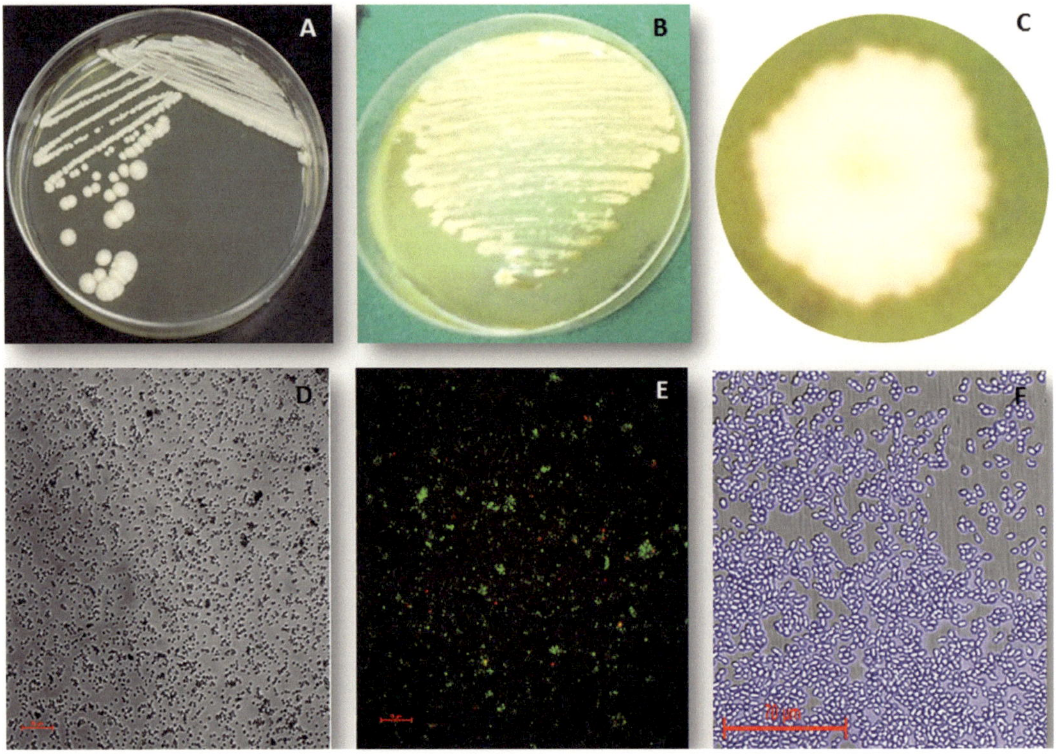

Figure 4. Macroscopic and microscopy of yeasts Saccharomyces cerevisiae. Macroscopies (A), (B), and (C) of PE-2 isolates, grown in YEPD (yeast extract-peptone-dextrose growth medium) solid with creamy and yellowish-white culture characteristics. Microscopies: (D) clear field microscopy of PE-2 isolate; (E) PE-2 isolate with FITC (fluorescein isothiocyanate fluoroforo) and Propidium Iodide cell tags; (F) PE-2 isolate with Calcofluor cellular target; Microscopy presence of oval yeasts with budding presence, with size of approximately 4–8 μm. Microscopies were performed in IN Cell Analyzer, objective of 20× and diameter 70 μm.

conditions of the growth medium, offering a primordial role in the processes of fermentation [29]. Some strains of *S. cerevisiae* have the capacity to be highly productive, dominating the entire fermentation process during the harvest period, allowing efficient and stable fermentations, which result in lower costs and higher fermentation performance (high production capacity of ethanol), and high viability throughout the process [30].

Studies indicate that *S. cerevisiae* is adaptable to different environments, revealing to be a rich source of phenotypic profiles in the *Saccharomyces* sp. species evolution [31, 32]. It has recently been shown that the interaction between environmental factors and organism may influence the identification of different specific characteristics of *S. cerevisiae* [33, 34]. *S. cerevisiae* is widely used and cultivated in industrial fermentation, due to the high capacity of the yeast adaptations to the variable conditions of the environment, such as sugar and ethanol concentrations, pH, oxygen concentrations, resistance to contaminants, salt stress, protein stress, temperature changes, and osmotic pressure [35].

According to Gao et al. [36], using thermophilic strains is interesting in the processes that involve simultaneous saccharification and fermentation (SSF), as this process may reach from

45 to 50°C, resulting in a greater bioethanol production. This occurs because the yeast suffers less damage with the temperature increase and there is a lower chance of microbial contaminations. It is therefore desirable that the thermostable yeast fermentation occurs at the optimum temperature of the enzyme, maximizing the ethanol production process. Thus, the increase in thermotolerance in yeast results in cost production, increases yield in the ethanol production with simultaneous saccharification and fermentation (SSF) system and reduces the possibility of contamination.

However, it is important to know the fermentative yeasts for the control and monitoring of alcoholic fermentation, especially in search of selected characteristics of dominance and resistance to bioethanol yeast production. The objective of this chapter is to distinguish the main phenotypic characteristics of *S. cerevisiae* yeasts in the alcoholic fermentation, for a possible selection of new strains with differentiated phenotypic characteristics, resistant and ideal for the production of bioethanol.

2. Phenotypic parameters of *S. cerevisiae* in alcoholic fermentation

2.1. Flocculation test

Cell flocculation of yeast strains such as *S. cerevisiae* is called cell aggregation and sedimentation in liquid media [37]. Cells have the characteristics of agglomeration at the end of each fermentation process, which makes it an interesting and divergent phenomenon in the industry, as shown in **Figure 5** [38]. Studies point out some divergences in flocculation, which can be a phenomenon of cooperative protection mechanism found in cells during adverse factors in the fermentation cycle [39]. It also facilitates the separation of yeasts at the end of the fermentation by sedimentation, thus helping the collection, centrifugation, and cellular treatment, contributing to the new inoculum stage, for a next cycle in the fermenter [40].

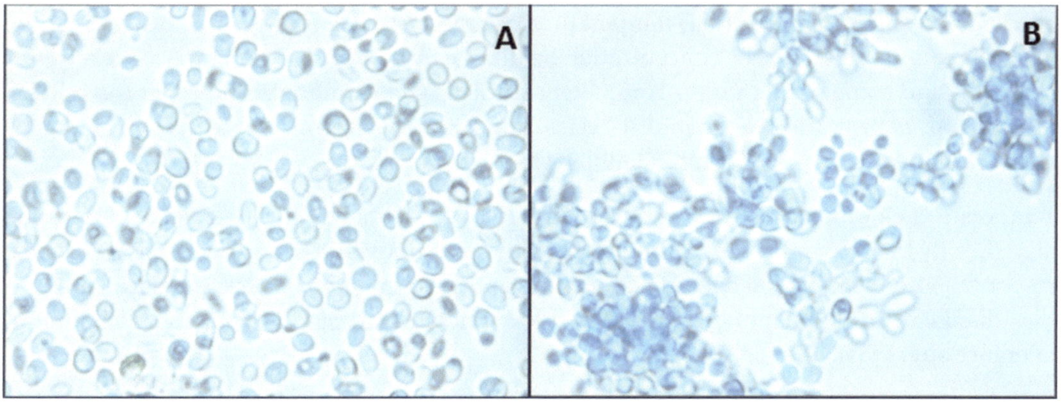

Figure 5. Characterization of aggregative power of *Saccharomyces cerevisiae* yeast cells, grown in YEPD liquid medium and stained with lactophenol. (A) Cells in normal condition without aggregative power. (B) Cells in condition of cellular aggregation.

The principle of cell adhesion is initiated through the recognition of mannose chains, located on cell surfaces, by lectin-like proteins, and agitation is necessary for the beginning of flocculation [41]. Two hypotheses for flocculation are well established: (I) sensitively to proteinases; (II) inhibited by saccharides, suggesting the existence of a protein that recognizes these sugars [42].

These two hypotheses classify two flocculation groups in yeasts that are distinguished by the inhibition of sugar: the first group, called New Flo phenotype, characterized by the inhibition of mannose, glucose, maltose, and sucrose with the exception of galactose; and the second group, the Flo1, is inhibited by mannose, but not by glucose, maltose, sucrose, and galactose, and its action is normally bound to a gene [43].

It is believed that these two distinct phenotypes are caused by two different proteins of the lectin type. Furthermore, the physical-chemical interaction in the cells surface may be involved in the aggregation process, where there is a correlation between flocculation and electrophoretic mobility of yeast cells on certain stress conditions. Other studies reported the correlation of hydrophobicity in the process of flocculation [44].

Another hypothesis for flocculation is the action of a dominant gene family (FLO1, FLO5, FLO8, FLO9, FLO10, and FLO11), where they encode a yeast cell wall protein that acts directly on the cell aggregation [45]. The proteins encoded by these FLO genes share a cellular/modular organization in three domains: an amino-terminal responsible for carbohydrate binding, a central domain, and a carboxyl-terminal domain containing a glycosylphosphatidylinositol anchor sequence [46]. However, the central domain contains tandem repeat regions of DNA sequence that can drive recombination reactions within and between FLO genes, resulting in new generations of FLO alleles, thus conferring yeast cells a wide diversity in the flocculation phenomenon [47].

2.2. Sensitivity test temperature and ethanol

The environmental adversities occurring in a fermentation cycle, such as the decrease of nutrients by sugar consumption, temperature changes, pH changes, risk of contamination, phenolic compounds, and the concentration of ethanol by its own production occurs in different forms and some of them were completely studied [48]. An understanding of the cellular mechanisms of protection to the multiphysical and chemical stresses that the yeast undergoes during fermentation cycle is fundamental for the selection of ideal yeast [49].

Temperature elevations result in reduced fermentation efficiency in *S. cerevisiae*, due to the high fluidity in the membranes, caused by the altered composition of fatty acids in the adverse response [50]. As one of the stress factors known in the fermentative cycle in yeast, temperature change restricts ethanol production and induces the accumulation of proteins bound to tolerance stress [51].

In the first-generation fermentation cycle, yeasts require a temperature of 30°C, whereas, in the production of second-generation bioethanol, where cellulose enzymes start the process by saccharification, yeasts require a higher temperature of 45–50°C [52]. The efficacy of the fermentation is decreased at high temperatures, because it causes damage to the yeast cell, such as the rupture of the protein structure or the loss of function, thus preventing cell proliferation, decreasing viability during the process, and leading to cell death [53]. This temperature control in the fermentative cycles is a problem for the plants in tropical countries, where the ambient temperature is already naturally high and cooling systems are necessary for the total control of this temperature [54].

In the bioethanol production, the process temperature must be stabilized at around 30°C (the cell growth temperature), which is reaching 40°C [55, 56]. Thus, thermotolerant yeast strains may be a promising approach to a profitable fermentation process, as is the case of simultaneous saccharification and fermentation that requires high temperatures to increase ethanol yield [36].

Osmotolerance can be an important factor in the production of ethanol for its adaptation strategy employed in all cell types by accumulating compatible solutes (sulfite), resulting in a decrease in the potential of intracellular water [57]. As sulphite and sulfite-generating compounds have long been used as antimicrobial agents in alcoholic fermentation, tolerance to sulfite in yeast is another desired characteristic for the production of bioethanol from sugarcane juice [58].

The high levels of ethanol in the fermentation medium are considered as negative parameters in the process conditions, because at the same time that the production is essential, the accumulation of ethanol by this production generates an acidification of the medium, leading to irreversible damages in the yeast membrane, thereby decreasing cell viability [59].

The true physiological and ecological relevance of ethanol tolerance in *S. cerevisiae* is its ability to generate mechanisms that protect the cell from chemical and physical damage at high levels of ethanol [60]; this is usually observed in a typical fermentation environment, where there is a large amount of sugars, leading later to ethanol production [30]. This stage generally occurs by stationary phase cells and its tolerance to the ethanol produced is only controlled by the integrity of the yeast membrane in contact with the ethanol accumulation, which is composed of chitin, glucans, glycoproteins, fatty acids, and ergosterol [61].

However, *S. cerevisiae* is resistant to ethanolic stress for its capacity of modifying the conformation of its membrane in the increase of fatty acids and ergosterol when coming in contact with the adverse environment, thus neutralizing the damages caused, mainly in relation to its viability [62, 63]. The accumulation of ethanol can also affect the structural compliance of the cellular proteins causing the inefficiency of its actions, such as the decrease of the activity of glycolytic enzymes: pyruvate kinase and hexokinase, besides altering the absorption of glucose, maltose, and amino acid. In some cases, there may occur cellular extravasation of essential cellular components [64].

In industrial fermentations, a high capacity of production is observed by the accumulation of ethanol in the medium, indicating a positive assimilation of residual sugar, which is measured by the visualization of cellular proliferation in the presence of the gradual levels of ethanol produced during the fermentative process [65]. Tolerance and ethanol characteristics of the main industrial strains of *S. cerevisiae* studied are described in **Table 1** [30, 66–69].

Strains	Group	Origin	Feedstocks	Temperature tolerance (°C)	Ethanol tolerance (%)
ZTW1	Industrial (fuel ethanol)	China	Grains	55	18
YJ5329	Industrial (fuel ethanol)	China	Grains	55	18
PE-2	Industrial (fuel ethanol)	Brazil	Sugarcane	40	15
CAT-1	Industrial (fuel ethanol)	Brazil	Sugarcane	40	15
AT-3	Industrial (fuel ethanol)	USA	Grains	40	14
ErOh red	Industrial (fuel ethanol)	USA	Grains	40	15

Table 1. Characteristics of temperature tolerance and ethanol for major industrial strains worldwide used for the production of bioethanol.

Although this assay is routinely used in industries as large-scale screenings, its actual importance in ethanol resistance in yeast is not elucidated, due to divergent of actions that this process can cause, for example, the negative side acting in the decrease of the cell viability, and positive the increase in resistance to contaminating microorganisms in the fermentation process [70].

The metabolic pathways correlated to the expression of genes responsive to high levels of heat stress and ethanol stress include heat shock proteins (HSPs) and also metabolic enzymes such as trehalose, which is directly involved in tolerance in *S. cerevisiae* [71]. HSPs play a role in folding and refolding, transport, and degradation of intracellular proteins, triggered by stress in fermentation process and located in the cytoplasm, nucleus, and mitochondria, acting immediately in response to an accumulation of denatured proteins, activating the transcription factors of thermal shock (HSF), and leading to a positive regulation of thermotolerance gene expression [72].

The interactions of multiple genes at loci for cellular functions under heat and ethanol stresses are essential [73]. HSPs are known as chaperones ensuring the functional and structural conformation of the yeast, on the action of genes such as SSA1, SSA2, SSA3, and SSA4 which are expressed together with the HSP genes HSP12, HSP26, HSP30, HSP31, and HSP150 which were also found active at high stress levels [74] and interactions between chaperones of different types are widely encountered [75].

However, the inference of several chaperones shows an effective activity in neutralizing the stress, with the activation of the functional chaperones specific to more complex structures in the yeast cell walls [76], which have as a main function to repair of these denatured proteins to maintain cell viability [77].

In addition to serving as chaperones, HSPs have numerous other functions, for example, Hsp30p is characterized as a hydrophobic plasma membrane protein that acts on the regulation of H^+-ATPase, Hsp31p, and Hsp32p functions as hydrolases and peptidase, and Hsp150 is characterized as a protein in supporting the cell wall stability and remodeling [78]. HSPs and chaperone-mediated genomic regulation are also linked to glucose metabolism, which are indispensable tools for stress tolerance in yeast metabolism, especially with storage of carbohydrates, such as trehalose [79].

Trehalose is a compound that acts to prevent the influx of excess salts resulting in irreversible dehydration of cells; therefore, yeasts are capable of accumulating trehalose up to 15% in a stress environment [80]. The trehalose acts by reducing the permeability of the membrane thereby rendering it hydrophobic, due to some regulatory genes such as TPS1, TPS2 and, TSL1, as well as acting in the remodeling of proteins under stress conditions [81]. Cells incapable of accumulating trehalose presented depreciated growth, leading to a significant decrease in cell viability during fermentation stresses [82].

2.3. Assimilation of sugars

Sugarcane juice is one of the main means used in the production of bioethanol, which is derived from the break of fermentable sugars such as sucrose, glucose, and fructose in contact with fermenting microorganisms such as yeast *S. cerevisiae* [83]. Yeast consumes the sugars in the medium in a complex and highly regulated manner, the principle of fermentation, where the sucrose is consumed first, followed by glucose and fructose, and finally maltose, this assimilation of sugars can occur simultaneously between the breaks of sugars, which is the standard process for sequential uptake of the glucose repression pathways or the catabolite repression pathway [84].

Glucose and sucrose may trigger beneficial effects on cells, including stimulation of cell proliferation, mobilization of storage compounds such as glycogen and trehalose, as well as decreased resistance to cell stress [85]. In contrast, negative impacts due to lack of glucose in the process can lead to several problems such as decreased or blocked fermentations, instability of cellular viability and low ethanol production [86], where the break of sugars, sucrose into simple sugars (glucose) occurs by an intracellular enzyme known as invertase, located in wall the yeast industries [87].

Microorganisms that possess the ability to assimilate the highest amount of sugars are indicated for the production of bioethanol, examples are shown in **Table 2** characterizing the main strains of *S. cerevisiae* worldwide used in industries for the production of bioethanol [30, 66–69].

Strains	Group	Origin	Feedstocks	Assimilation sugar (%)	Production bioethanol (%)
ZTW1	Industrial (fuel ethanol)	China	Grains	65	28
YJ5329	Industrial (fuel ethanol)	China	Grains	60	33
PE-2	Industrial (fuel ethanol)	Brazil	Sugarcane	51	22
CAT-1	Industrial (fuel ethanol)	Brazil	Sugarcane	52	26
AT-3	Industrial (fuel ethanol)	USA	Grains	42	18
ErOh red	Industrial (fuel ethanol)	USA	Grains	75	30

Table 2. Characteristics of the assimilation of residual sugars and ethanol production for large industrial strains of *Saccharomyces cerevisiae* yeasts used in the production of bioethanol.

The assimilation of sugars in the fermentation process is not exclusively the fermentation of sugarcane. Currently, new technologies are available to produce ethanol from vegetables such as potatoes, cassava, beets, cereals such as corn, and there are also studies showing the production of green bioethanol in algae fermentations [88].

This type of fermentation is due to the breakdown of starch, carried out by the action of the enzyme glucoamylase, acting directly on the conversion of starch to glucose, by breaking the successive bonds of the nonreducing end of the glucose finally producing straight chains [89]. The process of producing ethanol from starch involves two main steps: enzymatic hydrolysis as the main step and habitual fermentation as the second step [90].

2.4. Second-generation bioethanol

All adverse parameters studied for the first-generation fermentation process have been highly researched to reach an ideal model of production of second-generation bioethanol, which is characterized by being profitable and environmentally sustainable [91].

Second-generation bioethanol production starts from the lysis of the raw material (sugarcane bagasse, vinasse, and residues from the milling of grains). The main step is characterized by a pretreatment where the breakdown of the cellulose-hemicellulose-lignin complex allows the production of fermentable sugar levels for a subsequent fermentation, demonstrated in **Figure 6** [92].

The hydrothermal and lime pretreatments are the most used, known for making the method more effective in preparing the biomass bioconversion step [93], a strong advantage for the sugarcane bagasse. They can be carried out under conditions of low temperature and pressure, resulting in lower sugar degradation, whereas in the saccharification the pretreatment is observed with high temperatures and difficult breaks of carbohydrate chains, resulting in a lower amount of sugars [94].

The fermentation of lignocellulose hydrolysates for bioethanol production presents two main problems: first, the fermentation of xylose that requires a low and controlled oxygenation; second, the removal of microbial inhibitors, which can contaminate the process [95]. Furthermore, these yeasts present a certain tolerance limited to ethanol [96].

Genetic manipulation in the metabolism of xylose in yeast fermentation has advanced and pioneering studies on glucose transporters that mediate xylose uptake, allyl-xylitol-reductase genes, xylitol dehydrogenase and xylulokinase have been expressed, which allows a better assimilation and fermentation of xylose [97].

The main concern in this step is that the balanced supply of NADP (enzyme nicotinamide adenine dinucleotide phosphate) and NADPH (enzyme nicotinamide adenine dinucleotide phosphate oxidase) has to be constant to avoid the production of xylitol. The path is the reduction of NADPH production by blocking the oxidative pentose phosphate cycle in xylose assimilation [98]. Cellulose hydrolysates present different inhibitors from lignin derivatives and sugar degradation, resulting in high amounts of acetic acid, intrinsically necessary for the deconstruction of biomass [99].

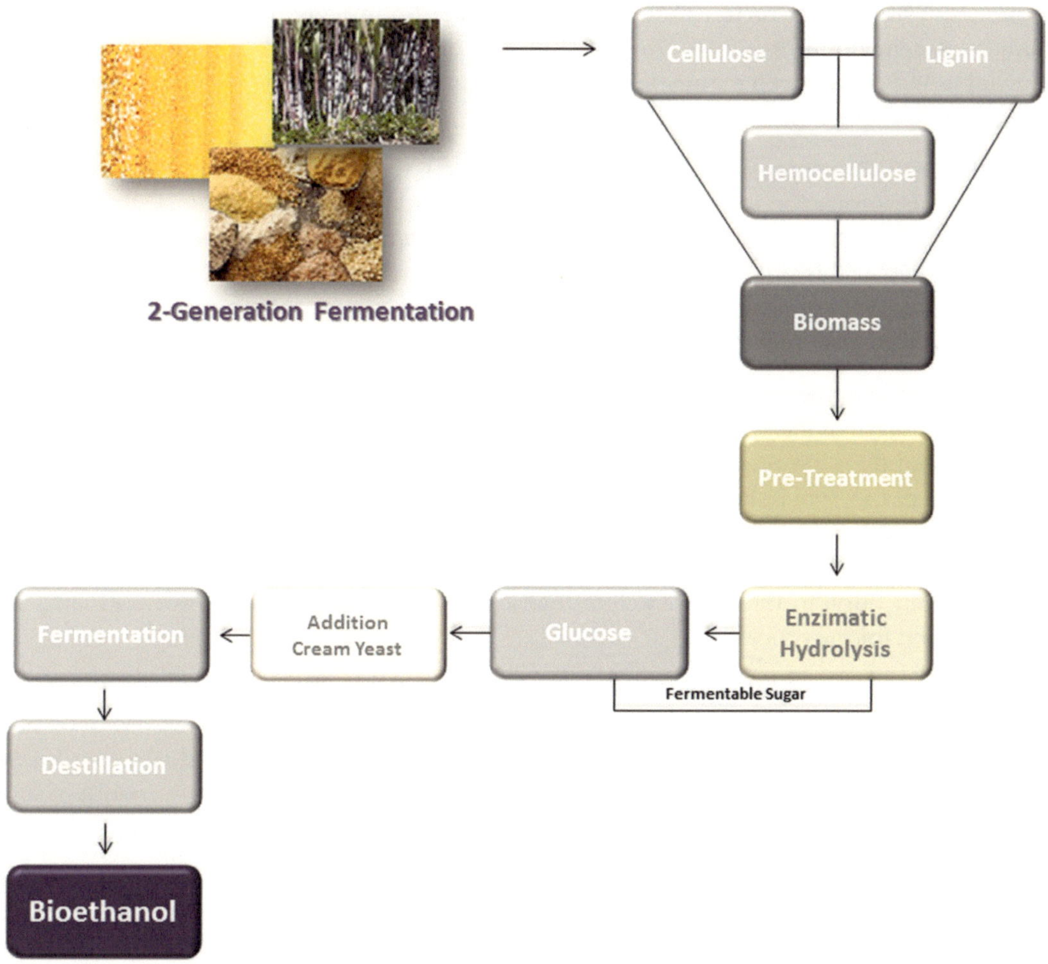

Figure 6. Simplified scheme of second-generation fermentation process with steps of biomass and hydrolysis for bioethanol production worldwide.

All adversities of typical fermentation first generation associated with pH, temperature, elevation of ethanol concentrations, and temperature among other stress factors present in large-scale fermentations are seen together in the adversity challenges of second-generation fermentation [100].

2.5. Advances and perspectives

First-generation fermentation over the years has mainly been used for large-scale industrial models. Although it is a well-established process, it is not definitively elucidated. Changes can be seen with each new process initiated presented for fundamental parameters and the behavior of the yeasts used.

S. cerevisiae has the characteristic of being adaptable to any environment, which leads to numerous behavioral responses during fermentation. As for each new cycle, changes are inevitable and checking all parameters of fermentation are of extreme importance for the success of the fermentation.

Flocculation is a divergent parameter, although it can have many advantages as a phenomenon of cellular protection to several stressors and contaminations in a process, it also presents disadvantages such as low yield in fermentation of fermenting tanks by their decanting. The question whether this phenomenon is beneficial or detrimental on flocculation is still uncertain; however, it is well-known and elucidated in its morphology or molecular action in yeast cells, and it contributes to the improvement of bioethanol production in the world industry.

For an alcoholic fermentation to be efficient, it is necessary and indispensable to know what happens throughout the process, the main steps and degrees that microorganisms go through for hours and days in order to remain viable and productive. For this reason, the study and knowledge of the two main parameters stress of fermentation (heat and ethanol) is of paramount importance for any beginning of the process, whether in small scale, as in laboratories, or large scale, as in industrial productions. The behavioral responses of the fermentation are measured through these parameters that are observed at all times, always aiming the improvement for the process.

The main step for a virtuous bioethanol production is the ability of the microorganism to breakdown the sugars and thus assimilate them to ferment. This detailed step has to be well studied so that there is no damage throughout the process, especially at industrial scales, so that both, a sufficient amount of microorganism concomitantly and adequate amount of sugars are essential to the start of the production.

Looking at the current scenario, the first-generation processes were modernized and studies and improvements resulted in second-generation fermentation, which aims to take advantage of all remaining residues and reaches to more sustainable processes. These processes are taking strides and improvements are being seen at all times to reach the ideal process.

In view of this profile, the search for yeasts with more robust characteristics in industrial lines is essential, and different strategies involving adaptation and functionality are highlighted by genetic engineering research. Advances in the area of a process and ideal yeast are positive, but the journey is still far from reaching perfection. The secrets and mysteries of fermentation are innumerable, but research is constantly revolutionizing and little by little these are being unraveled and the beginning of everything is the understanding of all the steps and all its parameters.

Acknowledgements

This study was financially supported by PADC-FCFAR UNESP and scientific agencies CAPES and CNPq.

Author details

Natália Manuela Strohmayer Lourencetti[1], Flávia Danieli Úbere[2], Maria Priscila Franco Lacerda[1], Maria José Soares Mendes-Giannini[1], Cleslei Fernando Zanelli[3], Ana Marisa Fusco-Almeida[1] and Edwil Aparecida de Lucca Gattás[2]*

*Address all correspondence to: edwilg@yahoo.com.br

1 Department Clinical Analysis, School of Pharmaceutical Sciences, São Paulo State University (UNESP), Araraquara, Brazil

2 Department Food and Nutrition, School of Pharmaceutical Sciences, São Paulo State University (UNESP), Araraquara, Brazil

3 Department Biological Sciences, School of Pharmaceutical Sciences, São Paulo State University (UNESP), Araraquara, Brazil

References

[1] Andrietta MGS, Andrietta SR, Steckelberg C, Stupiello ENA. Bioethanol – Brazil, 30 years of Proálcool. Inernational Sugar Journal. 2007;**109**:195-200

[2] Scholey DV, Burton EJ, Williams PE. The bio refinery; producing feed and fuel from grain. Food Chemistry. 2016;**197**:937-942. DOI: 10.1016/j.foodchem.2015.11.063

[3] Beltrade E, Vandenbergue LPS, Soccol CR, Sigoillot JC, FauldsC. First generation bioethanol. In: Soccol CR, Brar SK, Faulds C, Ramos LP, editors. Green Fuels Technology. Switzerland: Springer International Publishing; 2016. pp. 175-212

[4] Amorim HV, Lopes ML, Oliveira JVC, Buckeridge MS, Goldman GH. Scientific challenges of bioethanol production in Brazil. Applied Microbiology and Biotechnology. 2011;**91**:1267-1275. DOI: 10.1007/s00253-011-3437-6

[5] Macedo IC, Seabra JEA, Silva JE. Green house gases emissions in the productions and use of ethanol from sugarcane in Brazil: The 2005/2006 averages and a prediction for 2020. Biomass and Bioenergy. 2008;**32**:582-595. DOI: 10.1016/j.biombioe.2007.12.006

[6] Zanin GM, Santana CC, Bon EP, Giordano RC, Moraes FF, Andrietta SR, Carvalho CC, Macedo IC, Fo DL, Ramos LP, Fontana JD. Brazilian bioethanol program. Applied Biochemistry and Biotechnology. 2000;**86**:1147-1161. DOI: 10.1385/ABAB:84-86:1-9:1147

[7] Goldemberg J. The Brazilian biofuels industry. Biotechnology for Biofuels. 2008;**1**:6-15. DOI: 10.1186/1754-6834-1-6

[8] CONAB. Levantamento da safra 2016/17. 2017. Available from: www.conab.gov.br. [Accessed: 2017/01/20]

[9] Amorim HV, Lopes ML. Ethanol production in a petroleum dependent world: The Brazilian experience. Sugar Journal. 2005;**67**:11-14. DOI: 20053116848

[10] Brow NA, Castro PA, Figueiredo BCP, Savoldi M, Buckeridge MS, Lopes ML, Paullilo SCL, Borges EP, Amorim HV, Goldman MHS, Bonatto D, Malavazi I, Goldman GH. Transcriptional profiling of Brazilian *Saccharomyces cerevisiae* strains selected for semi-continuous fermentation of sugarcane must. FEMS Yeast Research. 2013;**13**:277-290. DOI: 10.1111/1567-1364.12031

[11] Paes LAD, Oliveira MA. Potential trash biomass of the sugar cane plant. 2005; p. 19-23. In: Hassuani SJ, Leal MRLV, Macedo IC. Biomass power generation: sugarcane, bagasse and trash. Piracicaba, SP, Brazil. 2005

[12] Canilha L, Chandel AK, Milessi TSS, Antunes FAF, Freitas WLC, Felipe MGA, Silva SS. Bioconversion of sugarcane biomass into ethanol: An overview about composition, pretreatment methods, detoxification of hydrolysates, enzymatic saccharification and ethanol fermentation. Journal of Biomedicine and Biotechnology. 2012;**2**:1-15. DOI: 10.1155/2012/989572

[13] Ensinas AV, Modesto M, Nebra SA, Serra L. Reduction of irreversibility generation in sugar and ethanol production from sugarcane. Energy. 2009;**34**:680-688. DOI: org/10. 1016/j.energy.2008.06.001

[14] Amorim HV. Ethanol Production in Brazil: a Successful History. Proceedings of the Sugar Processing Research Conference. 2006;**1**:44-47.

[15] Basílio ACM, Araújo PRL, Morais JOF, Silva-Filho EA, Morais MA, Simões DA. Detection and identification of wild yeast contaminants of the industrial fuel ethanol fermentation process. Current Microbiology. 2008;**56**:322-326. DOI: 10.1007/s00284-007-9085-5

[16] Barbrzadeh F, Jalili R, Wang C, Shokralla S, Pierce S, Robinson-Mosher A, Nyren P, Shafer RW, Basso LC, Amorim HV, Oliveira AJ, Davis RW, Ronaghi M, Gharizadeh B, Stambuk BU. Whole-genome sequencing of the efficient industrial fuel-ethanol fermentative *Saccharomyces cerevisiae* strain CAT-1. Molecular Genetics and Genomics. 2012;**287**:485-494. DOI: 10.1007/s00438-012-0695-7

[17] Christofoletti CA, Escher JP, Correia JE, Marinho, JF, Fontanetti CS. Sugarcane vinasse: Environmental implications of its use. Waste Management. 2013;**33**:2752-2761. DOI: 10.1016/j.wasman.2013.09.005

[18] Ingledew WD, Austin GD, Kelsall DR, Kluhspies C. The alcohol industry: How has it changed and mature? In: Ingledew WM, Austin GD, Kelsall DR, Kluhspies C, editors. The Alcohol Textbook. A Reference for the Beverage, Fuel and Industrial Alcohol Industries. Nottingham: Nottingham University Press; 2009. pp. 1-6

[19] Martinez-Amezcua C, Parsons CM, Singh V, Srinivasan R, Murthy GS. Nutritional characteristics of corn distillers dried grains with solubles as affected by the amounts of grains versus solubles and different processing techniques. Poultry Science. 2007;**86**:2624-2630. DOI: 10.3382/ps.2007-00137

[20] Godoy A, Amorim HV, Lopes ML, Oliveira AJ. Continuous and batch fermentation process: Advantages and disadvantages of these process in the Brazilian ethanol production. International Sugar Journal. 2008;**110**:175-181

[21] Deák T. Methods for the rapid detection and identification of yeasts in foods. Trends in Food Science & Technology. 1995;**6**:287-292. DOI: 10.1016/S0924-2244(00)89138-4

[22] Karabıçak N, Uludağ Altun H, Karatuna O, Hazırolan G, Aksu N, Adiloğlu A, Akyar I. Evaluation of common commercial systems for the identification of yeast isolates in microbiology laboratories: A multicenter study. Mikrobiyol Bul. 2015;**49**:210-220. DOI: 10.5578/mb.9370

[23] Bernardi T, Pereira G, Cardoso P, Dias E, Schwan R. *Saccharomyces cerevisiae* strains associated with the production of cachaca: Identification and characterization by traditional and molecular methods (PCR, PFGE and mtDNA-RFLP). Journal of Microbiology and Biotechnology. 2008;**24**:2705-2712. DOI: 10.1007/s11274-008-9799-y

[24] Song HT, Liu SH, Gao Y, Yang YM, Xiao WJ, Xia WC, Liu ZL, Li R, Ma XD, Jiang ZB. Simultaneous saccharification and fermentation of corncobs with genetically modified *Saccharomyces cerevisiae* and characterization of their microstructure during hydrolysis. Bioengineered. 2016;**7**:198-204. DOI: 10.1080/21655979.2016.1178424

[25] Muller H, Annaluru N, Schwerzmann JW, Richardson SM, Dymond JS, Cooper EM, Bader JS, Boeke JD, Chandrasegaran S. Assembling large DNA segments in yeast. Methods in Molecular Biology. 2012;**852**:133-150. DOI: 10.1007/978-1-61779-564-0_11

[26] de Melo Pereira GV, Ramos CL, Galvão C, Souza Dias E, Schwan RF. Use of specific PCR primers to identify three important industrial species of *Saccharomyces* genus: *Saccharomyces cerevisiae, Saccharomyces bayanus* and *Saccharomyces pastorianus*. Letters in Applied Microbiology. 2012;**51**:131-137. DOI: 10.1111/j.1472-765X.2010.02868.x

[27] Guillamón JM, Sabaté J, Barrio E, Cano J, Querol A. Rapid identification of wine yeast species based on RFLP analysis of the ribosomal internal transcribed spacer (ITS) region. Archives of Microbiology. 1998;**169**:387-392. DOI: 10.1007/s002030050587

[28] Oliveira VA, Vicente MA, Fietto LG, Castro IM, Coutrim MX, Schüller D, Alves H, Casal M, Santos JO, Araújo LD, da Silva PH, Brandão RL. Biochemical and molecular characterization of *Saccharomyces cerevisiae* strains obtained from sugar-cane juice fermentations and their impact in cachaca production. Applied and Environmental Microbiology. 2008;**74**:693-701. DOI: 10.1128/AEM.01729-07

[29] Hector RE, Mertens JA, Bowman MJ, Nichols NN, Cotta MA, Hughes SR. *Saccharomyces cerevisiae* engineered for xylose requires gluconeogenesis and the oxidate brach of the pentose phosphate pathway for aerobic xylose assimilation. Yeast. 2011;**28**:645-660. DOI: 10.1002/yea.1893

[30] Basso LC, Amorim HV, Oliveira AJ, Lopes ML. Yeast selection for fuel ethanol production in Brazil. FEMS Yeast Research. 2008;**8**:1155-1163. DOI: 10.1111/j.1567-1364.2008.00428.x

[31] Lidzbarrsky GA, Shkolnik T, Nevo, E. Adaptive response to DNA-damaging agents in natural *Saccharomyces cerevisiae* populations from "Evolution Canyon", Mt. Carmel, Israel. Plos One. 2009;**4**:1-8. DOI: 10.1371/journal.pone.0005914

[32] Liti G, Carter DM, Moses AM, Warringer J, Parts L, James SA, Davey RP, Roberts IN, Burt A, Koufopanou V, Tsai IJ, Bergman CM, Bensasson D, O'Kelly MJ, van Oudenaarden A, Barton DB, Bailes E, Nguyen AN, Jones M, Quail MA, Goodhead I, Sims S, Smith F, Blomberg A, Durbin R, Louis EJ. Populations genomics of domestic and wild yeasts. Nature. 2009;**458**:337-344. DOI: 10.1038/nature07743

[33] Camarasa C, Sanchez I, Brial P, Bigely F, Dequin S. Phenotypic landscape of *Saccharomyces cerevisiae* during wine fermentation: Evidence for origin-dependent metabolic traits. Plos One. 2011;**6**:1-12. DOI: 10.1371/journal.pone.0025147

[34] Warringer J, Cubillos FA, Zia A, Gjuvsland A, Simpson JD. Trait variation in yeast is defined by population history. Plos Genetics. 2011;**6**:1-15. DOI: 10.1371/journal.pgen.1002111

[35] Bai FW, Anderson WA, Moo-Young M. Ethanol fermentation technologies from sugar and starch feedstocks. Biotechnology. 2008;**26**:89-105. DOI: 10.1016/j.biotechadv.2007.09.002

[36] Gao L, Liu Y, Sun H, Li C, Zha Z, Liu G. Advances in mechanisms and modifications for rendering yeast thermotolerance. Journal of Bioscience and Bioengineering. 2015;**121**:599-606. DOI: 10.1016/j.jbiosc.2015.11.002

[37] Smit G, Straver MH, Lugtenberg BJJ, KIjne JW. Flocculence of *Saccharomyces cerevisiae* cells is induced by nutrient limitation, with cell surface hydrophobicity as a major determinant. Applied and Environmental Microbiology. 1992;**58**:3709-3714

[38] Alvarez F, CorreA LFM, Araújo TM, Mota BEF, Conceição LEFR, Castro IM, Brandão RL. Variable flocculation profiles of yeast strains isolated from cachaça distilleries. International Journal of Food Microbiology. 2014;**190**:97-104. DOI: 10.1016/j.ijfoodmicro.2014.08.024

[39] Soares EV. Flocculation in *Saccharomyces cerevisiae*: A review. Journal of Applied Microbiology. 2010;**110**:1-18. DOI:10.1111/j.1365-2672.2010.04897.x

[40] Smukalla S, Caldara M, Pochet N, Beauvais A, Guadagnini S, Yan C, Vinces MD, Jansen A, Prevost MC, Latgé JP, Fink GR, Foster KR, Verstrepen KJ. FLO1 is a variable green beard gene that drives biofilm-like cooperation in budding yeast. Cell. 2008;**135**:726-737. DOI: 10.1016/j.cell.2008.09.037

[41] Kobayashi O, Hayashi N, Kuroki R, Sone H. Region of FLO1 proteins responsible for sugar recognition. Journal of Bacteriology. 1998;**180**:6503-6510

[42] Ma K, Wakisaka M, Sakai K, Shirai Y. Flocculation characteristics of an isolated mutant flocculent *Saccharomyces cerevisiae* strain and its application for fuel ethanol production from kitchen refuse. Bioresource Technology. 2009;**100**:2289-2292. DOI: 10.1016/j.biortech.2008.11.010

[43] Stratford M, Assinder S. Yeast flocculation: Flo1 and NewFlo phenotypes and receptor structure. Yeast. 1991;**7**:559-574. DOI: 10.1002/yea.320070604

[44] Beavan MJ, Belk DM, Stewart GG, Rose AH. Changes in electrophoretic mobility and lytic enzyme activity associated with development of flocculating ability in *Saccharomyces cerevisiae*. Canadian Journal of Microbiology. 1979;**25**:88-95

[45] Govender P, Domingo JL, Bester MC, Pretorius IS, Bauer FF. Controlled expression of the dominant flocculation genes FLO1, FLO5, and FLO11 in *Saccharomyces cerevisiae*. Applied and Environmental Microbiology. 2008;**74**:6041-6052. DOI: 10.1128/AEM.00394-08

[46] Goossens KV, Willaert RG. The N-terminal domain of the Flo11 protein from *Saccharomyces cerevisiae* is an adhesin without mannose-binding activity. FEMS Yeast Research. 2012;**12**:78-87. DOI: 10.1111/j.1567-1364.2011.00766.x

[47] Goossens KV, Ielasi FS, Nookaew I, Stals I, Alonso-Sarduy L, Daenen L, Van Mulders SE, Stassen C, van Eijsden RG, Siewers V, Delvaux FR, Kasas S, Nielsen J, Devreese B, Willaert RG. Molecular mechanism of flocculation self-recognition in yeast and its role in mating and survival. MBio. 2015;**6**:1-16. DOI: 10.1128/mBio.00427-15

[48] Della-Bianca BE, de Hulster E, Pronk JT, van Maris AJ, Gombert AK. Physiology of the fuel ethanol strain *Saccharomyces cerevisiae* PE-2 at low pH indicates a context-dependent performance relevant for industrial applications. FEMS Yeast Research. 2014;**14**:1196-1205. DOI: 10.1111/1567-1364.12217

[49] Kitichantaropas Y, Boonchird C, Sugiyama M, Kaneko Y, Harashima S, Auesukaree C. Cellular mechanisms contributing to multiple stress tolerance in *Saccharomyces cerevisiae* strains with potential use in high-temperature ethanol fermentation. AMB Express. 2016;**6**:1-14. DOI: 10.1186/s13568-016-0285-x

[50] Henderson CM, Zeno WF, Lerno LA, Longo ML, Block DE. Fermentation temperature modulates phosphatidylethanolamine and phosphatidylinositol levels in the cell membrane of *Saccharomyces cerevisiae*. Applied and Environmental Microbiology. 2013;**79**:5345-5356. DOI: 10.1128/AEM.01144-13

[51] Ma M, Liu ZL. Mechanisms of ethanol tolerance in *Saccharomyces cerevisiae*. Applied Microbiology and Biotechnology. 2010;**87**:829-845. DOI: 10.1007/s00253-010-2594-2593

[52] Liu ZH, Qin L, Zhu JQ, Li BZ, Yuan YJ. Simultaneous saccharification and fermentation of steam-exploded corn stover at high glucan loading and high temperature. Biotechnology for Biofuels. 2014;**7**:1-16. DOI: 10.1186/s13068-014-0167-x

[53] Mejía-Barajas JA, Montoya-Pérez R, Salgado-Garciglia R, Aguilera-Aguirre L, Cortés-Rojo C, Mejía-Zepeda R, Arellano-Plaza M, Saavedra-Molina A. Oxidative stress and antioxidant response in a thermotolerant yeast. Brazilian Journal of Microbiology. 2017;**205**:1-7. DOI: 10.1016/j.bjm.2016.11.005

[54] Wallace-Salinas V, Gorwa-Grauslund MF. Adaptive evolution of an industrial strain of *Saccharomyces cerevisiae* for combined tolerance to inhibitors and temperature. Biotechnology for Biofuels. 2013;**6**:1-9. DOI: 10.1186/1754-6834-6-151

[55] Őhgren K, Bura R, Lesnicki G, Saddler J, Zacchi G. A comparison between simultaneous saccharification and fermentation and separate hydrolysis and fermentation using steam-pretreated corn stover. Process Biochemistry. 2007;**42**:834-839. DOI:10.1016/j.procbio.2007.02.003

[56] Basso LC, Basso TO, Rocha SN. Ethanol production in Brazil: The industrial process and its impact on yeast fermentation. Biofuel Production. 2011;**5**:85-100. DOI: 10.5772/17047

[57] Rep M, Krantz M, Thevelein JM, Hohmann S. The transcriptional response of *Saccharomyces cerevisiae* to osmotic shock. Hot1p and Msn2p/Msn4p are required for the induction of subsets of high osmolarity glycerol pathway-dependent genes. Journal of Biological Chemistry. 2000;**275**:290-300. DOI: 10.1074/jbc.275.12.8290

[58] Linderholm AL, Findleton CL, Kumar G, Hong Y, Bisson LF. Identification of genes affecting hydrogen sulfide formation in *Saccharomyces cerevisiae*. Applied and Environmental Microbiology. 2008;**74**:1418-1427. DOI: 10.1128/AEM.01758-07

[59] Stanley D, Bandara A, Fraser S, Chambers PJ, Stanley GA. The ethanol stress response and ethanol tolerance of *Saccharomyces cerevisiae*. Journal of Applied Microbiology. 2010;**109**:13-24. DOI: 10.1111/j.1365-2672.2009.04657.x

[60] Klis FM, Boorsma A, De Groot PW. Cell wall construction in *Saccharomyces cerevisiae*. Yeast. 2006;**23**:85-202. DOI:10.1002/yea.1349

[61] Dong SJ, Yi CF, Li H. Changes of *Saccharomyces cerevisiae* cell membrane components and promotion to ethanol tolerance during the bioethanol fermentation. The International Journal of Biochemistry & Cell Biology. 2015;**69**:196-203. DOI: 10.1016/j.biocel.2015.10.025

[62] Gibson BR, Lawrence SJ, Leclaire JP, Powell CD, Smart KA. Yeast responses to stresses associated with industrial brewery handling. FEMS Microbiology. 2007;**31**:535-569. DOI: 10.1111/j.1574-6976.2007.00076.x

[63] Ding J, Huang X, Zhang L, Zhao N, Yang D, Zhang K. Tolerance and stress response to ethanol in the yeast *Saccharomyces cerevisiae*. Applied Microbiology and Biotechnology. 2009;**85**:253-263. DOI: 10.1007/s00253-009-2223-1

[64] Zhao XQ, Bai FW. Mechanisms of yeast stress tolerance and its manipulation for efficient fuel ethanol production. Journal of Biotechnology. 2009;**144**:23-30. DOI: 10.1016/j.jbiotec.2009.05.001

[65] Kotarska K, Czupryński B, Kłosowski G. Effect of various activators on the course of alcoholic fermentation. Journal of Food Engineering. 2006;**77**:967-971. DOI.org/10.1016/j.jfoodeng.2005.08.041

[66] Ruyters S, Mukherjee V, Verstrepen KJ, Thevelein JM, Willems KA, Lievens B. Assessing the potential of wild yeasts for bioethanol production. Journal of Industrial Microbiology and Biotechnology. 2015;**42**:39-48. DOI 10.1007/s10295-014-1544-y

[67] Zhang K, Zhang L, Fang Y, Jin X, Qi L, Wu X, Zheng D. Genomic structural variation contributes to phenotypic change of industrial bioethanol yeast *Saccharomyces cerevisiae*. FEMS. 2016;**16**:1-12. DOI: 10.1093/femsyr/fov118

[68] Santos RM,. Nogueira FCS, Brasil AA, Carvalho PC, Leprevost FV, Domont GB, Eleutherio ECA. Quantitative proteomic analysis of the *Saccharomyces cerevisiae* industrial strains CAT-1 and PE-2. Journal of Proteomics. 2017;**151**:114-121. DOI: 10.1016/j.jprot.2016.08.020

[69] Della-Bianca BE, Gombert AK. Stress tolerance and growth physiology of yeast strains from the Brazilian fuel ethanol industry. Antonie van Leeuwenhoek. 2013;**104**:1083-1095. DOI 10.1007/s10482-013-0030-2

[70] Breisha GZ. Production of 16% ethanol from 35% sucrose. Biomass Bioenergy. 2010; **34**:1243-1249. DOI:10.1016/j.biombioe.2010.03.017

[71] Jacob P, Hirt H, Bendahmane A. The heat-shock protein/chaperone network and multiple stress resistance. Plant Biotechnology Journal. 2016;**10**:1-10. DOI: 10.1111/pbi.12659

[72] Verghese J, Abrams J, Wang Y, Morano KA. Biology of the heat shock response and protein chaperones: Budding yeast (*Saccharomyces cerevisiae*) as a model system. Microbiology and Molecular Biology Reviews. 2012;**76**:115-158. DOI: 10.1128/MMBR.05018-11

[73] Zander G, Hackmann A, Bender L, Becker D, Lingner T, Salinas G, Krebber H. mRNA quality control is bypassed for immediate export of stress-responsive transcripts. Nature. 2016;**540**:593-596. DOI: 10.1038/nature20572

[74] Mackenzie RJ, Lawless C, Holman SW, Lanthaler K, Beynon RJ, Grant CM, Hubbard SJ, Eyers CE. Absolute protein quantification of the yeast chaperome under conditions of heat shock. Proteomics. 2016;**16**:2128-2140. DOI: 10.1002/pmic.201500503

[75] Mokry DZ, Abrahão J, Ramos CH. Disaggregases, molecular chaperones that resolubilize protein aggregates. Anais da Academia Brasileira de Ciências. 2015;**87**:1273-1292. DOI: 10.1590/0001-3765201520140671

[76] Aslam K, Hazbun TR. Hsp31, a member of the DJ-1 superfamily, is a multitasking stress responder with chaperone activity. Prion. 2016;**10**:103-111. DOI: 10.1080/19336896.2016.1141858

[77] O'Connell JD, Tsechansky M, Royall A, Boutz DR, Ellington AD, Marcotte EM. A proteomic survey of widespread protein aggregation in yeast. Molecular BioSystems. 2014;**10**:851-861. DOI: 10.1039/c3mb70508k

[78] Needham PG, Patel HJ, Chiosis G, Thibodeau PH, Brodsky JL. Mutations in the yeast Hsp70, Ssa1, at P417 alter ATP cycling, interdomain coupling, and specific chaperone functions. Journal of Molecular Biology. 2015;**427**:2948-2965. DOI: 10.1016/j.jmb.2015.04.010

[79] Mahmud SA, Hirasawa T, Shimizu H. Differential importance of trehalose accumulation in *Saccharomyces cerevisiae* in response to various environmental stresses. Journal of Bioscience and Bioengineering. 2010;**109**:262-266. DOI: 10.1016/j.jbiosc.2009.08.500

[80] Tapia H, Koshland DE. Trehalose is a versatile and long-lived chaperone for desiccation tolerance. Current Biology. 2014;**24**:2758-2766. DOI: 10.1016/j.cub.2014.10.005

[81] Tapia H, Young L, Fox D, Bertozzi CR, Koshland D. Increasing intracellular trehalose is sufficient to confer desiccation tolerance to *Saccharomyces cerevisiae*. Proceedings of the

National Academy of Sciences of the United States of America. 2015;**112**:6122-6127. DOI: 10.1073/pnas.1506415112

[82] Matsuda F, Kinoshita S, Nishino S, Tomita A, Shimizu H. Targeted proteome analysis of single-gene deletion strains of *Saccharomyces cerevisiae* lacking enzymes in the central carbon metabolism. PLoS One. 2017;**12**:1-20. DOI: 10.1371/journal.pone.0172742

[83] Inai T, Watanabe D, Zhou Y, Fukada R, Akao T, Shima J, Takagi H, Shimoi H. Rim15p-mediated regulation of sucrose utilization during molasses fermentation using *Saccharomyces cerevisiae* strain PE-2. Journal of Bioscience and Bioengineering. 2013;**16**:591-594. DOI: 10.1016/j.jbiosc.2013.05.015

[84] Souza CS, Thomaz D, Cides ER, Oliveira KF, Tognolli JO, Laluce C. Genetic and physiological alterations occurring in a yeast population continuously propagated at increasing temperatures with cell recycling. World Journal of Microbiology and Biotechnology. 2007;**23**:1667. DOI: 10.1007/s11274-007-9414-7

[85] Koschwanez JH, Foster KR, Murray AW. Sucrose utilization in budding yeast as a model for the origin of undifferentiated multicellularity. PLoS Biology. 2011;**9**:1-10. DOI: 10.1371/journal.pbio.1001122

[86] Lopes ML, Paulillo SC, Godoy A, Cherubin RA, Lorenzi MS, Giometti FH, Bernardino CD, Amorim Neto HB, Amorim HV. Ethanol production in Brazil: A bridge between science and industry. Brazilian Journal of Microbiology. 2016;**47**:64-76. DOI: 10.1016/j.bjm.2016.10.003

[87] Naumov GI, Naumova ES. Invertase overproduction may provide for inulin fermentation by selection strains of *Saccharomyces cerevisiae*. Microbiology. 2015;**84**:130-134. DOI: 10.1134/S0026261715020095

[88] Ivey M, Massel M, Phister TG. Microbial interactions in food fermentations. Annual Review of Food Science and Technology. 2013;**4**:141-162. DOI: 10.1146/annurev-food-022811-101219

[89] Favaro L, Viktor MJ, Rose SH, Viljoen-Bloom M, van Zyl WH, Basaglia M, Cagnin L, Casella S. Consolidated bioprocessing of starchy substrates into ethanol by industrial *Saccharomyces cerevisiae* strains secreting fungal amylases. Biotechnology and Bioengineering. 2015;**12**:1751-1760. DOI: 10.1002/bit.25591

[90] Fang Z, Deng H, Zhang X, Zhang J, Bao J. Evaluation of the cellulase cost during the cassava cellulose ethanol fermentation process. Sheng Wu Gong Cheng Xue Bao. 2013; **29**:312-324

[91] Li H, Shen Y, Wu M, Hou J, Jiao C, Li Z, Liu X, Bao X. Engineering a wild-type diploid *Saccharomyces cerevisiae* strain for second-generation bioethanol production. Bioresources and Bioprocess. 2016;**3**:1-17. DOI: 10.1186/s40643-016-0126-4

[92] Grimaldi MP, Marques MP, Laluce C, Cilli EM, Pombeiro Sponchiado SR. Evaluation of lime and hydrothermal pretreatments for efficient enzymatic hydrolysis of raw sugarcane bagasse. Biotechnology for Biofuels. 2015;**8**:205. DOI: 10.1186/s13068.015.0384.y

[93] Djajadi DT, Hansen AR, Jensen A, Thygesen LG, Pinelo M, Meyer AS, Jørgensen H. Surface properties correlate to the digestibility of hydrothermally pretreated lignocellulosic *Poaceae* biomass feedstocks. Biotechnology for Biofuels. 2017;**23**:1-15. DOI: 10.1186/s13068-017-0730-3

[94] Choudhary J, Singh S, Nain L. Bioprospecting thermotolerant ethanologenic yeasts for simultaneous saccharification and fermentation from diverse environments. Journal of Bioscience and Bioengineering. 2017;**123**:34-36. DOI: 10.1016/j.jbiosc.2016.10.007

[95] Miranda I, Masiero M. O, Zamai T, Capella M, and Laluce C. Improved pretreatments applied to the sugarcane bagasse and release of lignin and hemicellulose from the cellulose-enriched fractions by sulfuric acid hydrolysis. Journal of Chemical Technology and Biotechnology. 2016;**91**:476-482. DOI: 10.1002/jctb.4601

[96] Raghavi S, Sindhu R, Binod P, Gnansounou E, Pandey A. Development of a novel sequential pretreatment strategy for the production of bioethanol from sugarcane trash. Bioresource Technology. 2016;**63**:495-409. DOI: 10.1016/j.biortech.2015.08.062

[97] Alff-Tuomala S, Salusjärvi L, Barth D, Oja M, Penttilä M, Pitkänen JP, Ruohonen L, Jouhten P. Xylose-induced dynamic effects on metabolism and gene expression in engineered *Saccharomyces cerevisiae* in anaerobic glucose-xylose cultures. Applied Microbiology and Biotechnology. 2016;**100**:969-985. DOI: 10.1007/s00253-015-7038-7

[98] Jeffries TW, Jin YS. Metabolic engineering for improved fermentation of pentoses by yeasts. Applied Microbiology and Biotechnology. 2004;**63**:495-409. DOI: 10.1007/s00253-003-1450-0

[99] Ko JK, Um Y, Lee SM. Effect of manganese ions on ethanol fermentation by xylose isomerase expressing *Saccharomyces cerevisiae* under acetic acid stress. Bioresource Technology. 2016;**222**:422-430. DOI: 10.1016/j.biortech.2016.09.130

[100] Moysés DN, Reis VC, de Almeida JR, de Moraes LM, Torres FA. Xylose fermentation by *Saccharomyces cerevisiae*: Challenges and prospects. International Journal of Molecular Sciences. 2016;**17**:1-18. DOI: 10.3390/ijms1703020

Chapter 10

Physicochemical Characterization of the Yeast Cells and Lignocellulosic Waste Used in Cell Immobilization for Ethanol Production

Agudelo-Escobar Lina María, Solange I. Mussatto, Mariana Peñuela and José António Teixeira

Additional information is available at the end of the chapter

http://dx.doi.org/10.5772/intechopen.70129

Abstract

Ethanol is one of the leading alternative fuels. Efforts have increased the development of technologies for producing ethanol efficiently and economically. The continuous fermentation using yeast cells immobilized in low-cost materials is presented as an excellent alternative. We used four lignocellulosic wastes for the inmobilization process. The materials were characterized physicochemically. The composition was determined by the Van Soest method. Zeta potential was measured to establish the hydrophobic or hydrophilic character of the material surfaces. The contact angles measurements were used to confirm the hydrophobic or hydrophilic character and the free energies interaction was established. Images were obtained by scanning electron microscope, and determination of surface areas and volumes was performed by adsorption and desorption isotherms. It was established that cell surface properties are modified by the immobilization process to which they are subjected. It was evident that cell immobilization depended on the properties of the carrier, as well as cell surface properties. Thus, in order to improve the process of cell immobilization, it is essential to understand the type of carrier-cell interactions that occur during the immobilization process, making necessary the knowledge of the main surface characteristics of both the media and of cells that can affect the process.

Keywords: ethanol, immobilization, yeast, lignocellulosic waste, continuos fermentation

1. Introduction

Global instability in the cost and supply of petroleum and the growing need in reducing the environmental impact of burning fossil fuels have renewed interest in alternative fuels from renewable sources. Ethanol is one of the most important biofuels that can be implemented alone or in combination with gasoline in the current transport system. Because of this, efforts have increased the development and implementation of technologies for producing ethanol more efficient and economical.

One of the improvements proposed to the alcoholic fermentation is the realization of continuous process with immobilized cells. This implementation offers great advantages such as ease of product separation, reuse of biocatalysts, and high volumetric productivity [1]. The immobilization matrices studied have been hydrophilic polymer gels such as alginate, carrageenan, agarose, etc. In these matrices, the cells are immobilized by entrapment in the gel [2, 3]. However, this immobilization method is impractical on an industrial scale because of the cost of raw materials, the complexity of preparation of biocatalysts, and the short operational lifetime of these in fermentations [4].

In the last decade, raw materials for agro-industrial residues have been evaluated as promising carriers for the immobilization of cells. On these media the cells are immobilized mainly by adsorption, a simple and economical methodology. The cellulosic materials are re-generable, reusable, sterilizable heat, biologically and chemically stable under different fermentation conditions and with sufficient mechanical strength, key features for potential implementation at industrial level [5–8].

The study of the interactions cell-cell has been approached by many researches; nevertheless, these studies have been approximations and simplifications of the phenomenon, due to the fact that it is impossible bear in mind all the factors and forces that are involved in this type of interaction. A complete study should include factors such as the chemical composition and form of the cells, the positive and negative surface loads and his not homogeneous distribution, the permanent changes that happen in the components of the membrane cellular, the ionic permeability of the membrane, the modification of the cellular topology during the process, the active movement or locomotion that can present the cell, the movement random Brownian, the ionic force, the pH, the temperature, brownian, the ionic force, the pH, the temperature, the viscosity, etc. [9].

Physico-chemical treatment of the cellular adhesion problem can provide a thermodynamic description of the adhesion event by considering the cell as a colloidal particle with uniform surface properties, although the cell may modify their geometry and redistribute their membrane components. This approach generates useful information on the earliest stages of the adhesion process; in some cases, it fails to predict actual adhesion. From a physicochemical point of view, microbial adhesion is often seen as interplay of Van der Waals long-range forces, electrostatic forces, and various short-range interactions [10].

This study characterized physico-chemically four lignocellulosic residues—wood shavings, bagasse, corncobs, and corn leaves—which were used as carriers in the immobilization of yeast cells for ethanol production in packed bed bioreactors. The characterization was also performed to yeast cells that remained free and immobilized cells in the four lignocellulosic materials during the basal processes of immobilization. It is hypothesized that the modification of surface properties of yeast is caused by the specific conditions of process, characteristics that modify the immobilization process on lignocellulosic substrates. To improve the process of cell immobilization, it is essential to understand the type of support-cell interactions that occur during the process, and it is essential to establish the main surface characteristics of both the media and of cells that affect the process of immobilization.

2. Methodology

2.1. The organism and culture media

In this work, we used the commercial yeast *Saccharomyces cerevisiae* Ethanol Red supplied by Fermentis. The strain was kept lyophilized and stored at 4°C. The medium for inoculum and fermentation is composed of: glucose 100 g/L, peptone 3.6 g/L, $(NH_4)_2SO_4$ 3 g/L, yeast extract 4 g /L, KH_2PO_4 2 g/L, and $MgSO_4.7H_2O$ 1 g/L. The pH was adjusted to 5.0.

2.2. Composition material determination

To determine the content of the lignin, cellulose, hemicellulose, ash, and cell wall of lignocellulosic waste, we followed the methodology used by Van Soest [11–15]. This methodology consists primarily in the digestion of the materials in a sequence of detergent solutions, with changes in acidity to achieve solubilization of components due to their resistance or susceptibility to chemical reagents.

2.3. Surface area determination

The technique of Brunauer–Emmett–Teller (BET) adsorption isotherms was used for establishing the surface area for each material. It was also performed to determine the type of pore and pore-size distribution, using the methodology of adsorption isotherms barrett-joyner-halenda (BJH) desorption.

2.4. Sedimentation test

The technique consists in the measurement of cell sedimentation after contact cell suspension with solution of $CaCl_2$, which promotes the aggregates cells formation. This methodology is used to establish the flocculating ability of cells. And their performance taking into account the work of Domingues et al. [16] and Branyik et al. [17].

2.5. Microbial Adhesion To Hydrocarbons test (MATH tests)

The technique consists of determining the ability of cell to adhere to several solvents and establish their electron-donor/electron acceptor or Lewis acid-base properties. This method is based on the comparison between microbial cell affinity to a polar solvent and to a non-polar solvent by simply measuring the fraction of cell removal from the aqueous phase in the presence of these solvents. To perform the test MATH, we used two pairs of substances, chloroform is an electron acceptor solvent and hexadecane is a nonpolar solvent. We also used ethyl acetate as a solvent strongly electron donor and decane as a nonpolar solvent. Due to surface tension properties of these solvents, the differences between the results obtained with chloroform and hexadecane, and the results obtained with ethyl acetate and decane indicate that cell surface interactions are of type electron donor or electron acceptor and reveal hydrophobic and hydrophilic properties. Strains adhering well to the hydrocarbon are considered to be "hydrophobic" and strains adhering poorly are considered "hydrophilic." The removal is greatest when you have a pH value where the zeta potential of the organisms and/or hydrocarbons is zero. Under this condition, there is no static repulsion [10]. This methodology was developed taking into account the work of Domingues et al. [16], Branyik et al. [17], and Bellon-Fontaine et al. [18].

2.6. Zeta potential measuresment

Zeta potential describes the static electric field strength of the double layer at the boundary between a particle and the fluid in which is immersed. Measurements are performed on equipment Zeta meter Malvern Instruments that employed folded capillary cells (zetasizer nano series Malvern).

2.6.1. For cells

Cells recovered from fermentation are centrifuged and resuspended in solution 10 mM KNO_3. They are washed at least twice with this solution. Prepare a cell suspension of concentration 0.85 g/L dry cell weight with solution 10 mM KNO_3. Adjust the pH with HNO_3. We can prepare three different cells suspension with pH values of 3.0, 4.0, and 5.0. The cell suspension is filled into the electrophoresis cell, and after at least 40 electrophoretic mobility readings, the average zeta potential is calculated.

2.6.2. For lignocellulosic materials

An amount of 0.5 g in dry state was triturated and then suspended in 100 mL of 10 mM KNO_3. The suspension of carrier particles was filtered through a polyester mesh (PE 15 mm Seidengazefabrik AG Thal, Switzerland) with mesh openings of 15 × 15 mm and the pH is adjusted with HNO_3. The cell suspension is filled into the electrophoresis cell and after at least 40 electrophoretic mobility readings the average zeta potential was calculated.

2.7. Contact angles

The technique consist in the measurement of contact angles, formed by sessile drops of three different liquids (two polar and one apolar), to enable the calculation of the surface free energy and the degree of hydrophobicity of cells and lignocellulosic materials. This methodology was developed taking into account the work of Henriques et al. [19] and Branyik et al. [17]. The apparatus used was a model OCA 15 PLUS, DATAPHYSICS.

2.7.1. Contact angle measurement for cells

Cells were harvested by centrifugation at 6000 g and 4°C for 10 min and washed with increasing concentrations of ethanol in water (10, 20, and 50% v/v). The resulting pellet was resuspended in 50% (v/v) ethanol. The final concentration was adjusted to 1.2 g/L dry cell weight. An aliquot of 1 mL of cell suspension is spread over the solidified agar layer in order to cover the entire surface. This layer is let to dry; this step is repeated four times. Contact angles are measured by the sessile drop technique at room temperature using water, formamide, and α-bromonaphtalene. Each assay is performed in triplicate and at least 20 contact angles, per sample, are measured.

2.7.2. Contact angle measurement for carriers

To contact angle measurements, carrier particles are fixed on a microscopic slide by an adhesive tape. Contact angles are measured by the sessile drop technique, it is employed a drop volume of 3 μL and the same procedure for cells determination.

2.8. Scanning electron microscope (SEM)

For the realization of the images, we used the scanning electron microscope (FEG/SEM) FEI Nova 200 with EDAX, EDS/EBSD, and STEM. The images were performed on samples of materials before and after basal immobilization.

2.9. FT-IR and FT-Raman measurements

Fourier Transform Infrared Perkin Elmer, Spectrum one model, deuterated triglycine sulfate (DTGS) detector was employed to obtain the spectra. The samples were dried at 80°C for 30 min, dispersed in KBr to form a tablet, which is analyzed in the spectrometer. The analysis conditions were temperature 24°C, the number of sweeps of 8 with a resolution of 4 cm^{-1} and a range of wavelengths (ν) 4000–400 cm^{-1}.

2.10. Basal immobilization process

We used pieces of each material: wood shavings, sugarcane bagasse, corncobs, and corn leaves. A cell suspension prepared in isotonic solution with 15 g/L of biomass was continuously recirculated. The process was realized by a period of 12 hours at 30°C, and we used column bioreactors with a volume of 150 mL.

2.11. Dry weight technique modified

Immobilized biomass was determined by the difference in dry weight of support before and after cell removal protocol. The removal was performed with a solution of NaOH 0.1% by mechanical agitation at 150 rpm for 24 hours at room temperature. The drying was performed at 105°C by 12 hours [20].

2.12. Free energy determinations

The total surface tension (γ_{tot}) and its components (γ_{LW}, γ^{+}, γ^{-}, γ_{AB}), the values of the free energy of interaction between cells and water and the components ($\Delta Gcwc_{tot}$) ($\Delta Gcwc_{LW}$, and $\Delta Gcwc_{AB}$) are calculated according to van Oss and co-workers [21].

3. Results and discussion

Table 1 shows the results for determining the composition of lignocellulosic waste. The content of lignin, cellulose, hemicellulose, ash, and cell wall presented significant differences for each material. Wood shavings, corn leaves, and bagasse have between 20 and 24% cellulose content more that lignin, however, the corncob has only 8%. The corn leaves in the material had higher content of cell wall and hemicellulose, whose main function is to provide the bond between the cellulose and lignin [22]. Lignin is the substance that gives structural rigidity to the material because it is responsible for holding the fibers harden and polysaccharide [23]. The corn leaf contains a low amount of lignin, a characteristic that may influence the ability of immobilization and the stability of the material in fermentation. We can experimentally verify that this material in the operation presents less stability. It tends to gather and pile up by the top of the bioreactor, resulting in not homogeneous distribution of the bed and its obstruction. The corn leaf was the material that showed less stability during cell removal protocol; it is easily disintegrated when placed in contact with NaOH at high concentration.

Natural materials typically have very low porosities and surfaces, so the amount of nitrogen adsorbed at 77 K was very low, this corresponds to materials with a poorly developed mesoporosity and no micropores. The size distributions of mesopores determined by the BJH method (desorption isotherm) resulted in none of the samples has a significant amount of mesopores than 20 nm; most of them are between 3 and 12 nm. The corn leaves and wood shaving materials presented mesopores between 3.1 and 3.4 nm. The values are summarized in **Table 2**. There is good agreement between the values of the areas calculated by the BET method and BJH method from the desorption isotherm, with the exception of wood shaving material. It is important that the BJH method counts only the pores with diameters greater than 3.1 and it is possible that this sample has a significant amount (in relative terms) of these pores. The corresponding size distribution suggests this. We can establish a small difference between the

	Lignin (%)	Cellulose (%)	Hemicellulose (%)	Ash (%)	Cell wall (%)	Total
Wood shavings	27.09	46.79	16.62	0.45	9.05	100
Bagasse	29.68	52.18	7.24	0.01	10.89	100
Corncobs	38.15	45.95	9.95	0.34	5.61	100
Corn leaves	5.44	29.19	24.63	1.71	39.03	100

Table 1. Material's composition obtained using the Van Soest Method.

Materials	S_{BET} (m²/g)[a]	S_{BJH} (m²/g)[b]	S_{BJH} (m²/g)[c]	V_P (cm³/g)[d]	V_P (cm³/g)[b]	V_P (cm³/g)[c]
Wood shavings	2.6	2.5	2.1	0.005	0.005	0.005
Corn leaves	2.8	2.2	2.0	0.004	0.004	0.004
Bagasse	5.3	3.7	3.7	0.006	0.005	0.006
Corncobs	4.8	3.1	3.0	0.005	0.005	0.005

[a]Calculated by the BET method (adsorption isotherm).
[b]Calculated by the BJH (desorption isotherm).
[c]Calculated by the BJH (adsorption isotherm).
[d]Calculated by the maximum amount adsorbed (P/P0 = 0.99).

Table 2. Specific surface areas and volumes for lignocellulosic wastes, determined by the methods of BET and BJH.

samples of bagasse, with the corncob, bagasse can absorb a little more nitrogen than the cob, and this is consistent with the parameters. The bagasse has a higher adsorption at low relative pressure zone. Also size distribution of present mesopores is less than 20 nm, while most are distributed in sizes 3 to 12 nm. The top curves correspond to diameters between about 3.5 and 4.5 nm, values that do not differ significantly from the other samples. We can say with some certainty that all samples have a very low porosity, mainly relating to mesopores with diameters below 12 nm, with specific areas between 1 and 4 surface m²/g and pore specific volume between 0.002 and 0.006 cm³/g. The advantages of cell adhesion to nonporous carriers consist in lower mass transfer limitation of substrates and products due to direct contact between cells and bulk liquid and also in the simplicity of the immobilization. The main risk of this method is the biofilm detachment induced by changes in cell environment [24].

All the spectra show similar absorption features, although the intensities of the absorption bands are different. The assignment of the following absorbance bands is in accordance with the literature [25]. The methylene (CH_2) structural unit has symmetric and asymmetric vibrations. The asymmetric C–H stretching vibrations for CH_2 involve one C–H bond contracting, while the other bond is lengthening. This vibration is observed at 2926 ± 10 cm^{-1}. The symmetric methylene stretch involves both the C–H bonds lengthening or contracting at the same time. This band typically appears at 2855 ± 10. These bands are present in both blank cells and immobilized cells. The presence of band at 1375 ± 10 cm^{-1} is a strong indication of the presence of a metyl group in a sample. These bands correspond to the symmetric C–H bending vibrations of the methyl (CH_3) group and are observed in the wood shaving sample. The spectral signature bands due to the C=O stretching vibration, which appears as an intense band between 1800 and 1600 cm^{-1}. This vibration is observed in both blank cells and immobilized cells. The overriding spectral feature of a carboxylic acid is the broad, intense O–H stretching band typically found from 3500 to 2500 cm^{-1}, and often centered around 3000 cm^{-1}. This band almost by itself, tells you that sample contains a carboxylic acid. The sharper C–H stretching vibrations are superimposed upon the broad O–H stretching band. However, some times the O–H stretch masks the C–H stretching bands. Note that on the low wavenumber side of the O–H stretching band in the blank and immobilized cells of both samples, between 2500 and 2800 cm^{-1}, there are some broad features of medium intensity. The in-plane O–H bending band is found from 1440 to 1395 cm^{-1}. In the spectrum, this band is found at 1400 cm^{-1}. Carboxylic acids contain a C–O bond, and the C–O stretch of carboxylic acids appears between 1320 and 1210 cm^{-1}. The C=O stretch of carboxylic acids is in the same regions as the C=O stretch of ketones, aldehydes and esters. The C=O stretching band by itself is not sufficient evidence to identify a carboxylic acid. Other bands, such as the O–H stretch must be found to confirm the assignment.

The NH_2 group in a primary amide can bend in addition to stretching. The in-plane bending vibration involves the H–N–H bond angle getting bigger and smaller, like the opening and closing of a pair of scissors. This vibration is called the NH_2 scissors mode, and is found from 1650 to 1620 cm^{-1}. Note that the wavenumber range for the C=O stretch and the NH_2 scissors overlap. The NH_2 part of a primary amide can also bend out of the plane defined by the functional group. This out-of-plane bending band is typically very broad and is found from 750 to 600 cm^{-1}. This band is observed in all spectra for the materials. Primary amides contain one C–N bond, and the stretching of this bond gives rise to a band from 1430 to 1390 cm^{-1}. It is found well-defined in the spectrums of corncobs, corn leaves, and wood shaving, and it is less evident in the bagasse spectrum. The secondary amides are probably the most common and important type of amide. Proteins, nylons, and other polymers contain secondary amide linkages. This band appears between 3370 and 3170 cm^{-1}. In the cells spectrum, this band is overlapped by the O–H stretching band. The carbonyl stretch of secondary amides appears in the same range as other amides, from 1680 to 1630 cm^{-1}. The N–H moiety can bend as well as stretch and, like many other functional groups, there is an in-plane and out-of-plane bending vibration in secondary amides. The in-plane N–H bending vibration usually appears from 1570 to 1515 cm^{-1}. It is found in the spectrum of both cells samples in 1542 cm^{-1}. The primary amine C–N stretching vibration for saturated molecules occurs from 1250 and 1020 cm^{-1}. It is observed at 1088 cm^{-1} for cells spectrum and 1062 cm^{-1} for the materials spectrum. In summary, the FT-IR spectra obtained showed the presence of CH, CH_2, CH_3, NH, NH_2, NH_3, COOH, and CONH groups on the surface of all samples cells and materials as shown in **Figure 1**).

Images obtained by SEM for the materials before and after basal immobilization are presented in **Figure 2(a)** and **(b)**. In **Figure 2(a)**, we can see the superficial differences between the materials prior to immobilization. The corncob is the material that has a rough, irregular surface, and multiple pores are observed, corresponding to embryo sacs that make up the fruit of the corn. The wood shaving has a surface with two types of arrays, one consisting of a flat, smooth region, equivalent to the wood fibers and is the main constituent. The other region is composed of pores corresponding to the tracheae in the body secondary woody plants. The corn leaf is the material having a smooth and homogeneous surface; there is complete cellular structure and the presence

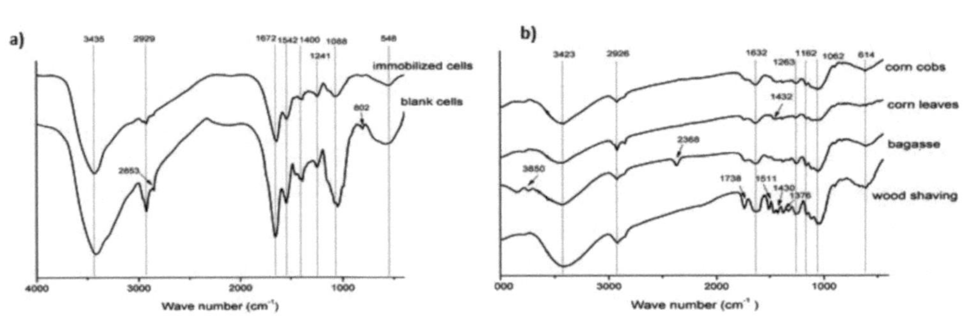

Figure 1. FT-IR results for (a) yeast cells and (b) lignocellulosic wastes.

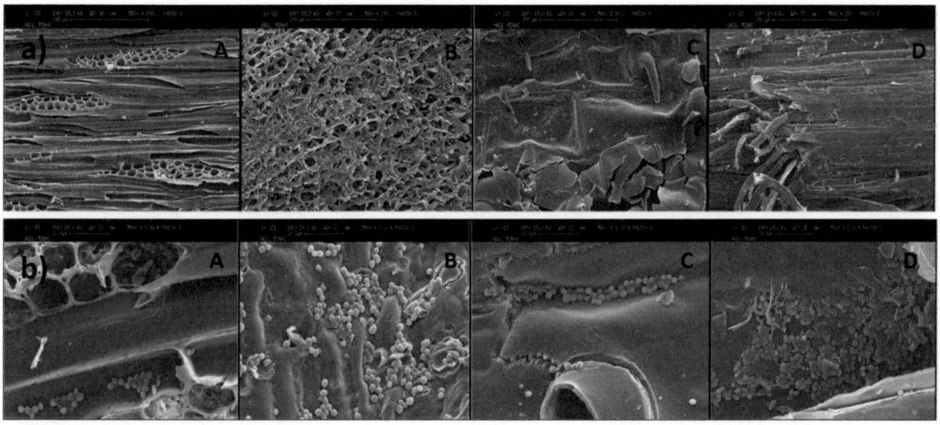

Figure 2. (a) Scanning Electronic Micrograph of lignocellulosic wasted, (b) Scanning Electronic Micrograph of lignocellulosic wasted after basal immobilization; (A) wood shavings, (B) corncobs, (C) corn leaves and (D) bagasse.

of trichomes or hairinesses characteristic of this tissue. The bagasse has an ordered structure in the form of overlapping plates; equivalent to longitudinal segment of cane stalk and due to mechanical treatment for the extraction of juice in sugar cane cannot be seen defined cellular structure.

Figure 2(b) shows the images of the material after basal immobilization. We can see that the yeast cells adhered to all surfaces of different materials. In the case of the wood shaving, immobilization was performed in both the pore region and the smooth region. In the corncob, the cells display more uniform adhesion across the surface of the material; however, in the corn leaf, the cells were immobilized preferentially in regions where there were creases. The bagasse laminar surface was covered uniformly by the cells. It can be seen that the cells are attached to chains and forming group. The highest number of immobilized cells was bagasse and corncob.

The superficial characteristic of cells before basal immobilization process, as shown in **Figure 3**, indicate an affinity for chloroform greater than hexadecane. The differences in affinity were due to interactions of Lewis acid-base, i.e., interactions donor / acceptor of electrons

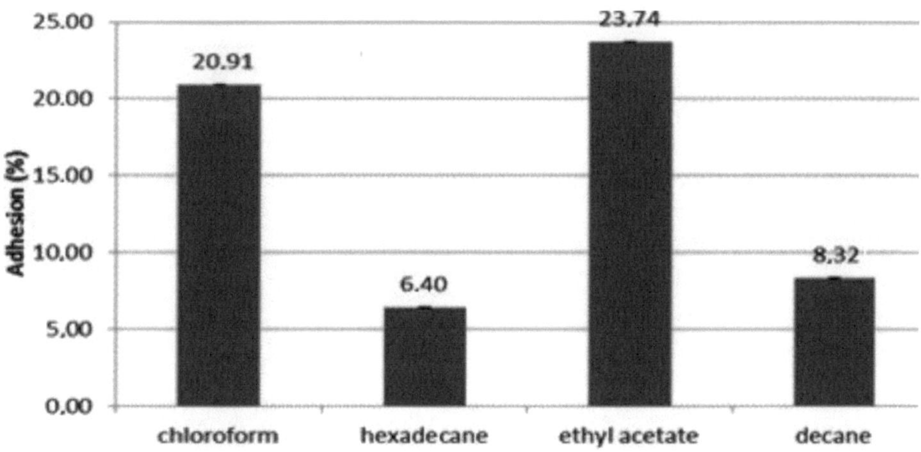

Figure 3. Test MATHS for free cells.

resulting from the electron donor nature of the yeast. These results are consistent with those obtained by Mercier-Bonin et al. [26]. From the results obtained with the other pair of solvents, we can establish that there is a higher affinity of the cells by the ethyl acetate solvent than the nonpolar solvent decane. It reveals an electron acceptor nature even slightly higher than its electron donor nature. The low affinity of cells to the nonpolar solvents indicates hydrophilic properties [10, 27]. We can conclude that yeast cells used in this study have an electron acceptor nature and hydrophilic properties.

Adhesion analysis for cellular samples obtained from the basal immobilization process on wood shaving, corncobs, and corn leaves carriers revealed an increase in the adhesion of immobilized cells over no immobilizes cells. Both free and immobilizes cells exhibited an electron donor nature, a result contrary to what was obtained with blank cells (free cells suspended in isotonic solution not subjected to immobilization). However, the results of adhesion to cells in the process with bagasse like carrier showed differences in this trend, i.e., retained their electron acceptor nature. It is important to note that all cell samples retained the hydrophilic character; it was evidenced by the poor adherence to the nonpolar solvents. These results suggest that the immobilization process modifies the surface characteristics of the cell and the immobilization by adhesion is influenced also by the surface characteristics of the carrier used. The results of cell adhesion after the basal immobilization process with each of the carriers are presented in **Figure 4**.

The results obtained for the sedimentation test performed on the cells used as blank and the free cells taken from the basal immobilization process with bagasse are presented in **Figure 5**. As shown no increased capacity flocculating or settling of the cells after being subjected to the immobilization process conducted by 12 hours was observed. It is possible that a change occurs with the time in cell physiology, allowing an increase in the capacity of sedimentation. This behavior has been reported by other researchers [17]; however, the period of time of immobilization process that the cells were subjected in these studies was superior to 100 hours; it was much higher than the evaluated in this study.

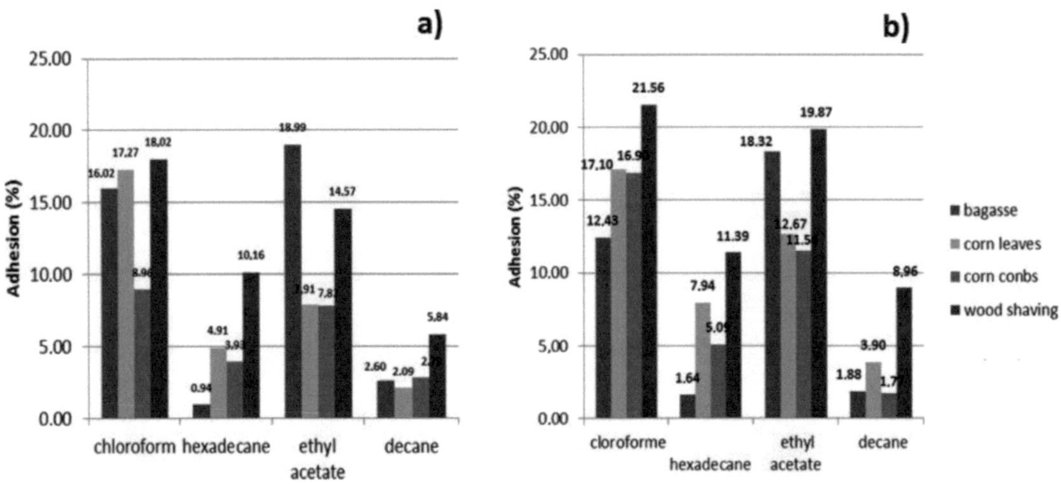

Figure 4. Test MATH for cells in basal immobilization process. (a) Free cells and (b) immobilized cells.

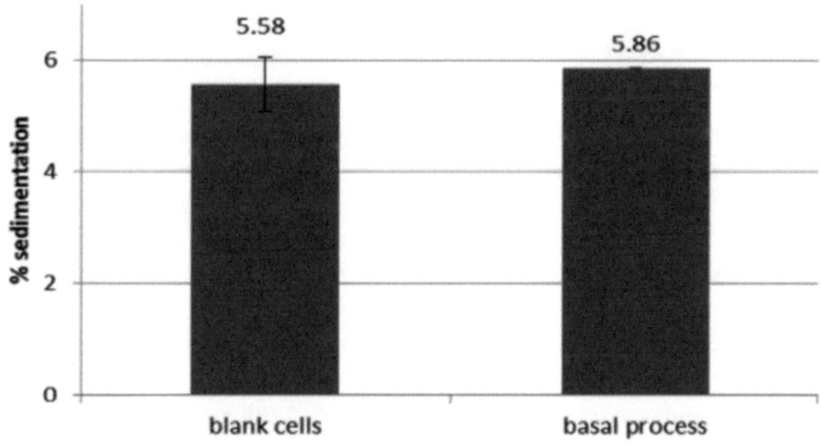

Figure 5. Sedimentation test for free cells obtained in different process.

An approach based on a balance of free energy of interaction between cell-liquid, liquid-medium, and medium-cell interface was used to estimate the physicochemical properties of the surface of cells and media, which could lead to adhesion. Electrodynamic forces (Lifshitz-Van der Waals) and hydrophobic forces (acids/Lewis bases) were determined using contact angle measurements and tests of electrophoretic mobility (zeta potential). The contact angle measurements were obtained from the formation of sessile drops of three different liquids; water and formamide as polar liquids and liquid α-bromonaphthalene as apolar. The calculation of surface free energy and the degree of hydrophobicity can be defined in terms of the change in free energy of interaction between two particles of the same material immersed in water (w). The free energy comprises a polar component (AB) and an apolar component (LW), $\Delta G_{1w1TOT} = \Delta G_{1w1AB} + \Delta G_{1w1LW}$. When the value of de ΔG_{1w1TOT} is negative, the interaction energy between molecules is attractive, which means that cells have less affinity for water than for themselves giving them a hydrophobic character. On the other hand, the cells are hydrophilic when this value is positive [27].

The value of contact angle with water can provide preliminary information on the hydrophobicity of the cells. If the value is greater than 50°, the surface is considered hydrophobic; if the value is less than 50°, the surface is hydrophilic [28]. In the results presented in **Table 3**, we can see values of contact angles above 50° for all lignocellulosic waste, indicating their

	Wood savings	Corncobs	Corn leaves	Bagasse	Free cells
Water (θ_w (°))	84.50 ± 06.40	84.40 ± 06.10	115.67 ± 15.28	104.20 ± 11.10	19.88 ± 1.87
α-Bromonaphthalene (θ_B (°))	32.27 ± 07.84	33.19 ± 09.63	44.63 ± 11.49	28.60 ± 10.00	76.50 ± 2.67
Formamide (θ_F (°))	74.25 ± 11.53	67.57 ± 10.15	75.72 ± 12.47	75.11 ± 11.37	13.80 ± 1.04

Table 3. Contact angle for lignocellulosic waste materials and yeast cells.

hydrophobic character. In the case of blank cells, the value of contact angle with water is less than 50°, indicating a hydrophilic character. These results are consistent with those previously obtained in the MATH test.

The cell surface is the site of the physicochemical interactions with the support leading to adhesion. Previous studies have suggested the relationship between the ability of adhesion and cell surface hydrophobicity [29]. The hydrophobicity of yeast cells is affected by temperature, nutrition, and growth phase [19]. While the nature of the substrate surface can be considered temporarily unchanging, the nature of the yeast surface is a function of their physiological state. In the **Table 4**, we shows the values of contact angles obtained for cell samples of immobilized process carried out with the four materials. The properties of the cell surface have varied. There is an increase in the value of contact angle with water for all samples of cells that were immobilized; this indicates an increase in cell surface hydrophobicity generated by changing the physiological state of the cell and the process conditions. The only material that did not show this behavior was the corn leaf. In this material, the contact angle value increased for the cells that remained free during the process, however, the immobilized cells showed a similar value to the blank cells.

In **Tables 5** and **6**, the results for surface tension and interaction energies are calculated. The high values obtained for γs^- are consistent with measurements of zeta potential, and also showed a negative surface charge to the cells (see results in **Figure 6(a)**. The result showed a pH of 3.0–5.0 and an ionic strength of 10 mM; the zeta potential has a range of −13 to 16 mV. These values reveal an electron donor nature of the yeast cell surface. Some authors suggest that it may be due to presence of carboxyl and phosphate groups [17]. **Figure 6(b)** shows that the zeta potential is negative for all lignocellulosic materials and there is a marked increase in negativity with the pH increase. For a pH range of 3.0 –5.0, the zeta potential range is between −7 and −21 mV.

Positive values obtained for ΔG_{1w1IF} (interaction free energy of cells) are positive, indicating the hydrophilic character of the cells. There were no changes to this feature during the process of cell immobilization, although there were differences between cells that were free and those that were immobilized. Cells that were immobilized had a value less hydrophilic, indicating a variation in the surface properties of yeast cells. For lignocellulosic material was obtained ΔG_{2w2IF} negative value, indicating its hydrophobicity. Also the high value of γs_{LW} confirms its apolar character. The interaction energy between the carrier and the cell when immersed in water is presented at the end of the table (ΔG_{1w2IF}), this value reveal the possibility of cell immobilization. Positive values of ΔG_{1w1AB} involve repulsion between the particles or molecules of the material 1 immersed or dissolved in the liquid (w). It is also a major factor in the stability of particles in suspensions.

It is theoretically possible for a given material to only have a value of $\gamma s+$ or presence of only one value for γs^-. So your γs_{AB} is zero and its surface tension (γs) is equal to total γs_{LW}. However, such substances, which are designated as single-pole, can interact strongly with bipolar materials and mono-polar materials of opposite polarity, despite the apparently non-polar nature of its surface tension [27]. This feature shows the lignocellulosic materials.

		Wood shaving		Corncobs		Corn leaf		Bagasse	
	Cells blank	Free cells	inmobilized cells	Free cells	Inmobilized cells	Free cells	inmobilized cells	Free cells	Inmobilized cells
θ_w (°)	19.88 ± 1.9	22.7 ± 3.4	27.3 ± 2.7	23.9 ± 3.7	19.5 ± 1.5	19.4 ± 1.7	29.5 ± 3.0	22.8 ± 2.0	29.0 ± 2.6
θ_B (°)	76.50 ± 2.7	75.6 ± 5.2	81.0 ± 2.9	80.1 ± 1.0	85.4 ± 1.3	85.6 ± 2.3	86.8 ± 2.1	84.5 ± 2.3	86.5 ± 3.2
θ_F (°)	13.80 ± 1.0	27.9 ± 4.0	29.4 ± 1.7	24.5 ± 2.6	22.6 ± 2.8	25.7 ± 1.9	27.2 ± 2.7	25.5 ± 2.1	27.1 ± 2.9

Table 4. Contact angles obtained for free cells and immobilized cells in basal immobilization process with each carried.

	Cells blank	Wood shaving		Corncobs		Corn leaf		Bagasse	
		Free cells	inmobilized cells	Free cells	Inmobilized cells	Free cells	Inmobilized cells	Free cells	inmobilized cells
Ys^{LW} (mJ/m^2)	16.898	17.345	14.860	15.247	12.955	12.880	12.383	13.341	12.518
Ys^{+} (mJ/m^2)	11.368	7.992	9.826	10.656	12.865	12.014	13.183	11.965	13.017
Ys^{-} (mJ/m^2)	46.504	51.619	47.823	47.869	50.270	52.211	43.519	49.271	44.039
s^{AB} (mJ/m^2)	45.985	40.622	43.355	45.170	50.862	50.090	47.905	48.560	47.886
s (mJ/m^2)	62.883	57.967	58.215	60.417	63.817	62.970	60.288	61.901	60.404
ΔG_{1w1}^{AB} (mJ/m^2)	11.929	19.022	14.292	13.343	11.942	13.791	8.795	12.543	9.175
ΔG_{1w1}^{LW} (mJ/m^2)	-0.658	-0.637	−1.371	−1.174	−2.298	−2.366	−2.672	−2.097	−2.619
ΔG_{1w1}^{IF} (mJ/m^2)	11.271	18.386	12.921	12.169	9.643	11.426	6.124	10.456	6.557
ΔG_{1w2}^{AB} (mJ/m^2)	−11.842	−13.574	−11.960	−10.128	−13.439	−3.734	0.304	−5.319	−2.813
ΔG_{1w2}^{LW} (mJ/m^2)	1.636	1.478	2.386	2.185	3.059	2.192	2.334	3.176	3.534
ΔG_{1w2}^{IF} (mJ/m^2)	−10.206	−12.096	−9.574	−7.943	−10.380	−1.541	2.638	−2.143	0.721

Table 5. Surface tensions and free energies of interaction between cells and water (1w1). Free energy of interaction between cells and substrates in water (1w2).

	Wood shaving	Corncobs	Corn leaves	Bagasse
Ys^{LW} (mJ/m^2)	37.631	37.195	32.305	38.829
Ys^{+} (mJ/m^2)	0.000	0.000	0.267	0.000
Ys^{-} (mJ/m^2)	10.582	6.760	0.000	0.220
s^{AB} (mJ/m^2)	0.000	0.000	0.000	0.000
s (mJ/m^2)	37.631	37.195	32.305	38.829
$\Delta G_{2w2}{}^{AB}$ (mJ/m^2)	−36.315	−49.113	−92.455	−95.837
$\Delta G_{2w2}{}^{LW}$ (mJ/m^2)	−4.354	−4.184	−2.299	−4.959
$\Delta G_{2w2}{}^{IF}$ (mJ/m^2)	−40.666	−53.297	−94.754	−100.796

Table 6. Superficial tensions and free energy of interaction between the carriers and water (2w2).

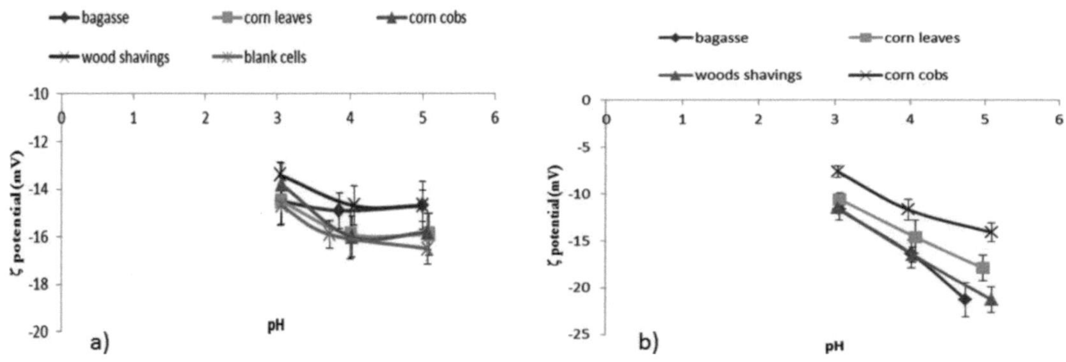

Figure 6. Zeta pontential (a) immobilized cells on basal process and (b) lignocellulosic wasted.

The implications when the interfacial free energy of interaction ΔG_{1w2} is greater than zero, a net repulsion must occur between particles or molecules of materials 1 and 2 immersed or dissolved in the liquid w (always that there is not a cancellation of the electrostatic attraction on materials 1 and 2), otherwise, i.e., ΔG_{1w2} less than zero, if there is an attraction. The energy of interaction between the carriers and cells is attractive for both wood shaving and corncobs free cells as the immobilized. Bagasse and the corn leaves have negative interaction energy (attraction) between the free cells and carriers, however, positive values are obtained for the energy of interaction with immobilized cells. Despite these results, the cell immobilization occurs on the supports, as shown in the images achieved in the SEM.

4. Conclusions

The characterization performed both lignocellulosic materials used as carriers, and yeast cells, gives relevant information about the nature of the surfaces. There are differences in the surfaces of four lignocellulosic wastes, as well as differences in their compositions. These variations can significantly influence cell adhesion and stability of biocatalysts in fermentation conditions used in ethanol production. It highlights the fact that the cells was immobilized on each of the four carrier evaluated. Yeast cells suspended in isotonic media was adhered to support, when they are placed into contact during 12 hours, corresponding to basal immobilization process. It was established that the yeast cells used in this study initially presented a hydrophilic surface and have electron acceptor nature.

After the basal immobilization process, the cell surface properties are modified. The cells have a less hydrophilic character and change their electron acceptor to electron donor character. Also, there are differences between free cells and the cells that adhere to the surface of the supports. These results confirm the variation of surface properties of cells, depending on their physiological state and the conditions under which the immobilization process was performed. To establish the potential presented by these materials in the development of functional biocatalysts for ethanol production in continuous packed-bed reactors, it is necessary to evaluate its performance under fermentative conditions, and it is necessary to establish that the yeast cells remain attached to the carriers during the process. It is necessary to establish the variations that exist in their surface properties. Nowadays, these studies are being developed in the group.

Acknowledgements

This research was in collaboration with the IBB Institute for Biotechnology and Bioengineering Centre of biological Engineering University of Minho, Campus de Gualtar, Braga, Portugal.

Author details

Agudelo-Escobar Lina María[1]*, Solange I. Mussatto[2], Mariana Peñuela[3] and José António Teixeira[4]

*Address all correspondence to: agudelo.linamaria@gmail.com and lina.agudelo@udea.edu.co

1 Research Group of Biotransformación, School of Microbiology, University of Antioquia, Medellín, Colombia

2 Novo Nordisk Foundation Center for Biosustainability, Technical University of Denmark, Kongens Lyngby, Denmark

3 Research Group of Bioprocesos, Faculty of Engineering, University of Antioquia, Medellín, Colombia

4 University of Minho, CEB-Centre of Biological Engineering, Braga, Portugal

References

[1] Wendhausen R, Fregonesi A, Moran PJS, Joekes ISJ, Rodrigues AR, Tonella E, Althoff K. Continuous fermentation of sugar cane syrup using immobilized yeast. Journal of Bioscience and Bioengineering. 2001;91(1):48-52

[2] Kurillová L, Gemeiner P, Vikartovská A, Mikov H, Rosenberg M, Ilavsky M. Calcium pectate gel beads for cell entrapment. 6. Morphology of stabilized and hardened calcium pectate gel beads with Immobilized cells for biotechnology. Journal of Microencapsulation. 2000;17:279-296

[3] Martins dos Santos VAP, Leenen EJTM, Rippoll MM, Sluis CVD, Vliet TV, Tramper J, Wijffels RH. Relevance of rheological properties of gel beads for mechanical their stability in bioreactors. Biotechnology and Bioengineering. 1997;56:517-529

[4] Serp D. From Whole Cell Immobilization: Practical Insight To Tools Applications. [thesis]. Switzerland: École Polytechnique Fédérale De Lausanne; 2002. p. 50

[5] Vasconcelos JN, Lopes CE, França FP. Continuous ethanol production using immobilized yeast on sugar-cane stalks. Brazilian Journal of Chemical Engineering. 2004;21(3):357-365

[6] Kourkoutas Y, Psarianos C, Koutinas AA, Kanellaki M, Banat IM, Marchant R. Continuous whey fermentation using kefir yeast immobilized on delignified cellulosic material. Journal of Agricultural of Food Chemistry. 2002;50:2543-2547

[7] Plessas S, Kourkoutas Y, Psarianos C, Kanellaki M, Koutinas AA. Continuous baker's yeast production using orange peel as promoting support in the bioreactor. Journal of the Science of Food and Agriculture. 2005;86(3):407-414

[8] Brányik T, Vicente AA, Machado JM, Teixeira JA. Spent grains – a new support for brewing yeast Immobilization. Biotechnology Letters. 2001 ;23:1073-1078

[9] Popescu AI. Cell-cell interactions. A physical approach. Bioelectrochemistry and Bioenergetics. 1996;40:153-157

[10] Van der Mei HC, Meinders JM, Busscher HJ. The influence of ionic strength and pH on diffusion of micro-organisms with different structural surface features. Microbiology. 1994;140:3413-3419

[11] Van Soest PJ. Use of detergents in the analysis of fibrous feeds. I. Preparation of fiber residues of low nitrogen content. Journal – Association of Official Analytical Chemists. 1963;46:825-829

[12] Van Soest PJ. Use of detergents in the analysis of fibrous feeds. II. A rapid method for the determination of fiber and lignin. Journal – Association of Official Analytical Chemists. 1963;46:829-835

[13] Van Soest PJ. Wine RH. Determination of lignin and cellulose in acid detergent fiber with permanganate. Journal – Association of Official Analytical Chemists. 1968;51:780-785

[14] Van Soest PJ. Collaborative study of acid-detergent fiber and lignin. Journal – Association of Official Analytical Chemists. 1973;56:781-784

[15] Van Soest PJ, Robertson JB, Lewis BA. Methods for dietary fiber, neutral detergent fiber and non-starch polysaccharides in relation to animal nutrition. Journal of Dairy Science. 1991;74:3583-3597

[16] Domingues L, Dantas M, Lima N, Teixeira J. Continuous ethanol fermentation of lactose by a recombinant flocculating *Saccharomyces cerevisiae* strain. Biotechnology and Bioengineering. 1999;64(6):692-697

[17] Branyik T, Vicente A, Oliveira R, Teixeira J. Physicochemical surface properties of brewing yeast influencing their immobilization onto spent grains in a continuous reactor. Biotecnolology and Bioengineering. 2004;88:84-83

[18] Bellon-Fontaine M-N, Rault J, van Oss CJ. Microbial adhesion to solvents: A novel method to determ ine the electron-donor/electron-acceptor or Lewis acid-base properties of microbial cells. Colloids and Surfaces B: Biointerfaces. 1996;7:147-143

[19] Henriques M, Gasparetto K, Azevedo J, Oliveira R. Experimental methodology to quantify Candida albicans cell surface hydrophobicity. Biotechnology Letters. 2002;24:1111-1115

[20] Branyik T, Vicente A, Dosta P, Teixeira A. Continuous beer fermentation using immobilized yeast cell bioreactor systems. Biotechnology Progress. 2005;21:653-663

[21] Van Oss CJ. Hydrophobicity of biosurfaces: Origin, quantitative determination and interaction energies. Colloids Surfaces: B Biointerfaces. 1995;5:91-110

[22] Sun RC, Fang JM, Rowlands P, Bolton J. Physicochemical and thermal characterization of wheat straw hemicelluloses and cellulose. Journal of Wood Chemistry and Technology. 1998;46:2804-2809

[23] Howard RL, Abotsi E, Rensburg J, Howard S. Lignocellulose biotechnology: Issues of bioconversion and enzyme production. African Journal of Biotechnology. 2003;2:602-619

[24] Mozes N, Rouxhet PG. Microbial hydrophobicity and fermentation technology. In: Doyle RJ, Rosenberg M, editors. Microbial cell Surface Hydrophobicity. Washington: American Society for Microbiology; 1990. pp. 76-105

[25] Smith B, editor. Infrared Spectral Interpretation. 1st ed. Boca Raton Florida: CRC Press LLC; 1999. p. 269

[26] Mercier-Bonin M, Ouazzani K, Schmitz P, Lorthois S. Study of bioadhesion on a flat plate with a yeast/glass model system. Journal of Colloid and Interface Science. 2004;271:342-350

[27] Van Oss CJ. Hydrophobicity and hydrophilicity of biosurfaces. Current Opinion in Colloid & Interface Science. 1997;2:503-512

[28] Van Oss CJ, Giese RF. The hydrophilicity and hydrophobicity of clay minerals. Clay Minerals. 1995;43:474-477

[29] Panagoda GJ, Ellepola AN, Samaranayake LP. Adhesion to denture acrylic surfaces and relative cell surface hydrophobicity of Candida parapsilosis and Candida albicans. APMIS. 1998;106:736-742

Chapter 11

Potential Production of Ethanol by *Saccharomyces cerevisiae* Immobilized and Coimmobilized with *Zymomonas mobilis*: Alternative for the Reuse of a Waste Organic

Alejandro Ruiz Marin, Yunuen Canedo Lopez,
Asteria Narvaez Garcia,
Juan Carlos Robles Heredia and
Jose del Carmen Zavala Loria

Additional information is available at the end of the chapter

http://dx.doi.org/10.5772/intechopen.69991

Abstract

Fermentation technologies have been developed to improve the production of ethanol and an alternative is the immobilization technology, which offers the possibility of efficiently incorporating symbiotic bacteria in the same matrix. This study analyzes the potential use of immobilized and coinmobilized systems on beads of calcium alginate for ethanol production used mango waste (*Mangifera indica*) by *Zymomonas mobilis* and *Saccharomyces cerevisiae* compared with free cells culture and evaluate the effect of glucose concentration on productivity in coimmobilized system using a Chemostat reactor Ommi Culture Plus. For free cell culture, the productivity was higher for *Z. mobilis* (5.76 g L^{-1} h^{-1}) than for *S. cerevisiae* (5.29 g L^{-1} h^{-1}); while in coimmobilized culture, a higher productivity was obtained (8.80 g L^{-1} h^{-1}) with respect to immobilized cultures (8.45 g L^{-1} h^{-1} - 8.70 g L^{-1} h^{-1}). The conversion of glucose to ethanol for coimmobilized system was higher (6.91 mol ethanol) with 50 g L^{-1} of glucose compared to 200 g L^{-1} of glucose (5.82 mol ethanol); suggesting the immobilized and coimmobilized cultures compared with free cells offer an opportunity for the reuse of organic residues and high alcohol production.

Keywords: *Mangifera indica*, immobilization, coimmobilization, ethanol, *Zymomonas mobilis*, *Saccharomyces cerevisiae*

1. Introduction

1.1. Use of agroindustrial waste in fermentation processes

Alcoholic fermentation is a process by which microorganisms convert hexoses, mainly glucose, fructose, mannose and galactose, in the absence of oxygen and get products as alcohol (ethanol), carbon dioxide and adenosine triphosphate (ATP) molecules. Approximately 70% of the energy is released as heat and the remainder is preserved in two terminal phosphate bonds of ATP, for use in transfer reactions, such as activation of glucose (phosphorylation) and amino acids before of the polymerization. In other words, fermentation is a set of chemical reactions carried out by microorganisms in which an organic compound is oxidized, partially in the absence of oxygen to obtain chemical energy and understood as a partial oxidation when all the carbon atoms of the compound are oxidized to form CO_2. It is a process known since antiquity and is currently the only industrial process for the preparation of ethyl alcohol in all countries. The glucose as raw material is not only used, but other types of raw material much cheaper. However, the process of alcoholic fermentation occurs naturally, originated by the activity of some microorganisms through its anaerobic energy cell metabolism; for a large-scale production process, it is necessary for microorganisms (bacteria, fungi and yeasts) to accelerate the process of alcoholic fermentation and increase the conversion rate [1]. During the twentieth century and until the beginning of the twenty-first century, alcoholic fermentation has focused exclusively on the improvement of fermentation processes and specifically on the optimization of industrial performance through a good selection of yeast strains, which are the most used microorganisms for the production of ethanol by fermentation, due to its high productivity in the conversion of sugars and better separation of the biomass after fermentation. Yeasts are unicellular (usually spherical) microorganisms of size 2–4 μm and are present naturally in some products such as fruits, cereals and vegetables. Different species of fermentative microorganisms have been identified, among which are mainly *Saccharomyces cerevisiae, Kluyveromy cesfragilis, Torulaspora* and *Zymomonas mobilis* [2].

S. cerevisiae is a unicellular organism that is able to follow two metabolic routes to obtain the energy necessary to carry out its vital processes: alcoholic fermentation and aerobic respiration. The first is characterized by the evolution of CO_2 and the production of ethanol out of contact with oxygen, obtaining the energy necessary to carry out its vital processes from metabolizing carbohydrates. The yeast requires glucose to be catalyzed by the glycolysis or Embden-Meyerhof pathway, to obtain pyruvate that is then converted by anaerobically into ethanol and CO_2 by the action of specific enzymes. Its optimal temperature of growth varies between 22 and 29°C and does not survive more than 53°C. It ferments a sugar solution with a concentration of less than 12% and is inactivated when the sugar concentration exceeds 15% due to the osmotic pressure of medium on the cell. On the other hand, *Z. mobilis* is a facultative anaerobic gram-negative bacterium that can ferment certain sugars through a metabolic pathway producing bioethanol, sometimes, more efficiently than yeasts. It has an incomplete Krebs cycle, but it has characteristics to perform the pyruvic synthesis pathways from glucose or glyceraldehyde-3-phosphate. This organism also shows a high rate of sugar uptake and a yield of ethanol as fuel of the 97% [3].

The alcoholic fermentation processes using agroindustrial products present a great challenge given the inconveniences that could arise when using raw material for human consumption or edible vegetable crops for the production of ethanol, and, on the other hand, the change in the use of land destined for the cultivation of vegetables that will be used to produce ethanol and bioethanol, which would sometimes lead to deforestation, food shortages, increase of desert regions and greater inability of soils to retain water, thus disrupting the balance of the hydrological cycle [4].

On a global scale, the use of energy raw materials for energy purposes and in the production of ethanol has led to higher prices for products such as maize or barley, as well as making ethanol production economically unviable. Therefore, it is important to use raw materials that do not compete with food products and that are low cost in the production of biofuels, must also ensure a good profitability and are environmentally sustainable projects. In the energy sector, it has been estimated that the use of all world food surpluses could only produce bioethanol to replace 1% of the oil currently used. Concluding that if food crops were used to produce ethanol, a chain of food imbalances would be generated, which would be unsustainable [5].

An alternative to producing ethanol is through the use of other nontraditional raw materials, which arise as by-products and/or waste from industrial processes. Propose new technologies that allow the production of ethanol from cane residues, solid waste and those materials containing cellulose and hemicellulose, which allows the revalorization of waste from various industries, converting them into raw material for the production of ethanol.

At present, efforts have been made mainly in the search for cheap raw materials, which replace the traditional ones, in order to achieve greater efficiency in the processes of fermentation, recovery and purification of alcohol produced. The importance of the production of bioethanol has as main interest to compete with the use of fossil fuels since ethanol can be used as fuel for motor vehicles increasing the octane number, and therefore the reduction of consumption and contaminants (10–15% less carbon monoxide and hydrocarbons). Ethanol can be mixed with unleaded gasoline from 10 to 25% without difficulty, although some engines have been able to incorporate 100% alcohol as fuel. Thus, ethanol could substitute for methyl tert-butyl ether (MTBE), an oxygenation product with which gasolines have been reformulated in Mexico since 1989, which has reduced CO_2 emissions. This action is very important since MTBE, being a very stable compound, with low degradation and very soluble in water.

The production of bioethanol lost importance at the end of the first half of the XX century, being replaced by the production of synthetic ethanol, from petroleum derivatives, which is cheaper, but cannot be used in food preparation, alcoholic beverages or medications. The rise in oil prices turned our eyes toward the fermentation route of ethanol production, and today, we work mainly in the search for cheap raw materials, replacing the traditional sugary materials. Studies carried out by different researchers suggest that the by-products of mango juice, cane juice and molasses are an efficient alternative for the production of ethanol without affecting the food item, besides increasing the productivity and concentration of ethanol in the fermentation medium, and therefore reduce the costs of ethanol production [6].

Historically, the sugar industry in Mexico is one of the most important, characterized by sugarcane harvests throughout the year, with a production of cane of 46,231,229 tons per year, and the remaining residue derived of sugarcane has been exploited as energy biomass and for the production of different biotechnological products by fermentation. Other alternatives of raw material are mango juice and its residues, and the fruit is grown in all the countries of Latin America, Mexico being the main exporting country of this fruit, with an annual production of approximately 1 million 452 thousand tons of mango and of which more than 60% of this production is given to the South-Southeast region of our country.

The alternative of using residues or products that can replace the raw materials normally used in ethanol production is now a highly promising possibility, because the cost of production of ethanol is closely related and dependent on the cost of the raw material, the volume and the composition of the same. The existing economy in Mexico related to cane cultivation (experience and sugar tradition) and the export of mango types offers technological alternatives that allow the fermentation of cane juice, molasses and mango juice through *S. cerevisiae* and *Z. Mobilis* as viable sources for the production of ethanol, whether in the manufacture of alcoholic beverages or for the production of biofuels.

1.2. Tecnología de inmovilización

Research has been developed in order to increase the productivity of alcoholic fermentation processes. The productivity, expressed as grams of ethanol produced per hour per unit of fermentation volume, can be increased by optimizing the composition of the culture medium, by the selection of an appropriate microorganism strain or through the adaptation of the design of reactors [7]. One challenge today is to reduce ethanol production costs, and an alternative is to reduce the cost of the culture media, which can represent about 30% of the final production costs of ethanol [8].

Some fermentation technologies have been developed to improve the production of ethanol and its concentration in the culture media [9, 10]. Among these, the immobilization technology offers advantages in contrast to free cell cultures, such as increased retention time in bioreactors, high cellular metabolic activity, high cell load and protection for cells from stress [10, 11]. The immobilization cell technologies have been applied for different purposes as for the production of hydrogen [12] and compounds commercially used in the food industry [13]. Other studies have been developed with immobilized algal cells to remove nutrients (N and P) from wastewater, phenol and hexavalent chromium [14–17]. Similarly, the immobilization of *Zymomonas* and *Saccharomyces* have been used for the bio-ethanol production from waste materials [7–10, 18–20].

On the other hand, the immobilization technology provides the possibility of efficiently incorporating symbiotic bacteria [21, 22]. The interaction between two microorganisms in the same matrix is called coimmobilization, and this association can be positive with higher growth and production. However, there are relatively less applications in the ethanol production involving the immobilization of mixed-culture systems and/or coimmobilized cultures.

In a petroleum deficiency situation, bioethanol from yeast and bacterial fermentation has become a promising alternative source for fuel. Agricultural and industrial waste containing sugar, starch and cellulose, such as cassava peels, fruit bunches, and the effluents from sugar and pineapple cannery productions have been successfully applied for the bioethanol production [23, 24]. In this context, the municipality of Ciudad del Carmen, Campeche, Mexico, has an annual production of about 2.868 ha mango (*Mangifera indica*), obtained through various forms of cultivation and orchard-based technology, but the lack of local market and the poor product distribution to other locations cause much of the product be wasted, with significant losses in the locality. Hence, the need to seek alternatives to use these wastes and generate added value in the economy of the region.

This study was to determine whether the association between *S. cerevisiae* coimmobilized with *Z. mobilis* improved growth, and ethanol production using a culture medium equivalent to mango juice (*M. indica*) creates an opportunity for a regional fruit for exploitation in the production of ethanol. In this study, both microorganisms were confined in small alginate beads, a practical means of using microorganisms for environmental applications.

2. Materials and methods

2.1. Microorganism and medium

The yeast strain *S. cerevisiae* (ATCC® 2601) and bacteria *Z. mobilis* (ATCC® 8938) were obtained from the laboratory Microbiologis® and used for fermentation in coimmobilized and immobilized systems. Both microorganisms were cultured in a medium containing composition (g L^{-1}), as described by Demirci et al. [25]: 20 g glucose, 6 g yeast extract, 0.23 g $CaCl_2{\bullet}2H_2O$, 4g $(NH_4)_2SO_4$, 1 g $MgSO_4{\bullet}7H_2O$ and 1.5 g of KH_2PO_4, previously sterilized by autoclave. Strains were maintained in 250 mL of culture at 30°C and pH 4.5 with manual shaking three times a day. Transfers of fresh medium were made every 24 h for three consecutive days prior to use in experiments.

2.2. Preparation of immobilized and coimmobilized cells

For the preparation of immobilized cells, we used the technique described by Tam and Wong [26]. Both microorganisms were harvested by centrifugation at 3500 rpm for 10 min. The bacteria and yeast cells were resuspended in 50 mL of distilled water to form a concentrated cell suspension. The suspension was then mixed with a 4% sodium alginate solution in 1:1 volume ratio to obtain a mixture of 2% microorganism–alginate suspension. The mixture was transferred to a 50-mL burette, and drops were formed when "titrated" into a calcium chloride solution (2%). This method produced approximately 6500 uniform algal beads of approximately 2.5 mm in diameter with biomass content for *Z. mobilis*-alginate of 0.0055 g $bead^{-1}$ and for *S. cerevisiae* of 0.00317 g $bead^{-1}$ for every 100 mL of the microorganism–alginate mixture (**Figure 1**).

The beads were kept for hardening in the $CaCl_2$ solution for 4 h at 25 ± 2°C and then rinsed with sterile saline solution (0.85% NaCl) and subsequently with distilled water. A concentration of 2.6 beads mL^{-1} of medium (equivalent to 1:25 bead: medium v/v) was placed in a Chemostat

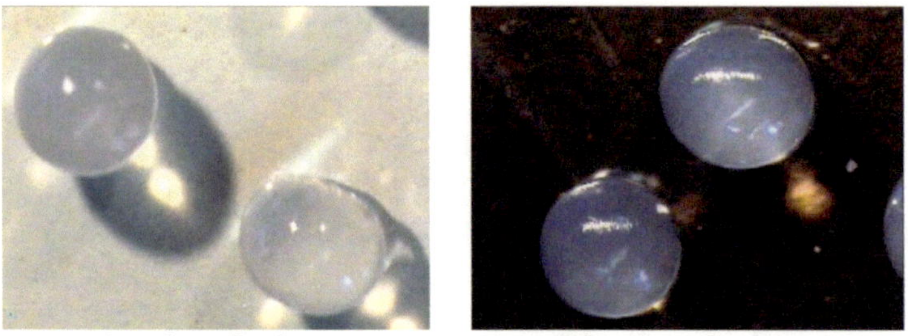

One day of culture Five day of culture

Figure 1. Microorganism–alginate beads suspension.

Ommi Culture Plus (Virtis) containing 2 L of culture medium. The reactor was maintained under stirring at 120 rpm and 30°C (**Figure 2**). A similar procedure was used for coimmobilization, with the difference that the concentrate of bacteria (25 mL) and yeast (25 mL) was mixed and then mixed with 50 mL of alginate; this procedure allowed retaining the same concentration of cells in all experiments.

Figure 2. Chemostat Ommi Culture Plus (Virtis) used for fermentation experimental process.

2.3. Experimental setup and procedure

This study was divided into two parts: (1) the batch experiment consisted of evaluating the growth and ethanol production in a medium equivalent to mango juice in cultures with free cells, immobilized and coimmobilized, and (2) evaluate the effect of glucose concentration in the production of ethanol in the system previously selected in the first experimental part based on the ethanol productivity obtained. Fermentation was performed in a Chemostat Ommi Culture Plus (Virtis) with a volume of 2 L operation, adjusting stirring at 120 rpm and maintaining a temperature of 30°C. The medium equivalent to mango juice was similar to that described by Demirci et al. [25] by adjusting the composition of the medium to a concentration of 200 g L^{-1} glucose, equivalent to that observed in the mango juice (*M. indica*).

The experimental design consisted of triplicate cultures in a Chemostat reactor Ommi Culture Plus for *S. cerevisiae* and *Z. mobilis* in free cell culture, immobilized and coimmobilized. For each experiment, the biomass was collected, as well as samples of the culture medium to the end of the logarithmic phase every 20 h. For the determination of ash-free dry weights, five beads were dissolved and filtered through a GF-C glass fiber filter (2.5 cm diameter), previously rinsed with distilled water, and incinerated at 470°C for 4 h. The samples were dried at 120°C and put to constant weight for 2 h in a conventional oven and then in a muffle furnace at 450°C for 3 h. The soluble solids of each fermenting medium were determined every 20 h by taking 1 mL aliquot from each reactor and testing for the Brix level in an refractometer.

Ethanol content (% v/v) was obtained using the Anton Paar DMA 4100M instrument, which determines the density of the mixture in relation to the standard OIML-STD-90, which can determine the content of distillate ethanol (% v/v); according to the ethanol density recorded, it was possible to obtain an approximate of ethanol content (grams of ethanol per liter of culture) produced for each experiment. Prior to the determination of the ethanol content, a distillation of cultures was conducted with a plate column distiller PS-DA-005/PE of four plates, at small scale. The cooling water flow was 3 L h^{-1} at 15°C. An aliquot of 3 L was distilled for 4 h, maintaining the operating conditions at atmospheric pressure, without reflux and with a temperature ramp in the heating jacket of 30°C up to 80°C.

The STATISTICA 7.0 software for statistical analysis and calculated mean and standard deviation for each treatment were used. The covariance analysis (ANCOVA) with $P \leq 0.05$ was used to evaluate the growth in free cell cultures, immobilized and coimmobilized. The Tukey test ($P \leq 0.05$) was used when significant differences were observed.

3. Results

3.1. Growth

In free cell cultures, the growth was observed immediately after being inoculated in the reactor of 2 L. Growth kinetics shows an exponential phase for *S. cerevisiae* and *Z. mobilis* of 120 h. After this period of cultivation, both species showed a decline in the production of biomass, finalizing treatment after 200 h of culture. The maximum values of biomass concentration were 14.18 and 11.80 g L^{-1} dry weight for *S. cerevisiae* and *Z. mobilis*, respectively. Both microorganisms grew satisfactorily under the culture conditions used in this study (**Figure 3A**), with a higher growth rate (μ) for *S. cerevisiae* (0.0547 d^{-1}) with respect to *Z. mobilis* (0.0418 d^{-1}). Growth rates in free cell cultures for both microorganisms *S. cerevisiae* and *Z. mobilis* were not significantly different ($P \geq 0.05$).

For immobilized cells, both yeast and bacteria presented immediate growth after adding the beads to the culture medium; in both treatments, the exponential phase of growth reached a maximum of 80 h. It is noteworthy that although both microorganisms were immobilized under the same procedure, the content of biomass per bead at the beginning of treatment was lower for *Z. mobilis* (0.0031 g $bead^{-1}$) compared to *S. cerevisiae* (0.0039 g $bead^{-1}$). Despite these differences, both microorganisms were able to tolerate immobilization (**Figure 3B**), reaching maximum biomass content values of 0.0055 and 0.0047 g $bead^{-1}$ for *S. cerevisiae* and *Z. mobilis*, respectively. In relation to growth, *Z. mobilis* showed a higher growth rate (0.142 d^{-1}) with respect to *S. cerevisiae* (0.106 d^{-1}), but there were no significant differences ($P \leq 0.0001$).

3.2. Glucose-substrate removal

The decrease of substrate showed significant differences ($P \leq 0.0001$) between treatments with free and immobilized cells for both species (**Figure 4**). However, the Tukey test analysis showed that the two species in free culture were not significantly different ($P > 0.05$) in 200 h of treatment. While for the immobilized and coimmobilized cell cultures, only the immobilized *Z. mobilis* bacteria showed no significant differences ($P = 0.245$) during removal of the substrate with the coimmobilized system during 140 h of culture (**Figure 4B**).

It is a fact that consumed substrate was greater in free culture for *S. cerevisiae* and *Z. mobilis* from 200 to 80 g L^{-1} (60% removal) after the 200-h treatment period (**Figure 4A**), compared to the immobilized system with 40% removal for *S. cerevisiae* (from 200 to 120 g L^{-1}) and 30% removal for *Z. mobilis* (from 200 to 140 g L^{-1}), while in those cultures of coimmobilized cells consumption ranged from 200 to 130 g L^{-1} (35% removal) (**Figure 4B**).

The average consumption analysis based on removal rates determined during the exponential growth for both species showed that free culture *S. cerevisiae* and *Z. mobilis* reached removal rates of 2.0 and 2.7 g-substrate per g-biomass d, respectively. This suggested greater productivity for the bacteria (5.76 g h^{-1}) with respect to yeast (5.29 g h^{-1}) (**Table 1**).

In cultures with immobilized cells, the removal rate in the exponential phase (80 h) was greater for *S. cerevisiae* (0.165 g-substrate per g-biomass d) with respect to *Z. mobilis* (0.056 g-substrate per g-biomass d), but in coimmobilized culture it was greater (0.235 g-substrate per g-biomass d) since both species contribute to reducing glucose and increasing the removal rate. Similar results were observed in the productivity, where the coimmobilized cell culture showed higher values (8.80 g L^{-1} h^{-1}) with respect to the immobilized cells of *S. cerevisiae* (8.45 g L^{-1} h^{-1}) and *Z. mobilis* (8.70 g L^{-1} h^{-1} (**Table 1**). In general, the highest productivity levels were recorded in coimmobilized and immobilized cultures with respect to free cell cultures because shorter ethanol production time (80 h) compared to free cultures (120 h).

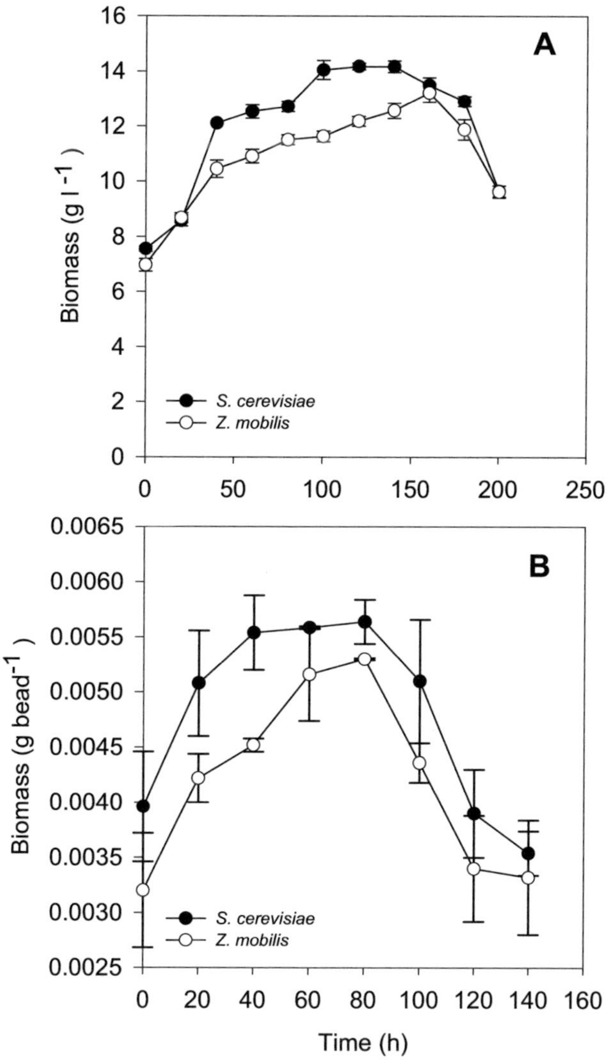

Figure 3. Average increase of biomass for *Saccharomyces cerevisiae* and *Zymomonas mobilis* in free culture (A) and immobilized cells (B).

3.3. Effect of initial concentration of glucose on ethanol production

It is a fact that most cultures from fruits may contain a high concentration of fiber solids that cause problems of mixture in the reactor, and consequently a low contact between cells and substrate. Mango juice is no exception. In this study, we evaluated the growth and alcohol production of *Z. mobilis* coimmobilized with *S. cerevisiae* in cultures with dilutions of 200 and 50 g L^{-1} of substrate in equivalent medium.

For the coimmobilized of *Z. mobilis* and *S. cerevisiae* within alginate beads, an immediate increase in biomass content was observed. Although the biomass content for both treatment

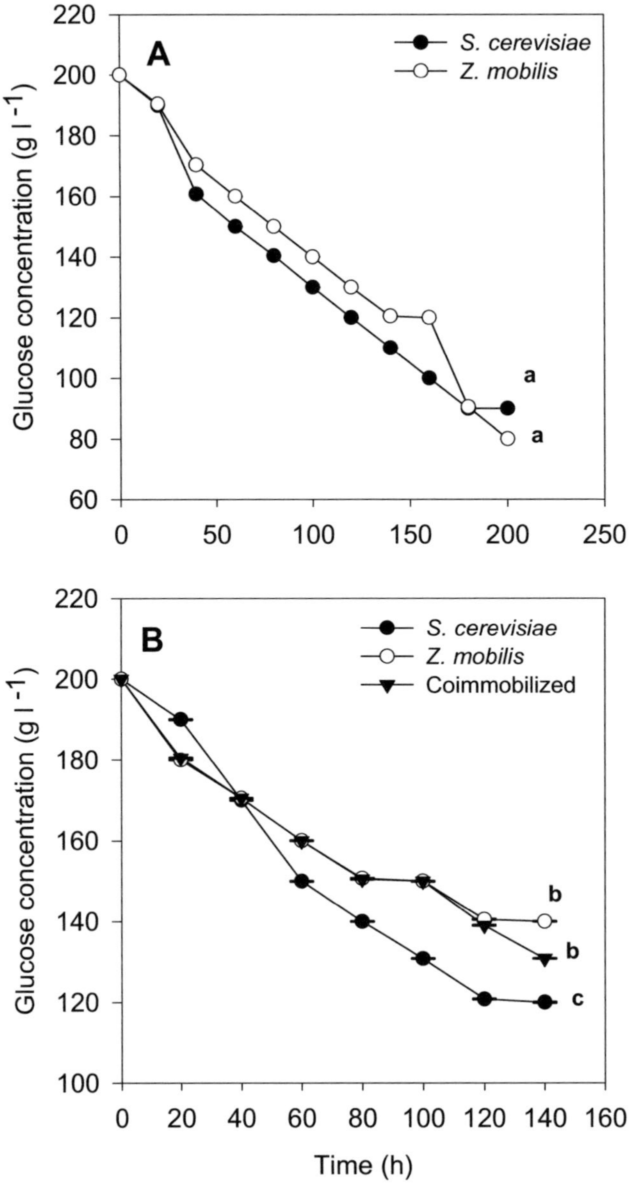

Figure 4. Average reduction of glucose (g L^{-1}) for *Saccharomyces cerevisiae* and *Zymomonas mobilis* in free culture (A) and immobilized culture (B). Different letters indicate significant differences ($P \leq 0.05$).

showed significant differences ($P \leq 0.0024$), the results suggest that the concentration of substrate was not a limiting factor for the growth for bacteria and yeast (**Figure 5**). The maximum biomass content in the treatment of glucose to 50 g L^{-1} (Gl_{50}) was obtained in the first 100 h of culture with about 0.0063 g $beads^{-1}$; while for the treatment of 200 g L^{-1} glucose (Gl_{200}) was of 0.053 g $beads^{-1}$ during a period of 80 h (**Figure 5**).

Culture	Cells	Ethanol formed (% v/v)	Y ($g\ L^{-1}\ h^{-1}$)	Glucose consumed ($g\ L^{-1}$)	Uptake rate (g-substrate removed per g-biomass d)	$\frac{moles\ Eth}{mole\ Glc}$
Free cells	*S. cerevisiae*	80.6	5.29[a]	90	2.71[a]	2.76
	Z. mobilis	87.7	5.76[a]	80	2.0[a]	3.38
Immobilized cells	*S. cerevisiae*	86.0	8.45[b]	60	0.165[b]	4.42
	Z. mobilis	87.7	8.70[b]	43	0.056[b]	6.29
Coimmobilized cells	*Z. mobilis* and *S. cerevisiae*	88.7	8.80[c]	47	0.235[c]	5.82

[a,b,c] Indicate significant differences ($P \leq 0.05$).

Table 1. Uptake rate, productivity (Y) and ethanol mole produced per glucose mole for *Saccharomyces cerevisiae* and *Zymomonas mobilis* in free culture, immobilized and coimmobilized.

The content of alcohol produced had no significant differences ($P \leq 0.05$) with respect to glucose concentration. However, uptake rates exhibit a decline as the glucose content in the reactor decreases (**Table 2**). The highest uptake rate occurred at a concentration of 200 $g\ L^{-1}$ glucose (0.235 g-substrate per g-biomass d) with a 76.5% removal, compared to 50 $g\ L^{-1}$ glucose (0.08 g-substrate per g-biomass d). Although the production of alcohol was similar in both treatments, the ratio mol-ethanol produced per consumed mol-glucose was higher in cultures of 50 $g\ L^{-1}$ glucose with a value of 6.91, with respect to 200 $g\ L^{-1}$ glucose with a ratio of 5.82 mol-ethanol produced per consumed mol-glucose (**Table 2**). Similarly, higher productivity was obtained (8.85 $g\ L^{-1}\ h^{-1}$) at a lower glucose concentration compared to a medium with high glucose content (8.80 $g\ L^{-1}\ h^{-1}$).

C_0	Ethanol formed (% v/v)	Y ($g\ L^{-1}\ h^{-1}$)	Glucose consumed ($g\ L^{-1}$)	Uptake rate (g-substrate removed per g-biomass d)	$\frac{moles\ Eth}{mole\ Glc}$
50	89.63	8.85	40.0	0.08	6.91
200	88.70	8.80	47.0	0.235	5.82

C_0: Initial concentration of glucose ($g\ L^{-1}$).

Table 2. Productivity (Y), uptake rate and ethanol mole produced per glucose with minimum content of glucose for coimmobilized *Z. mobilis* and *S. cerevisiase*.

4. Growth and productivity

To evaluate the capacity of growth in free and immobilized culture, two microorganisms, yeast and bacteria, were subjected to the same culture conditions (**Figure 3A**). In free culture, the yeast *S. cerevisiae* showed a higher cell density and specific growth rate (0.0547 d^{-1}) with respect to

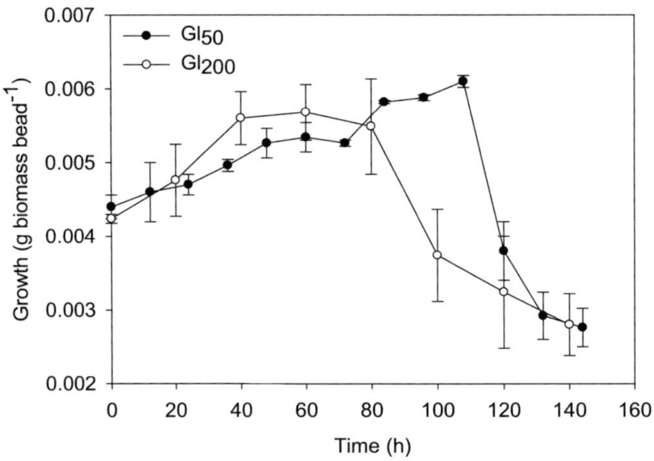

Figure 5. Growth (g biomass bead^{-1}) in coimmobilized system at different substrate concentrations.

bacteria *Z. mobilis* (0.0418 d^{-1}) in a treatment time of 120 h. The immobilized systems are known to have a greater capacity of cell growth and high metabolic activity [27, 28], which is consistent with the results obtained in this study. The result showed a high growth rates for immobilized *Z. mobilis* (0.142 d^{-1}) and *S. cerevisiae* (0.106 d^{-1}) with respect to free cell cultures, suggesting that immobilization did not affect growth in both microorganisms and increased biomass content favorably. Furthermore, the high activity in immobilized cell was observed with in a decrease of substrate in a shorter time of treatment (80 h) compared to free cell cultures (120 h). The short time of treatment for immobilized cell could be attributed to the increase of biomass within the beads and consequently an immediate decay of the substrate; however, this indicates that increasing cell population within the beads can cause a limited effects of nutrients on the cells located at the center of the beads, causing a decrease in cellular activity [28, 29].

Another factor that probably favors the rapid decline in cell density is attributed to the production of CO_2 as result of fermentation activity. Studies suggest the adverse effect of CO_2 gas, because if the diffusion of CO_2 is lower compared to its production, it will accumulate inside of alginate bead [30]. In this study, the CO_2 is observed in the reactor as bubbles attached on the surface of the beads, suggesting that the spread of CO_2 gas in the first 80 h was not a factor that inhibited growth and the production of alcohol; however, after this time, gas saturation in the reactor was probably high, affecting the diffusion of CO_2. This coupled with a limitation in the transport of nutrients and subsequent inhibition of microorganisms and may have caused glucose consumption to be lower compared to free cells (**Table 1**).

In particular, in free cell culture, the lower percentage of alcohol obtained by yeast during the increasing fermentation culture commonly relates to the fact that this is affected by the high concentration of ethanol in the solution, which may inhibit metabolism and decrease efficiency [31], unlike bacteria *Z. mobilis* [30]. In the present study, the lowest biomass produced by the bacteria (0.0047 g L^{-1}) with respect to yeast (0.0055 g L^{-1}) may be practical from the standpoint of waste generation. Similarly, observations were reported by Amin and Verachtert [9] for *Z. mobilis* and *Saccharomyces bayanus* immobilized in carrageenan with 5.6 and 9.9 g L^{-1}, respectively.

http://dx.doi.org/10.5772/intechopen.69991

It is evident that ethanol production was not inhibited in immobilized or coimmobilized systems, and even showed higher productivity with respect to free cells (**Table 1**), suggesting that they are more efficient in the conversion of sugar with respect to time. Krishnan et al. [19] reported lower productivity for *Z. mobilis* immobilized in carrageenan (1.6 g L^{-1} h^{-1}) compared to that obtained in this study (8.7 g L^{-1} h^{-1}); this difference may be attributed to the lower amount of glucose content in the culture medium of 32 g L^{-1} with respect to that used in the present study of 200 g L^{-1}.

Interestingly, the immobilized systems showed a higher conversion of substrate of 4.42 and 6.29 mole of ethanol per mole of glucose for yeast and bacteria, respectively, compared to the obtained by free cells, from 2.7 to 3.3 mole of ethanol per mole of glucose. In general, treatments with immobilized cells showed a higher output of ethanol per mole of glucose with respect to that reported by Amin and Verachtert [9] for *Z. mobilis* and *S. bayanus* immobilized in carrageenan with values of 1.8–1.9 mole of ethanol produced per mole of consumed glucose. Gunasekaran et al. [32] and Krishnan et al. [19] suggest that *Z. mobilis* is a good candidate to obtain alcohol with approximately 1.9 mole ethanol per mole of glucose; similarly, Rogers et al. [33] reported that specific productivity of ethanol (g ethanol g^{-1} biomass dry weight) is greater for *Zymomonas* than for *Saccharomyces uvarum*.

According to the results, immobilization and coimmobilization exhibited a lower uptake rate compared to free cells; this shows that there was less consumption of substrate (**Table 1**). Nevertheless, there was greater productivity, which indicates that it is possible to obtain high alcohol content with a lower requirement of substrate, but with the disadvantage of residual glucose in the medium; this problem can be solved with sequenced systems, as suggested by Demirci et al. [25]. Another alternative of solving this problem is to increase the cell number or inoculum size within the reactor. This is reasonable because a high number of cells could create a greater sorption of substrate (glucose) into the cell and eventually consumed substrate. However, Siripattanakul-Ratpukdi [10] suggested that with different cell yeast loads, the same reduction (>90%) of substrate is obtained at the end of a treatment period of 10 h.

The low glucose reduction observed in this study in alginate beads can be attributed to the decline in cell density, but it is likely that the diffusion of substrate could have been prevented. Studies have reported that the adsorption of substrate by the matrix was observed in the first hours of treatment, with a possible decrease of substrate diffusion within the matrix in a continuous process [34].

On the other hand, Robinson et al. [35] suggest that the diffusion rate within the alginate matrix depends on the concentration gradient between the culture medium and matrix; this is, when the nutrient concentration in the culture medium decreases, the diffusion rate occurs within the matrix and therefore the removal rate. In this study, during the first hours of treatment, the matrix is probable a partial saturation with substrate (glucose), because the substrate is decreased during the culture time, and cell growth for both microorganisms was continuous. Clearly, the immobilized cell system successfully decreased glucose by adsorption of the matrix (immobilized glucose) and biodegradation (bioconversion of glucose), being the main process the biodegradation. This suggests that the main factor that could limit glucose removal may have been the high concentration of CO_2 in the reactor.

4.1. Coimmobilization of *Z. mobilis* and *S. cerevisiae* at different glucose concentrations

The biomass content in alginate beads shows that a high concentration of glucose (200 g L^{-1}) leads to a rapid decrease compared to cultures with a low glucose concentration (50 g L^{-1}). This confirms the fact that the high concentration of glucose saturates beads faster, reducing the diffusion between beads and the culture medium; consequently, the diffusion of CO_2 produced can be reduced and remains trapped inside the bead, causing a decrease in growth and substrate consumption.

Conversely, the low concentration of substrate of the culture medium indicates the presence of a soft transport and substrate accumulation within the matrix, allowing a proper consumption and growth of bacteria and yeast. Therefore, a low concentration of substrate may actually increase the production of alcohol with minimal residual glucose, reaching values of 6.91 mol of ethanol per mole of glucose, with respect to a high glucose concentration (**Table 2**).

Previous studies in our laboratories have shown that the fermentation process of mango juice for a coimmobilized system can produce a production ratio of 1.4 L of alcohol (79% v/v ethanol) for every 3 L of mango juice.

5. Conclusions

The present study has shown the existing potential of using coimmobilized systems in the production of ethanol. The association of *Z. mobilis* and *S. cereviase* was positive, obtaining a higher ethanol content and high conversion of substrate compared to free and immobilized cells.

In general, the immobilization technology offers an alternative by increasing productivity and conversion of substrate compared to culture systems with free cells. In the present study, the immobilized systems showed high conversion capacity to obtain high alcohol content with a lower requirement of substrate.

The possible substrate inhibition was not a factor affecting cell growth in both organisms; it is clear that the immobilized cell system successfully reduced glucose by the matrix adsorption (immobilized glucose) and biodegradation (bioconversion of glucose), being biodegradation the main process. This suggests that the main factor that could limit further growth was the high concentration of CO_2 in the reactor. Furthermore, although no significant differences were detected in the alcohol content in immobilized culture in diluted medium, the conversion from glucose to ethanol is greater in those media with a glucose concentration of 50 g L^{-1}. For practical purposes, it is desirable that the fermentation of waste organic be performed through dilutions to increase the homogeneity of alginate beads within the reactor and consequently allow the diffusion of CO_2 and substrate through the beads.

http://dx.doi.org/10.5772/intechopen.69991

Author details

Alejandro Ruiz Marin*, Yunuen Canedo Lopez, Asteria Narvaez Garcia, Juan Carlos Robles Heredia and Jose del Carmen Zavala Loria

*Address all correspondence to: aruiz@pampano.unacar.mx

Environmental Sciences Laboratory, Faculty of Chemistry, Autonomous University of Carmen, Ciudad del Carmen, Mexico

References

[1] Mathews KC, Van Holde KE, Kevin GA. Biochemistry, San Francisco, California, editorial Addison Wesley, 3er edition, 2002: P.1368

[2] Beltran G, Torija MJ, Novo M, Ferrer N, Poblet M, Guillamon JM, Rozes N, Mas A. Analysis of yeast populations during alcoholic fermentation: A six year follow-up study. Systematic and Applied Microbiology. 2002;**25**(2):287-293

[3] Doelle HW, Kirk L, Crittenden R, Toh H, Doelle MB. *Zymomonas mobilis*—science and industrial application. Critical Reviews in Biotechnology. 1993;**13**(1):57-98

[4] Sheehan J, Himmel M. Enzymes, energy, and the environment: A strategic perspective on the U.S. Department of Energy's research and development activities for bioethanol. Biotechnology Progress. 1999;**15**(5):817-827

[5] Christen P, Minier M, Renon H. Ethanol extraction by supported liquid membrane fermentation. Biotechnology & Bioengineering. 1990;**36**:116-123

[6] Mejía-Giraldo LF, Martínez-Correa HA, Betancourt-Gutiérrez JE, Castrillón Castaño CE. Usage of the common mango agroindustrial waste (*Mangifera indica L.*) in the destruction of fermentable sugars. Ingeniería y Ciencia. 2007;**3**(6):41-62

[7] Amin GV, Eynde D, Verachtert E. Determination of by-products formed during the ethanolic fermentation, using batch and immobilized cell systems of *Zymomonas mobilis* and *Saccharomyces bayanus*. European Journal of Applied Microbiology and Biotechnology. 1983;**18**:1-5

[8] Sivers MV, Zacchi G, Olsson L, Hahn-Hagerdal B. Cost analysis of ethanol production from willow using recombinant *Escherichia coli*. Biotechnology Progress. 1994;**10**:555-560

[9] Amin G, Verachtert H. Comparative study of ethanol production by immobilized-cell systems using *Zymomonas mobilis* or *Saccharomyces bayanus*. European Journal of Applied Microbiology and Biotechnology. 1982;**14**:59-63

[10] Siripattanakul-Ratpukdi S. Ethanol production potential from fermented rice noodle wastewater treatment using entrapped yeast cell sequencing batch reactor. Applied Water Science. 2012. DOI: 10.1007/s13201-011-0024-z

[11] Tam NFY, Lau PS, Wong YS. Wastewater inorganic N and P removal by immobilized *Chlorella vulgaris*. Water Science & Technology. 1994;**30**:369-374

[12] Wu SY, Lin CN, Chang JS. Hydrogen production with immobilized sewage sludge in three-phase fluidized-bed bioreactors. Biotechnology Progress. 2003;**19**:828-832

[13] Kawaguti HY, Manrich E, Sato HH. Production of isomaltulose using *Erwinia sp*. D12 cells: Culture medium optimization and cell immobilization in alginate. Biochemical Engineering Journal. 2006;**29**:270-277

[14] Bandhyopadhyay K, Das D, Bhattacharyya P, Maiti BR. Reaction engineering studies on biodegradation of phenol by *Pseudomonas putida* MTCC 1194 immobilized on calcium alginate. Biochemical Engineering Journal. 2001;**8**:179-186

[15] Park DH, Chab JM, Ryua HW, Lee GW, Yu EY, Rheea JI, Park JJ, Kimd SW, Joe YI, Ryu YW, Hur BK, Park JK, Park K. Hydrogen sulfide removal utilizing immobilized *Thiobacillus sp*. IW with Ca-alginate bead. Biochemical Engineering Journal. 2002;**11**:167-173

[16] Humphries AC, Nott KP, Hall LD, Macaskie LE. Reduction of Cr (VI) by immobilized cells of *Desulfovibrio vulgaris* NCIMB 8303 and *Microbacterium sp*. NCIMB 13776. Biotechnology & Bioengineering. 2005;**90**(5):597-605

[17] Ruiz-Marin A, Mendoza-Espinosa GL, Stephenson T. Growth and nutrient removal in free and immobilized green algae in batch and semi-continuous cultures treating real wastewater. Bioresource Technology. 2010;**101**:58-59

[18] Vallejo-Becerra V, Vazquez-Bahena JM, Santiago-Hernández JA, Hidalgo-Lara ME. Immobilization of the recombinant invertase INVB from *Zymomonas mobilis* on nylon-6. Journal of Industrial Microbiology and Biotechnology. 2008;**35**:1289-1295

[19] Krishnan MS, Blanco M, Shattuck KC, Nghiem PN, Davison BH. Ethanol production from glucose and xylose by immobilized *Zymomonas mobilis* CP4 (pZB5). Applied Biochemistry and Biotechnology. 2000;**84-86**:525-541

[20] Yu J, Yue G, Zhong J, Zhang X, Tan T. Immobilization of *Saccharomyces cerevisiae* to modified bagasse for ethanol production. Renewable Energy. 2010;**35**:1130-1134

[21] Travieso L, Benitez F, Weiland P, Sánchez E, Dupeyrón R, Dominguez AR. Experiments on immobilization of microalgae for nutrient removal in wastewater treatments. Bioresource Technology. 1996;**55**:181-186

[22] De-Bashan LE, Moreno M, Hernández JP, Bashan Y. Removal of ammonium and phosphorus ions from synthetic wastewater by the microalgae *Chlorella vulgaris* coimmobilized in alginate beads with the microalgae growth promoting bacterium *Azospirillum brasilense*. Water Research. 2002;**36**:2941-2948

[23] Nigan JN. Continuous ethanol production from pineapple cannery waste using immobilized yeast cells. Journal of Biotechnology. 2000;**80**:189-193

[24] Kassim MA, Kheang LS, Bakar NA, Aziz AA, Som RM. Bioethanol production from enzymatically saccharified empty fruit bunches hydrolysate using *Saccharomyces cerevisiae*. Research Journal of Environmental Sciences. 2011;**5**:573-586

[25] Demirci A, Pometto AL, Ho KLG. Ethanol production by *Saccharomyces cerevisiae* in biofilm reactors. Journal of Industrial Microbiology & Biotechnology. 1997;**19**:299-304

[26] Tam NFY, Wong YS. Effect of immobilized microalgal bead concentrations on wastewater nutrient removal. Environmental Pollution. 2000;**107**(1):145-151

[27] Cohen Y. Biofiltration—the treatment of fluids by microorganisms immobilized into the filter bedding material. Bioresource Technology. 2000;**77**:257-274

[28] Mallick N. Biotechnological potential of immobilized algae for wastewater N, P and metal removal: A review. BioMetals. 2002;**15**:377-390

[29] Uemoto H, Saiki H. Distribution of *Nitrosomonas europaea* and *Paracoccus denitrificans* immobilized in tubular polymeric gel for nitrogen removal. Applied and Environmental Microbiology. 2000;**66**(2):816-819

[30] Kim BG, Choi CY. A study on the ethanol production by immobilized cells of *Zymomonas mobilis*. Korean Journal of Chemical Engineering. 1984;**1**(1):13-19

[31] Nuwamanya E, Chiwona-Karltun L, Robert SK, Yona B. Bio-ethanol production from non-food parts of Cassava (*Manihot esculenta* Crantz). Ambio. 2011;**41**(3):262-270

[32] Gunasekaran P, Karunakaran T, Kasthuribai M. Fermentation pattern of *Zymomonas mobilis* strains on different substrate: A comparative study. Journal of Biosciences. 1986;**10**(2):181-186

[33] Rogers PL, Jon JL, Tribe DE. Kinetics of alcohol production by *Zymomonas mobilis* at high sugar concentration. Biotechnology Letters. 1979;**1**:165

[34] Siripattanakul S, Khan E. Fundamentals and applications of entrapped cell bioaugmentation for contaminant removal. In: Shah V, editor. Emerging Environmental Technologies. Vol. 2. Berlin: Springer; 2010. pp. 147-169

[35] Robinson PK, Reeve JO, Goulding KH. Phosphorus uptake kinetics of immobilized *Chlorella* in bath and continuous-flow culture. Enzyme and Microbial Technology. 1989;**11**:590-596

Section 4

Bioremediation

Chapter 12

Yeast Biomass: An Alternative for Bioremediation of Heavy Metals

Wifak Bahafid, Nezha Tahri Joutey, Meryem Asri,
Hanane Sayel, Nabil Tirry and Naïma El Ghachtouli

Additional information is available at the end of the chapter

http://dx.doi.org/10.5772/intechopen.70559

Abstract

Heavy metal pollution has become one of the most serious environmental problems throughout the world. Among the innovative solutions for treatment of contaminated water and soil, bioremediation that use biological materials like living or dead microorganisms is a promising, safe and economical technology. One of the most ubiquitous biomass types available for bioremediation of heavy metals is yeast. Yeast cells represent an inexpensive, readily available source of biomass that retains its removal ability for a broad range of heavy metals to varying degrees. Furthermore, yeasts exhibit the ability to adapt to extreme conditions such as temperature, pH and high levels of organic and inorganic contaminants. To understand the different mechanisms of interactions between metals and yeast strains in the environment, this paper will give an overview on the role that yeasts play in the immobilization/mobilization of toxic metals and factors affecting these processes. Biotechnological applications in the bioremediation of heavy metal such as bioaugmentation using degradation abilities of yeasts will also be discussed.

Keywords: bioremediation, heavy metals, yeast, interaction mechanisms, bioaugmentation

1. Introduction

The industrialization has long been accepted as a hallmark of civilization. However, million tons of contaminating compounds such as toxic heavy metals (Cd, Cu, Hg, Pb, Mn, As, Ni, Zn, etc.) are produced and directly or indirectly released into the environment [1]. Unlike organic contaminants, these pollutants are not biodegradable and can be transferred through the food chain via bioaccumulation [2]. Actually, the build-up of dangerous concentrations of toxic metals in water sources and in grains and vegetables grown in contaminated soils is critically alarming due to the harmful effects of metals on human life and aquatic biota [1]. The actual challenge is to develop innovative and cost-effective solutions to decontaminate polluted environments and to protect the functioning of the ecosystems. Volesky [3] and Domenech [4] shortlisted some available conventional methods for removing the dissolved heavy metals including chemical precipitation, filtration, ion exchange, oxidation or reduction, reverse osmosis, evaporation, membrane technology, and electrochemical treatment. But most of these techniques become ineffective when the concentrations of heavy metals are less than 100 mg/L [5]. Additionally, strong and contaminating reagents are used for desorption, resulting in toxic sludge and secondary environmental pollution [1].

In order to minimize the effects of environmental pollution, the biological methods of metal removal such as bioremediation were considered. Different kinds of organisms isolated from contaminated soil, waste waters, compost and extreme environments are proved useful for bioremediation, from plants to microbes [6, 7]. Their success to survive in such a harsh environment can be attributed to metabolic possibilities allowing biological organisms to explore, detoxify and survive in exotic and complex substrates. The microbial metabolic diversity and versatility are two of the reasons why they are suitable as agents for remediation among many living organisms [8]. These microorganisms have evolved various measures to respond to heavy-metal stress via processes such as transport across the cell membrane, biosorption to cell walls and entrapment in extracellular capsules, precipitation, complexation and oxidation-reduction reactions [9–14].

One of the most ubiquitous biomass types available for bioremediation of heavy metals is yeast cells that retain their ability to accumulate a broad range of heavy metals to varying degrees under a wide range of external conditions [15]. Yeast can serve as a suitable model for studying physiological and molecular mechanisms of eukaryotic cell interactions with heavy metals. Studies on bacteria, yeast, fungi and microalgae showed that yeasts are better biosorbent for the removal of heavy-metal ions from wastewater due to their high growth rate and cell wall structure [16]. Indeed, cell wall phosphates and carboxyl groups have been reported to be the major determinant of negative yeast cell surface charge which enhances the ability of yeast cells to bind heavy-metal cations. This is likely due to electrostatic interactions [17]. It was also reported that the various metal-binding groups, viz., amine, imidazole, phosphate, sulfate, sulfhydryl and hydroxyl, are present in the polymers of the cell wall of fungi [18]. In addition, yeasts are larger than bacteria and can physically intertwine the mycelia/pseudomycelia to form floes. The net-structured yeast floc facilitates oxygen diffusion and eliminates the necessity of excessive dissolution of oxygen in water. Therefore using yeasts can give high efficient oxygen supply and reduce energy consumption by reducing the supplied air flow [19].

On the other hand, yeast biomass is an inexpensive, readily available source of biomass. It can be easily cultivated in cheap growth media, and it can be cheaply available in good quantities from fermentation industries for wastewater remediation [20, 21]. Many industrial waste-biomass types were investigated for their biosorptive potential. These include the yeasts, *Saccharomyces cerevisiae*, from the food and beverage industry and *Candida albicans*, a clinical isolate [14].

2. Bioremediation

Bioremediation is considered as an alternative processing method to reduce the environmental pollutants into less toxic forms [2]. It is defined as the process by which organic or inorganic wastes are biologically degraded or transformed usually to innocuous materials [22]. Mueller et al. also defined bioremediation as a process where organic wastes are biologically degraded under controlled conditions to an innocuous state or to levels below concentration limits established by regulatory authorities [23].

The major strategies for implementing bioremediation processes include biostimulation and bioaugmentation [22]. Biostimulation is the bioremediation process that can be enhanced by adding an electron acceptor, nutrient or other factors to a contaminated site with the objective of stimulating growth of the microbial population already present there [2]. When microorganisms are imported to a contaminated site to enhance degradation, we have a process known as bioaugmentation. Thus, the microorganisms used in bioremediation may be indigenous to a contaminated area, or they may be isolated from elsewhere. Recent studies show that microorganisms isolated from contaminated sites present high-tolerance adaptation of multiple environmental conditions and have excellent capability of removing significant amounts of metals [2, 24–26].

It was reported that yeasts and fungi are able to grow in matrices that have high concentrations of metal compounds compared to other microorganisms. In addition to their resistance, Ksheminska et al. [27] reported that yeast strains are capable of removing significant quantities of these pollutants. Other studies with yeasts showed also that, upon metal exposure, the main goal of the yeast cell is to protect and detoxify the environment by rendering the metal ions unavailable to promote cytotoxic effects. Furthermore, in comparison with bacteria which only use the active metabolizing capabilities, yeasts have the property of being used whether they are alive (metabolically active) or dead (metabolically inactive/passive) to remove these contaminants [24, 28, 29].

3. Yeast bioremediation mechanisms

Information about metal detoxification mechanisms in yeasts is considerably less available when compared to prokaryotes. The most studied yeasts mainly belong to the ascomycetous group, such as *S. cerevisiae*, *Schizosaccharomyces pombe* and *Candida* sp. [30]. The studies showed that yeasts evolved several different detoxifying mechanisms by which they can mobilize, immobilize or transform metals. The immobilization mechanisms include (i) biosorption, interaction of metals with the cell membrane via different processes such as ion exchange, complexation, crystallization, adsorption and precipitation; (ii) biotransformation, toxic metals are reduced to less toxic forms; and (iii) bioaccumulation, intracellular uptake of metal ions by living microorganisms [2]. The mobilization mechanisms involve mainly bioleaching through production and excretion of some acids which interact with metal ions to produce insoluble complex. Thus, in general the immobilization and mobilization are the two main techniques used for the bioremediation of metals by yeast and fungi (**Figure 1**) [31–33].

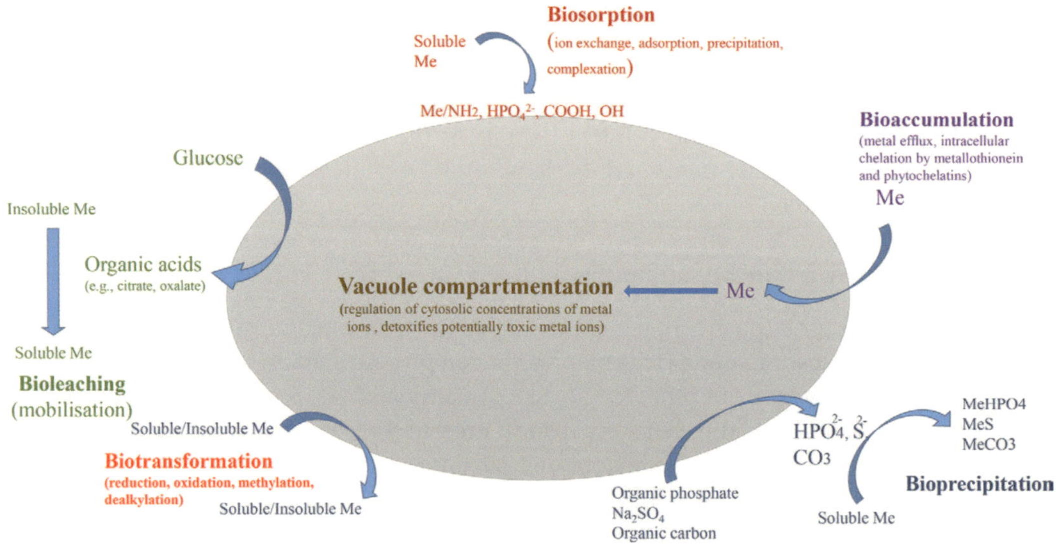

Figure 1. Interactions of metal and fungi cells (adapted from Refs. [31–33]).

3.1. Biosorption

Biosorption is a nondirected physico-chemical interaction that may occur between metal and cellular compounds of biological species [34]. It consists of the ability of biological materials to bind and concentrate heavy metals through metabolically mediated or physico-chemical pathways [35].

Among the promising types of biosorbent studied, yeast and fungal biomass seems to be good sorption materials with a sufficiently high metal-binding capacity and selectivity for heavy metals [36].

3.1.1. Yeast cell wall properties

The first stage of metal ion binding in microorganism cells does not depend on their metabolism and consists in ion chemisorption into cell wall components. Thus, the biosorption efficiency of heavy metals by microbial biomass is mainly connected with the structure of the microorganism cell wall and consequently with cell surface properties in which structure determines the interaction nature between micro-organism and metal cation [36–38].

Yeast cell walls are negatively charged, and the ability of yeast cells to bind heavy-metal cations is likely due to electrostatic interactions [39]. Indeed, heavy metals can be biosorbed by microbes at binding sites present in cellular structure without the involvement of energy. Among the various reactive compounds associated with cell walls, the extracellular polymeric substances such as exopolysaccharide (EPS) are well known to have a considerable effect on acid–base properties and a great ability to complex heavy metals [40]. The structure and the distribution of homopolysaccharides (mannans and glucans), single saccharides and acid components, which are good binding agents, also dictate the cell wall's biosorption capacity [41].

3.1.2. Mechanisms of biosorption

It is likely that various mechanisms can be involved in biosorption and can operate simultaneously to various degrees. Interaction of metal with yeast cell wall involves a complex mechanism that includes several processes such as ion exchange, complexation, adsorption and precipitation [2, 42]. Many evidences have proved that ion exchange mechanism exists in biosorption system [38, 43, 44]. However, it was suggested by many researchers that ion exchange is neither the sole nor the main mechanism for metal biosorption [45]. Ion exchange is the replacement of an ion in a solid phase in contact with a solution by another ion. More specifically, it is the replacement of an absorbed, readily exchangeable ion by another [46]. Rapid release of 70% of cellular K^+, followed by a slower release of approximately 60% of cellular Mg^{2+}, but little loss of Ca^{2+}, was observed in Cu^{2+} removal by *S. cerevisiae* [47], indicating the existence of an ion exchange mechanism. Chen and Wang [48] also reported that *S. cerevisiae* acts as a biosorbent for the removal of Zn (II) and Cd (II) through the ion exchange mechanism. According to Vasudevan et al. [49], the release of Ca^{2+}, Mg^{2+} or H^+ was also observed in the biosorption process of heavy-metal ions (Sr^{2+}, Mn^{2+}, Zn^{2+}, Cd^{2+}, Cu^{2+}, Ti^+) by living, non-metabolizing cells of *S. cerevisiae,* which also confirmed the existence of ion exchange. Although it is a simple concept, in reality, ion exchange can be a mechanistically highly complex process depending on the system [50, 51].

Metal precipitation is also involved in biosorption. The term precipitation in most cases refers to the formation of insoluble inorganic metal precipitates [52]. This may be more easily understood when metals are bound to extracellular polymeric substances excreted by eukaryotic microorganisms such as yeast and fungi. The precipitates may be formed and remain in contact with or inside the microbial cells or may be independent from the solid phase of the microbial cell. Some researches proved that purified biomolecule products from isolated cells such as glucan, mannan, and chitin accumulate greater quantities of cations than the intact cells and can form metal precipitates [53].

Several yeast species such as *S. cerevisiae, Pichia anomala, Candida tropicalis, C. albicans,* and *Cunninghamella elegans* emerged as a promising sorbents against heavy metals [54]. **Table 1** summarizes some of the important results of metal biosorption using these yeast biomasses.

3.1.3. Factors affecting yeast biosorption capacity

It is noted that the same yeasts species have different biosorption capacities for the same metal ion. It was shown that different yeasts species present different cell surface properties and cell wall compositions, which brings about a differentiation in biosorption ability, affinity and interaction specificity. Moreover, biosorption depends on many factors that are related to the biomass or to environmental conditions. Indeed, Nguyen et al. [68] studied the polysaccharide composition of the cell walls of several yeast species, such as *Debaryomyces hansenii, Zygosaccharomyces bailii* and *S. cerevisiae,* and results indicated that the cell wall composition varied over the species and strains. Growth, nutrition and age of the biomasscan also influence the variability of cell size, cell wall composition and extracellular product formation [69]. It was found that the cell wall composition and polysaccharide content could vary by more than 50% with the nature of the carbon source, nitrogen limitation, temperature and aeration

Biomass type	Metal studied	Initial concentration of metal studied (mg/L)	Biosorption capacity (mg/g) or % removal	Biomass concentration (g/L)	References
S. cerevisiae	Hg	25–200	76.20	2	[55]
S. cerevisiae	Pb		67–82%		[56]
	Cd		73–79%		[56]
S. cerevisiae	Cd	169	5.96		[57]
Schizosaccharomyces pombe	Ni	400	33.8		[58]
S. cerevisiae	Cr(III)	200	35.00		[59]
S. cerevisiae	Cr(VI)	150	120	0.5	[60]
S. cerevisiae	Pb		270.30	10	[61]
S. cerevisiae	Zn		23.40	1	[62]
S. cerevisiae	Hg		64.20	1	[62]
S. cerevisiae	Cd		15.60		[63]
S. cerevisiae	Pb		17.50		[63]
S. cerevisiae	Pd		40.60		[64]
S. cerevisiae	Cr(VI)		32.60	10	[61]
S. cerevisiae	Ni		46.30	10	[61]
S. cerevisiae sp.	Zn		100%	0.4	[65]
S. cerevisiae	Cd		95%	0.4	[65]
C. pelliculosa	Cu	100	95.04%	13.39	[66]
C. utilis	Cd	50	81.46	3	[67]
C. tropicalis	Cr	100	29.10	2	[24]
W. anomalus	Cr	100	28.14	2	[24]
C. fabianii	Cr	100	18.90	2	[24]

Table 1. Some data on the biosorptive capacities of yeasts for different metal ions reported in literatures.

and pH and with the mode of cell cultivation [70]. Temperature is considered as one of the important factors in the biosorption process. It was reported that adsorption reactions are mostly exothermic and the extent of adsorption augment with decreasing temperature [32]. Thus, a better biosorption capacity for Ni and Pb by *S. cerevisiae* was observed at a low temperature (25°C) and found to diminish as the temperature was increased to 40°C [71].

Metal biosorption was also frequently shown to be strongly pH dependent. Sulaymon et al. [35] indicated that pH of solution affects the solution chemistry of the metals, the activity of the functional groups in the biomass and the competition of metallic ions. Some studies showed that yeast cells of *S. cerevisiae* are able to remove heavy-metals between pH 5.0 and

9.0, being pH close to 5–6 as the optimal for Cu^{2+}, Cd^{2+}, Pb^{2+}, Ni^{2+}, Zn^{2+} and Cr^{3+} biosorption by yeast cells [47, 72]. However, other studies demonstrated that biosorption of metals like Cu, Cd, Ni, Co and Zn is often reduced at low pH values [73, 74]. Generally, the heavy metal uptake for most of the biomass types declines significantly when pH of the metal solutions is decreased from pH 6.0 to 2.5 [72]. The cell surface hydrophobicity may also affect biosorption capacity, facilitating hydrophobic bonds. Edyta [75] has reported that a lower relative hydrophobicity and a higher negative surface charge can be related to better availability of polar/charged groups such as carboxyls and mannosylphosphates on yeast cell surface.

3.1.4. *Industrial application*

Biosorption process has not only been used in laboratory scale (batch and column) but also in pilot plant-scale studies. Tigini et al. [76] tested this process using *C. elegans* biomass in 200 L of pilot plant installation for dye removal. Obtained results showed that biosorption was an effective process for the removal of different pollutants from spent baths and wastewaters [76].

The apparent exploitation potential of biosorption is often cited in the literature, and the commercial applications of biosorption are currently available on the market. The biological materials can be commercialized as powdered biosorbents, e.g. BIO-FIX® (sphagnum, peat moss, algae, yeast, bacteria and aquatic flora immobilized in polysulfone), MetaGeneR and RAHCO Bio-Beads [77]. All the preparations are suitable for the treatment of wastewaters from metallurgical industry or mining operations. The future possible applications may concern using biosorption in the separation and purification of high value molecules, e.g. high-value proteins, steroids, pharmaceuticals or antibodies [77].

3.2. Bioaccumulation

Bioaccumulation is defined as the uptake of toxicant by living cells and their transport into the cell [78]. It is a growth-dependent process mediated only by living biomass [79]. The mechanism of intracellular uptake is more complex than biosorption itself and is not fully understood yet. Generally, the process is regarded as a two-step process [80]. The first step called passive biosorption proceeds rapidly within several minutes by any one or a combination of the following metal-binding mechanisms: coordination, complexation, ion exchange or physical adsorption (e.g. electrostatic). The metal ions are adsorbed to the surface of cells by interactions between metal-functional groups displayed on the surface of cells. The second step which slowly takes place is that metal ions penetrate the cell membrane and enter into the cells. Raspor et al. [81] pointed out that after initial rapid biosorption on cell walls, active transport mechanisms into the cells take place and metal ions penetrate the cell membrane and enter into the cells. Metal ions transported across the cell membrane, are transformed to other species or precipitated within the cell by active cells, including transportation.

Metal accumulation strategies for essential and non-essential metal ions may be different. Literature in model yeasts indicates that upon non-essential metal exposure, cytoplasmic detoxification is the major strategy [30]. This cytoplasmic detoxification can be achieved by metal transport to the outside of the cell or to less sensitive cellular compartments, making the metal ions unavailable to promote cytotoxic effects [82]. Thus, microorganisms developed different mechanisms including cell membrane metal efflux [83], intracellular chelation by metallothionein proteins and glutathione-derived-peptides called phytochelatins [84, 85] as well as metal compartmentalization in vacuoles [3], but the exact mechanism of intracellular accumulation is not elucidated.

In general, metals enter yeast cells by dedicated transporters, which are often target of negative regulation when the specific metal is present in excess intracellularly. In the past decades, a large number of references on metal ion transport in the baker's yeast *S. cerevisiae* were published [86, 87]. Baker's yeast was found to possess two or more substrate-specific transport systems to accumulate any single metal ion, and a large number of yeast genes that function in metal ion transport or its regulation were confirmed [87]. After entering into the cell, the metal ions react with thiol groups present in Cys residues such thiolated peptides include the glutathione (GSH), phytochelatins (PC), and the metallothioneins (MT) [82]. The resulting metal-thiolated peptide complexes may be used as a substrate for metal(loid) extrusion to the outside of the cell, or for accumulation in cellular compartments such as the vacuole. However, thiolated peptides can be produced to chelate metals, reducing their reactivity and availability to the cells [30].

The most described yeast MT is the Cup1 of *S. cerevisiae*, which is mainly associated with Cu detoxification [88]. Several authors attributed Cd, in addition to Cu, detoxification to chelation by this MT [88, 89]. In *S. cerevisiae*, GSH synthesis was also described. In comparison with a wild-type strain of *S. cerevisiae*, the mutant strain displayed a high sensitivity to arsenite, selenite and cadmium which confirms the role of GSH on detoxification of these metals.

The role of the vacuole in the detoxification of metal ions was also investigated and a large number of researches signaled that fungal vacuole plays an important role in molecular degradation, storage of metabolites and regulation of cytosolic concentrations of metal ions and detoxifies potentially toxic metal ions. The results showed that vacuole-deficient strain displayed much higher sensitivity and decreased large uptake of As, Zn, Mn, Co and Ni [90, 91]. Avery and Tobin [92] also confirmed that Sr^{2+} accumulated mainly stays in the vacuole of the living yeast cell of *S. cerevisiae*.

The active mode of metal accumulation by living cells is dependent on structural properties, physiological and genetic adaptation, environmental modification of metal specification, availability and toxicity [93]. The process also depends of several factors (which are almost identical to the factors influencing the cultivation of an organism): the composition of the growth medium, pH, temperature, the presence of other pollutants or other inhibitors, surfactants, etc. [94]. Metabolic activities such as respiration, nutrient uptake, and metabolite release will alter the microenvironment around the cells which, in turn, may affect mechanisms involved in bioaccumulation (adsorption, ion exchange, complexation and precipitation) [69].

In yeasts, heavy metals can be accumulated by bioaccumulation process more than biosorption. But the biosorption process seems to be more feasible for large scale application compared to the bioaccumulation process, because microbes will require addition of nutrients for their active uptake of heavy metals [27]. Thus, the biosorptive capacity of yeast was studied extensively in comparison to bioaccumulation.

3.3. Bioreduction

The detoxification of metal ions can also be realized by oxidation or reduction. When reduction of a metal to a lower redox state occurs, mobility and toxicity can be reduced, thus offering potential bioremediation applications [2].

Information available for metal detoxification, such as reduction mechanisms in yeasts, only considers neutrophilic yeasts and there is considerably less information available when compared to prokaryotes. Indeed, for eukaryotic microbial cells and primarily yeasts, the data on the metal-reducing systems are more ambiguous. It is generally unknown what system enzymatic or non-enzymatic and intracellular or extra-cellular plays a leading role in the chromate detoxification process [9].

Tamás and Wysocki [82] proved that one mechanism of detoxification of As(V) was the reduction of As(V) to As(III), a process catalysed by arsenate reductase enzymes. The removal of toxic hexavalent chromium from aqueous solution by biosorption by different biomass types was as well extensively reported [9]. Cr(VI) can be reduced as a powerful oxidative agent to Cr(III) by cellular-reducing systems that can include enzymatic and non-enzymatic pathways.

The intracellular reduction of Cr(VI) to Cr(III) is known to be the main detoxification mechanism of chromium. In aerobic condition, microbial reduction of Cr^{6+} is catalysed by soluble enzymes (chromate reductase) [95]. Many yeasts like *Cyberlindnera fabianii, P. anomala, Rhodotorula pilimanae* D-76, and *Pichia guilliermondii* ATCC 201911 were known for their enzymatic reduction ability of Cr^{6+} to Cr^{3+} [27, 28, 96, 97].

Cr(VI) removal may also be associated with its simultaneous reduction to Cr(III). It was reported that Cr(VI) can be reduced to Cr(III) through a redox reaction unrelated to any enzyme activity [98, 99]. Thus inactivated biomass, e.g. *S. cerevisiae, C. tropicalis, P. anomala* and *Penicillium chrysogenum*, removes Cr(VI) from aqueous solutions by reduction to Cr(III) when contacted with the biomass [96, 100]. Glutathione and cysteine can be considered as the most powerful non-enzymatic chromate reductants for microbial cells and ascorbate for higher organisms [2, 101]. Therefore, the removal of Cr(VI) from aqueous solution by dead cells may involve two mechanisms: (i) direct reduction, in the aqueous phase, Cr(VI) is directly reduced to Cr(III) by contact with the electron-donor groups of the biomass which has lower reduction potential values than that of Cr(VI) (+1.3 V), and (ii) indirect reduction, which includes three steps—(a) adsorption of Cr(VI) anionic species to the biomass surface containing the positively charged groups, (b) reduction of Cr (VI) to Cr(III) by adjacent electron-donor groups and (c) release of Cr (III) ions in the aqueous phase due to the electronic repulsion between the Cr (III) ions and positively charged groups, or adsorption of Cr(III) with adjacent groups.

3.4. Bioleaching

In the context of bioremediation, immobilization processes such as biosorption, accumulation and precipitation may enable metals to be transformed in situ into insoluble forms. They are particularly applicable to remove metals from mobile phases such as ground waters and leachates. In contrast, the process of metal solubilization provides a way to remove metals from soils and sediments by leaching. Generally bioleaching is a process described as being "dissolution of metals from their mineral source by certainly and naturally occurring microorganisms". However, there are some slight differences in definition: Usually, "bioleaching" is described as the conversion of solid metal values into their water soluble forms using microorganisms [102].

In the soil environment, metals can be held on inorganic soil constituents through various sorption or ion exchange reactions, complexed with soil organic materials or precipitated as pure or mixed solids [103]. However, in most acidic soils, metals may be speciated into more mobile forms [104]. It was reported that in such locations, fungi are the most predominating and often comprise the largest biomass suggesting its intervention in metal solubilization [33, 104]. Gadd [33] confirmed that microbes interact with metal and mineral in natural and synthetic environments, altering their physical and chemical state. Therefore, the biochemical activity of fungi and other microorganisms can affect metal speciation and mobility in the soil, modifying their biogeochemical cycles. The most important mechanisms of metal and minerals solubilization by fungi are acidolysis and complexolysis. It was revealed that some excreted metabolites with metal-complexing properties, e.g. phenolic compounds, and organic acids may be involved in metals solubilization [105]. Gadd [104] also indicated that low molecular weight organic acids, such as citric and oxalic acid, are the most important chemical fungal and yeast products and they were used for heavy metal solubilization. In fact, previous studies showed that many metal citrates are highly mobile. They also proved that oxalic acid can act as a leaching agent for those metals, which forms soluble oxalate complexes, including Al and Fe [106]. It was shown that organic acids provide both source of protons for solubilization and metal-chelating anion to complex the metal cation [107]. They have the double function: (i) to acidify the substrate, thus enhancing ions solubility, and (ii) to form complexes with solubilized ions, which leads into mobilizing them [104, 105].

Yeast doesn't show a broad area of examples in the field of bioleaching and the available information about the metal extraction from solid substrates using yeasts is limited. Some yeasts associated with bioleaching are *Rhodotorula rubra* and *Rhodotorula mucilaginosa*. Studies of yeast *R. mucilaginosa sp. lm*9 isolated from Kupferschiefer black shale showed that organic acids (malic and oxalic) produced by this strain can effectively mobilize copper from sedimentary rock [108]. Marcinčáková et al. [109] also demonstrated that 0.17% of lithium can be recovered from lepidolite by microbial leaching using the heterotrophic microorganism of *R. rubra*.

The bioleaching process is carried out at the mine site. It is indicated that the bioleaching process can be used to process low-grade ores and arsenic-containing ores that could not be processed effectively by high temperature smelting. Two types of bioleaching processes exist: bioleaching in stirred tank reactors (STR), which is dedicated to high-grade ores due to the relatively high costs of investments, and heap leaching, which is dedicated to low-grade ores. Bench scale columns were also used with ore in Australia in the period from 1964 to 1968 [110].

There are many and various applications of bioleaching. It can be applied to a wide variety of base-metal sulfides, mainly in large operations located in many countries and several interesting projects were conducted in order to develop this technology. The earliest commercial applications of the process involved in situ leaching of uranium in Canada, heap bioleaching of copper in Toromocho and dump leaching of copper in the United States (bioleach, Chili). Nowadays, the production of copper from low-grade ores is the most important industrial application. Early application of bioleaching to copper mining was centered in its recovery from heap or dump/stockpile [111, 112]. In these operations, copper is extracted from ores containing minerals: secondary copper sulfides such as covellite (CuS), chalcocite (Cu_2S) and bornite (Cu_5FeS_4) and the primary copper sulfide, chalcopyrite ($CuFeS_2$).

As well as detoxification of pollutant metals and copper production, the recovery of precious metals such as gold is also a potential area for exploitation. Industrial-scale bioleaching of refractory gold concentrates was practiced in South Africa, Brazil, Australia, Ghana, Peru, China and Kazakhstan. In this case, the process is used to leach sulfide minerals such as pyrite (FeS_2) and arsenopyrite ($FeAsS$), which encapsulate microscopic and sub-microscopic gold particles. By dissolving these sulfide minerals, the gold particles are exposed and can be recovered by further treatment [113].

Although yeasts have a high potential for bioleaching, there are no studies on the use of these microorganisms in bioleaching projects. So, it would be interesting to develop this technology using yeasts by testing their capacity on a large scale.

4. Biotechnological approaches: bioaugmentation

Bioaugmentation is one of the promising techniques of bioremediation; it is referring to the process of adding selected microbial strains or mixed cultures to biological waste treatments or contaminated sites in order to enhance effectively the removal of specific pollutants [114]. The rationale for this approach is that indigenous microbial populations may not be able to degrade the wide range of potential substrates present in complex mixtures [115] or when the indigenous pollutant-degrading population is low. Thus, the acceleration of decontamination is the primary advantage for the introduction of microorganisms.

The approach of this technology is taking advantage of microbial consortia designed for specific physico-chemical properties of the bioprocess [116]. This remediation way was shown to be more efficient than using undefined inocula [114]. However, one of the difficulties of bioaugmentation processes is the presence of many uncharacterized organisms that enter in competition with introduced populations. It was reported that some foreign microorganisms (those in inocula) were applied successfully in laboratory, but their efficiency depends on their ability to compete with indigenous microorganisms, predators and various abiotic factors. Thus, successful bioaugmentation treatments depend on the use of inocula consisting of microbial strains or microbial consortia well adapted to the site to be decontaminated [117].

Bioaugmentation was carried out using different microorganisms. The yeast-based bioaugmentation was specifically shown to be an advantageous soil and water clean-up approach for contaminated sites, containing heavy metals and/or organic pollutants [118–120].

Many isolated and laboratory-qualified microorganisms were reviewed, but they are not valid in situ yet. The soil bioaugmentation with *C. fabianii* removed more than 60% of a soil's chromium contamination of 40 mg.Kg^{-1}. It has reduced by the way its phytotoxic effects on *Phaseolus vulgaris L.* and promoted its growth under chromium stress [121]. In another work, both alive and dead *C. tropicalis* biomass showed a great ability to significantly reduce the bioavailability of 40 mg.Kg^{-1} of Cr(VI) in soils (up to 58.7 and 72.25% of reduction, respectively) [120]. In that work, clover plants were used as bioindicator where significant increase in seed germination and growth of seedlings was detected in the inoculated soils by *C. tropicalis* cells. In a further work, the bioaugmentation with the specific yeast strain *C. tropicalis SK21*showed a great efficiency for the bioremediation of petroleum-contaminated soil [119] where 96 and 42% of total petroleum hydrocarbons (TPH) were degraded by the strain at the initial diesel oil concentrations of 0.5 and 5% (v/v), respectively. The bioaugmentation of acidic oily sludge-contaminated soil with *Candida digboiensis* showed its great ability to degrade the acidic petroleum hydrocarbons under laboratory and field conditions [118]. From an application perspective, the bioaugmentation using microbial consortia rather than pure cultures is surely more advantageous [122–125]. It provides divers metabolic pathways and robustness required for field applications.

The application of bioaugmentation in different countries around the world was extensively reported using bacteria. Numerous works suggested the use of yeast strains as potential tools for bioaugmentation process [117]. Nevertheless, an impending gap between laboratory trials and on-field studies was detected. Hence, further efforts should be deployed by scientific community for a higher-scale application. A successful industrial application of bioaugmentation requires the application of strategies customized for the specific environmental conditions of the contaminated sites [124]. Thus, many studies were carried out to take advantages of this treatment technology for industrial application, as it represents an economical and environmental friendly remediation way [126].

Acknowledgements

The authors are grateful for the financial and scientific support rendered by Microbial Biotechnology Laboratory of Faculty of Sciences and Innovation City, Sidi Mohamed Ben Abdellah University, Fes, Morocco.

Author details

Wifak Bahafid, Nezha Tahri Joutey, Meryem Asri, Hanane Sayel, Nabil Tirry and Naïma El Ghachtouli*

*Address all correspondence to: naima.elghachtouli@usmba.ac.ma

Microbial Biotechnology Laboratory, Faculty of Sciences and Techniques, Sidi Mohamed Ben Abdellah University, Fez, Morocco

References

[1] Anushree M. Metal bioremediation through growing cells. Environmental International. 2004;**30**:261-278

[2] Siddique S, Rovina K, Al Azad S, Naher L, Suryani S, Chaikaew P. Heavy metal contaminants removal from wastewater using the potential filamentous fungi biomass: A review. Journal of Microbial and Biochemical Technology. 2015;**7**:384-393. DOI: 10.4172/1948-5948.1000243

[3] Volesky B. Advances in biosorption of metals: Selection of biomass types. FEMS Microbiology Review. 1994;**14**:291-302

[4] Domenech X. Quimica Ambiental, El Impacto Ambiental de los Residous. Madrid: Miraguano; 1998. p. 127

[5] Ahluwalia SS, Goyal D. Microbial and plant derived biomass for removal of heavy metals from wastewater. Bioresource Technology. 2007;**98**:2243-2257

[6] Kolvenbach BA, Helbling DE, Kohler HP, Corvini PF. Emerging chemicals and the evolution of biodegradation capacities and pathways in bacteria. Current Opinion in Biotechnology. 2014;**27**:8-14

[7] McGenity TJ. Hydrocarbon biodegradation in intertidal wetland sediments. Current Opinion in Biotechnology. 2014;**27**:46-54

[8] Trama B, Fernandes JDS, Labuto G, Franco de Oliveira JC, Viana-Niero C, Pascon RC, Vallim MA. The evaluation of bioremediation potential of a yeast collection isolated from composting. Advances in Microbiology. 2014;**4**:796-807

[9] Tahri Joutey N, Sayel H, Bahafid W, El Ghachtouli N. Mechanisms of hexavalent chromium resistance and removal by microorganisms. Reviews of Environmental Contamination and Toxicology. 2015;**233**:45-69. DOI: 10.1007/978-3-319-10479-9_2

[10] Avery SV, Tobin JM. Mechanism of adsorption of hard and soft metal ions to *Saccharomyces cerevisiae* and influence of hard and soft anions. Applied and Environmental Microbiology. 1993;**59**:2851-2856

[11] Brady D, Duncan JR. Bioaccumulation of metal cations by *Saccharomyces cerevisiae*. Applied Microbiology and Biotechnology. 1994;**41**:149-154

[12] Brady D, Stoll A, Duncan JR. Biosorption of heavy metal cations by non-viable yeast biomass. Environmental Biotechnology. 1994;**15**:429-438

[13] Krauter P, Martinelli R, William K, Martins S. Removal of Cr^{4+} from ground water by *S. cervisiae*. Biodegradation. 1996;**7**:277-286

[14] Veglio F, Beolchini F. Removal of metals by biosorption: A review. Hydrometallurgy. 1997;**44**:301-316

[15] Abbas BA, Badr SQ. Bioremediation of some types of heavy metals by *Candida* spp. International Journal of Engineering and Technical Research. 2015;**3**:2454-4698

[16] Ram CB, Nelly G, Nevena L. Bioremediation of chromium ions with filamentous yeast *Trichosporon cutaneum* R57. Journal of Biology and Earth Sciences. 2012;**2**:70-75

[17] Kordialik-Bogacka E. Studies on surface properties of yeast cells during heavy metal biosorption. Central European Journal of Chemistry. 2011;**9**:348-351

[18] Crist RH, Oberholser K, Shank K, Nguyen M. Nature of bonding between metallic ions and algal cell walls. Environmental Science and Technology. 1981;**15**:1212-1217

[19] Wongkarnka M. The application of aerobic yeast for treatment of high strength food processing wastewater containing furfural. [thesis]. Iowa State University Ames; 2005

[20] Ertugrul S, San NO, Dönmez G. Treatment of dye (Remazol blue) and heavy metals using yeast cells with the purpose of managing polluted textile wastewaters. Ecological Engineering. 2009;**35**:128-134

[21] Sanghi R, Sankararamakrishnan N, Dave BC. Fungal bioremediation of chromates: Conformational changes of biomass during sequestration, binding, and reduction of hexavalent chromium ions. Journal of Hazardous Materials. 2009;**169**:1074-1080

[22] Huang JP, Huang CP, Morehart AL. Removal of heavy metals by fungal (*Aspergillus oryzae*) adsorption. In: Vernet JP, editor. Heavy Metals in the Environment. London: Elsevier; 1991

[23] Mueller JG, Cerniglia CE, Pritchard PH. Bioremediation of environments contaminated by polycyclic aromatic hydrocarbons. In: Bioremediation: Principles and Applications. Cambridge University Press: UK; 1996. p 125-194.

[24] Bahafid W, Tahri Joutey N, Sayel H, Iraqui-Houssaini M, El Ghachtouli N. Chromium adsorption by three yeast strains isolated from sediments in Morocco. Geomicrobiology Journal. 2013;**30**:422-429

[25] Sayel H, Bahafid W, Tahri Joutey N, Derraz K, Fikri Benbrahim K, Ibnsouda Koraichi S, El Ghachtouli N. Cr(VI) reduction by enterococcus gallinarum isolated from tannery waste-contaminated soil. Annals of Microbiology. 2012;**62**:1269-1277

[26] Tahri Joutey N, Bahafid W, Sayel H, El Abed S, El Ghachtouli N. Remediation of hexavalent chromium by consortia of indigenous bacteria from tannery waste contaminated biotopes in Fez, Morocco. International Journal of Environmental Studies. 2011;**686**:901-991

[27] Ksheminska H, Fedorovych D, Honchar T, Ivash M, Gonchar M. Yeast tolerance to chromium depends on extracellular chromate reduction and Cr(III) chelation. Food Technology and Biotechnology. 2008;**46**:419-426

[28] Bahafid W, Tahri Joutey N, Sayel H, EL Ghachtouli N. Mechanism of hexavalent chromium detoxification using *Cyberlindnera fabianii* yeast isolated from contaminated site in Fez Morocco. Journal of Materials and Environmental Science. 2013;**4**:840-847

[29] Brierley CL. Bioremediation of metal-contaminated surface and groundwater. Geomicrobiology Journal. 1990;**8**:201-223

[30] Assis Fidalgo CI. Heavy metal resistance in extremophilic yeasts: A molecular and physiological approach. [thesis]. Faculdade de Ciências: Universidade de lisboa; 2011

[31] Nancharaiah YV, VenkataMohan S, Lens PNL. Biological and bioelectrochemical recovery of critical and scarce metals. Trends in Biotechnology. 2016;**34**:137-155

[32] Valls M, de Lorenzo V. Exploiting the genetic and biochemical capacities of bacteria for the remediation of heavy metal pollution. FEMS Microbiology Review 2002;**26**:327-338.

[33] Gadd GM. Interactions of fungi with toxic metals. The New Phytologist. 1993;**124**:25-60

[34] Shumate ES, Strandberg WG. Accumulation of metals by microbial cells. Comprehensive Biotechnology. 1985;**13**:235-247

[35] Abbas SH, Ismail IM, Mostafa TM, Sulaymon AH. Biosorption of heavy metals: A review. Journal of Chemical Science and Technology. 2014;**3**:74-102

[36] Wang JL, Chen C. Biosorption of heavy metals by *Saccharomyces cerevisiae*: A review. Biotechnology Advances. 2006;**24**:427-451

[37] Wang JL, Chen C. Biosorbents for heavy metals removal and their future. Biotechnology Advances. 2009;**27**:195-226. DOI: 10.1016/j.biotechadv.2008.11.002

[38] Ahluwalia SS, Goyal D. Microbial and plant derived biomass for removal of heavy metals from wastewater. Bioresource Technolology. 2007;**98**:2243-2257

[39] Fukudome K, Sato M, Takata Y, Kuroda H, Watari J, Takashio M. Evaluation of yeast physiological state by Alcian blue retention. Journal of the American Society of Brewing Chemists. 2002;**60**:149-152

[40] Comte S, Guibaud G, Baudu M. Biosorption properties of extracellular polymeric substances (EPS) towards Cd, Cu and Pb for different pH values. Journal of Hazardous Materials. 2008;**151**:185-193

[41] Raspor P, Zupan J. Yeasts in extreme environments. Yeast Handbook, Biodiversity and Ecophysiology of Yeasts. Springer-Verlag: Berlin; 2006. 371-417p

[42] Ruchita D, Wasiullah DM, Kuppusamy P, Udai BS, Asha S, Renu S, Bhanu PS, Jai PR, Pawan KS, Harshad L, Diby P. Bioremediation of heavy metals from soil and aquatic environment: An overview of principles and criteria of fundamental processes. Sustainability. 2015;**7**:2189-2212. DOI: 10.3390/su7022189

[43] Schneider IAH, Rubio J, Smith RW. Biosorption of heavy metals onto plant biomass: Exchange adsorption or surface precipitation? International Journal of Mineral Processing. 2001;**62**:111-120

[44] Davis TA, Volesky B, Mucci A. A review of the biochemistry of heavy metal biosorption by brown algae. Water Research. 2003;**37**:4311-4330

[45] Brady JM, Tobin JM. Binding of hard and soft metal ions to *Rhizopus arrhizus* biomass. Enzyme and Microbial Technology. 1995;**17**:791-796

[46] Sposito G. The Chemistry of Soils. New York: Oxford University Press; 1989

[47] Brady D, Duncan JR. Cation loss during accumulation of heavy metal cations by *Saccharomyces cerevisiae*. Biotechnology Letters. 1994;**16**:543-548

[48] Chen C, Wang JL. Characteristics of Zn^{2+} biosorption by *Saccharomyces cerevisiae*. Biomedical and Environmental Sciences. 2007;**20**:478-482

[49] Vasudevan P, Padmavathy V, Dhingra SC. Biosorption of monovalent and divalent ions on baker's yeast. Bioresource Technology. 2002;**82**:285-289

[50] Stumm W, Morgan JJ. Aquatic Chemistry. Chemical Equilibria and Ratesin Natural Waters. NewYork: Wiley; 1996

[51] Crist DR, Crist RH, Martin JR, Watson JR. Ion exchange systems in proton-metal reactions with algal cell walls. FEMS Microbiology Review. 1994;**14**:309-314

[52] Macaskie LE, Bonthrone KM, Rauch DA. Phosphatase-mediated heavy metal accumulation by *Citrobacter* sp. and related enterobacteria. FEMS Microbiology Letters. 1994;**121**:141-146

[53] Remoundaki E, Hatzikioseyian A, Kousi P, Tsezos M. The mechanism of metals precipitation by biologically generated alkalinity in biofilm reactors. Water Research. 2003;**37**:3843-3854

[54] Tigini V, Prigione V, Giansanti P, Mangiavillano A, Pannocchia A, Varese GC. Fungal biosorption, an innovative treatment for the decolourisation and detoxification of textile effluents. Water. 2010;**2**:550-565

[55] Anaemene IA. The use of *Candida* sp. in the biosorption of heavy metals from industrial effluent. European Journal of Experimental Biology. 2012;**2**:484-488

[56] Damodaran D, Suresh G, Mohan BR. Bioremediation of soil by removing heavy metals using *Saccharomyces cerevisiae*. In: . 2nd International Conference on Environmental Science and Technology. IPCBEE; 2011

[57] Breierova E, Vajczikova I, Sasinkova V, Stratilova E, Fiserac M, Gregorc T, Sajbidor J. Biosorption of cadmium ions by different yeast species. Zeitschrift für Naturforschung. 2002;**57**:634-639

[58] Selcen DS, Alpagu SA, Topal-Sarikaya A, Alp S. Biosorption of Ni (II) by *Schizosaccharomyces pombe*: Kinetic and thermodynamic studies. Bioprocess and Biosystems Engineering. 2011;**34**:997-1005

[59] Abd-Elsalam IS. Factorial design for some parameters affecting on chromium (iii) uptake by *Saccharomyces cerevisiae*. International Journal of Applied Biology and Pharmaceutical Technology. 2011;**2**:33-40

[60] Bingol A, Ucun H, Bayhan YK, Karagunduz A, Cakici A, Keskinler B. Removal of chromate anions from aqueous stream by a cationic surfactant-modified yeast. Bioresource Technology. 2004;**94**:245-249

[61] Ozer A, Ozer D. Comparative study of the biosorption of Pb(II), Ni(II) and Cr(VI) ions onto *S. cerevisiae*: Determination of biosorption heats. Journal of Hazardouz Material. 2003;**27**(1-3):219-229

[62] Al-Saraj MR, Monzir A, El-Nahhal IM, Baraka R. Bioaccumulation of some hazardous metals by sol–gel entrapped microorganisms. Journal of Non-Crystalline Solids. 1999;**248**(2-3):137-140. DOI: 10.1016/S0022-3093(99)00306-3

[63] Göksungur Y, Uren S, Güvenç U. Biosorption of cadmium and lead ions by ethanol treated waste baker's yeast biomass. Bioresource Technology. 2005;**96**(1):103-109

[64] Xie DD, Liu YY, CL W, JK F, Xue R. Studies on biosorption of Pd^{2+} by the immobilized *Saccharomyces cerevisiae* waste biomass. Microbiology. 2003;**30**:29-34

[65] Mapolelo M, Torto N. Trace enrichment of metal ions in aquatic environments by *Saccharomyces cerevisiae*. Talanta. 2004;**64**:39-47

[66] Andréa HA, Luciana VP, Vanete ST, Masaki R, Ernani FF, Alexsandro G, Carlo SR. Optimization of biomass production with copper bioaccumulation by yeasts in submerged fermentation. Brazilian Archives of Biology and Technology. 2011;**54**:1027-1034

[67] Zu Y, Zhao XH, Hu M, Ren Y, Xiao P, Zhu L, Cao Y, Zhang Y. Biosorption effects of copper ions on *Candida* utilis under negative pressure cavitation. Journal of Environmental Sciences. 2006;**18**:1254-1259

[68] Nguyen T, Fleet G, Rogers P. Composition of the cell walls of several yeast species. Applied Microbiology and Biotechnology. 1998;**50**:206-212

[69] Gadd GM. Biosorption: Critical review of scientific rationale, environmental importance and significance for pollution treatment. Journal of Chemical Technology and Biotechnology. 2009;**84**:13-28

[70] Aguilar-Uscanga B, François JM. A study of the yeast cell wall composition and structure in response to growth conditions and mode of cultivation. Letters in Applied Microbiology. 2003;**37**:268-274

[71] White C, Sayer JA, Gadd GM. Microbial solubilization and immobilization of toxic metals: Key biogeochemical processes for treatment of contamination. FEMS Microbiology Reviews. 1997;**20**:503-516

[72] Machado MD, Soares EV, Soares HMV. Removal of heavy metals using a brewer's yeast strain of *Saccharomyces cerevisiae*: Application to the treatment of real electroplating effluents containing multielements. Chemical Technology and Biotechnology. 2010;**85**:1353-1360

[73] Greene B, Darnall DW. Microbial oxygenic photoautotrophs (cyanobacteria and algae) for metal-ion binding. In: Ehrlich HL, Brierley CL, editors. Microbial Mineral Recovery. New York: McGraw-Hill; 1990. p. 227-302

[74] Deng X, Wang P. Isolation of marine bacteria highly resistant to mercury and their bioaccumulation process. Bioresource Technology. 2012;**121**:342-347

[75] Edyta K-B. Surface properties of yeast cells during heavy metal biosorption. Central European Journal of Chemistry. 2011;**9**(2):348-351. DOI: 10.2478/s11532-011-0008-8

[76] Tigini V, Prigione V, Anastasi A, Spina F, Varese GC. Scale up of biosorption process for the treatment of the textile wastewaters with fungal biomasses. In: 7th International Conference on Polymer and Textile Biotechnology; 2011; Milano. Italy

[77] Volesky B. Biosorption and me. Water Research. 2007;**41**:4017-4029

[78] Malik A. Metal bioremediation through growing cells. Environment International. 2004;**30**:261-278

[79] Karna RR, Sajani LS, Mohan PM. Bioaccumulation and biosorption of CO^{2+} by Neurospora crassa. Biotechnology Letters. 1996;**18**:1205-1208

[80] Chojnacka K. Biosorption and Bioaccumulation in Practice. New York: Nova Science Publishers; 2009 137 p

[81] Raspor P, Batic M, Jamnik P, Josic D, Milacic R, Pas M, Recek M, Rezic-Dereani V, Skrt M. The influence of chromium compounds on yeast physiology (a review). Acta Microbiologica et Immunologica Hungarica. 2000;**47**:143-173

[82] Tamás MJ, Wysocki R. How *Saccharomyces cerevisiae* copes with toxic metals and metalloids. FEMS Microbiology Reviews. 2010;**34**:925-951

[83] Kamizono A, Nishizawa M, Teranishi Y, Murata K, Kimura A. Identification of a gene conferring resistance to zinc and cadmium ions in the yeast *Saccharomyces cerevisiae*. Molecular & General Genetics. 1989;**219**:161-167

[84] Kneer R, Kutchan TM, Hochberger A, Zenk MH. *Saccharomyces cerevisiae* and *Neurospora crassa* contain heavy metal sequestering phytochelatin. Archives of Microbiology. 1992;**157**:305-310

[85] Presta A, Stillman MJ. Incorporation of copper into the yeast *Saccharomyces cerevisiae*. Identification of cu(I)-metallothionein in intact yeast cells. Journal of Inorganic Biochemistry. 1997;**66**:231-240

[86] Portnoy ME, Schmidt PJ, Rogers RS, Culotta VC. Metal transporters that contribute copper to metallochaperones in *Saccharomyces cerevisiae*. Molecular Genetics and Genomics. 2001;**265**(5):873-882

[87] Eide DJ. The molecular biology of metal ion transport in *Saccharomyces cerevisiae*. Annual Review of Nutrition. 1998;**18**:441-469

[88] Tamás MJ, Wysocki R. Mechanisms involved in metalloid transport and tolerance acquisition. Current Genetics. 2001;**40**:212

[89] Perego P, Howell SB. Molecular mechanisms controlling sensitivity to toxic metal ions in yeast. 1997;**147**:312-318.

[90] Ramsay LM, Gadd GM. Mutants of *Saccharomyces cerevisiae* defective invacuolar function confirm a role for the vacuole intoxic metalion detoxification. FEMS Microbiology Letters. 1997;**152**(2):293-298

[91] Thorsen M, Jacobson T, Vooijs R, Navarrete C, Bliek T, Schat H, Tamás MJ. Glutathione serves an extracellular defence function to decrease arsenite accumulation and toxicity in yeast. Molecular Microbiology. 2012;**84**(6):1177-1188

[92] Avery SV, Tobin JM. Mechanisms of strontium uptake by laboratory and brewing strains of *Saccharomyces cerevisiae*. Applied and Environmental Microbiology. 1992;**58**:3883-3889

[93] Blackwell KJ, Singleton I, Tobin JM. Metal cation uptake by yeast: A review. Applied Microbiology and Biotechnology. 1995;**43**:579-584

[94] Kujan P, Votruba J, Kamenik V. Substrate-dependent bioaccumulation of cadmium by growing yeast *Candida utilis*. Folia Microbiologica. 1995;**40**(3):288-292

[95] Cheung KH, Gu J. Mechanism of hexavalent chromium detoxification by microorganisms and bioremediation application potential: A review. International Biodeterioration and Biodegradation. 2007;**59**:8-15

[96] Bahafid W, Sayel H, Tahri Joutey N, EL Ghachtouli N. Removal mechanism of hexavalent chromium by a novel strain of *Pichia anomala* isolated from industrial effluents of Fez (Morocco). Journal of Environmental Science & Engineering. 2011;**5**:980-991

[97] Muneer B, Rehman A, Shakoori FR, Shakoori AR. Evaluation of consortia of microorganisms for efficient removal of hexavalent chromium from industrial wastewater. Bulletin of Environmental Contamination and Toxicology. 2009;**82**(5):597-600

[98] Ackerley DF, Gonzalez CF, Park CH, Blake R II, Keyhan M, Matin A. Chromate-reducing properties of soluble flavoproteins from *Pseudomonas putida* and *Escherichia coli*. Applied and Environmental Microbiology. 2004;**70**:873-882

[99] Cardenas-Gonzalez JF, Acosta-Rodriguez I. Hexavalent chromium removal by a *Paecilomyces* sp. fungal strain isolated from environment. Bioinorganic Chemistry and Applications. 2010;1-6

[100] Park D, Yun YS, Park JM. Use of dead fungal biomass for the detoxification of hexavalent chromium: Screening and kinetics. Process Biochemistry. 2005;**40**:2559-2565

[101] Elangovan R, Abhipsa S, Rohit B, Ligy P, Chandraraj K. Reduction of Cr(VI) by a *Bacillus* sp. Biotechnology Letters. 2006;**28**:247-252

[102] Hamidian H. Microbial leaching of uranium ore. In: Tsvetkov P, editor. Nuclear Power – Deployment, Operation and Sustainablity. InTech; 2011. p. 291-304

[103] Knox AS, Seaman JC, Mench MJ, Vangronsveld J. Remediation of metal- and radionuclide-contaminated soils by in situ stabilization techniques. In: Iskandar IK, editor. Environmental Restoration of Metals-Contaminated Soil. BocaRaton, FL: Lewis Publishers; 2000. pp. 21-61

[104] Gadd GM. Fungal production of citric and oxalic acid: Importance in metal speciation, physiology and biogeochemical processes. Advances in Microbial Physiology. 1999;**41**:47-92

[105] Gadd GM. Mycotransformation of organic and inorganic substrates. Mycologist. 2004;**18**:60-70. DOI: 10.1017/S0269915XO4002022

[106] Strasser H, Burgstaller W, Schinner F. High yield production of oxalic acid for metal leaching purposes by *Aspergillus niger*. FEMS Microbiology Letters. 1994;**119**:365-370

[107] Devevre O, Garbaye J, Botton B. Release of complexing organic acids by rhizosphere fungi as a factor in Norway spruce yellowing in acidic soils. Mycological Research. 1996;**100**:1367-1374

[108] Rajpert L, Sklodowska A, Matlakowska R. Biotransformation of copper from Kupferschiefer black shale (fore-Sudetic monocline, Poland) by yeast *Rhodotorula mucilaginosa* LM9. Chemosphere. 2013;**91**:1257-1265

[109] Marcinčáková R, Kaduková J, Mražíková A, Velgosová O, Vojtko M. Lithium bioleaching from lepidolite using the yeast *Rhodotorula rubra*. Journal of the Polish Mineral Engineering Society. 2015;**16**:1-6

[110] Allman MB, Harris JA. An experiment in heap leaching. Mining Congress Journal. 1969;28-31

[111] Dreisinger D. Copper leaching from primary sulfides: Options for biological and chemical extraction of copper. Hydrometallurgy. 2006;**83**:10-20

[112] Watling HR. The bioleaching of sulphide minerals with emphasis on copper sulphides - a review. Hydrometallurgy. 2006;**84**:81-108

[113] Neale J. Bioleaching Technology in Minerals Processing. Randburg, South Africa: Mintek; 2006

[114] Herrero M, Stuckey DC. Bioaugmentation and its application in wastewater treatment: A review. Chemosphere. 2015;**140**:119-128

[115] Leahy JG, Colwell RR. Microbial degradation of hydrocarbons in the environment. Microbial Reviews. 1990;**53**(3):305-315

[116] Van der Gast CJ, Whiteley AS, Thompson IP. Temporal dynamics and degradation activity of a bacterial inoculum for treating waste metal-working fluid. Environmental Microbiology. 2004;**6**:254-263

[117] Godleads OA, Prekeyi TF, Samson EO, Igelenyah E. Bioremediation, biostimulation and bioaugmention: A review. International Journal of Environmental Bioremediation & Biodegradation. 2015;**3**:28-39

[118] Sood N, Sonali P, Banwari L. Bioremediation of acidic oily sludge-contaminated soil by the novel yeast strain *Candida digboiensis* TERI ASN6. Environmental Science and Pollution Research. 2010;**17**:603-610

[119] Fan M, Rui-Jie X, Gang Q. Bioremediation of petroleum contaminated soil by a combined system of biostimulation-bioaugmentation with yeast. Environmental Technology. 2013;**35**:391-399

[120] Bahafid W, TahriJoutey N, Sayel H, Boularad I, El Ghachtouli N. Bioaugmentation of chromium-poluted soil of microcosms with *Candida tropicalis* diminishes phytoavailable chromium. Journal of Applied Microbiology. 2013;**115**:727-734

[121] Bahafid W, Tahri Joutey N, Sayel H, Asri M, Laachari F, EL Ghachtouli N. Soil bioaugmentation with *Cyberlindnera fabianii* diminish phytotoxic effects of chromium (VI) on *Phaseolus vulgaris* L. Journal of Materials and Environmental Sciences. 2017;**8**:438-443

[122] Gallego JL, García-Martínez MJ, Llamas JF, Belloch C, Peláez AI, Sánchez J. Biodegradation of oil tank bottom sludge using microbial consortia. Biodegradation. 2007;**18**:269-281

[123] Li X, Xin L, Peijun L, Wan L, Li W, Fang M, Chukwula KS. Biodegradation of the low concentration of polycyclic aromatic hydrocarbons in soil by microbial consortitum during incubation. Journal of Hazardous Materials. 2009;**172**:601-605

[124] Tyagi M, Da Fonseca MR, Carvalho CCCR. Bioaugmentation and biostimulation strategies to improve the effectiveness of bioremediation processes. Biodegradation. 2011;**22**:231-241

[125] He Z, Yao Y, Lu Z, Ye Y. Dynamic metabolic and transcriptional profiling of *Rhodococcus* sp. strain YYL during the degradation of tetrahydrofuran. Applied and Environmental Microbiology. 2014;**80**:2656-2664. DOI: 10.1128/AEM.04131-13

[126] Chai L, Huang S, Yang Z, Peng B, Huang Y, Chen Y. Cr(VI) remediation by indigenous bacteria in soils contaminated by chromium-containing slag. Journal of Hazardous Materials. 2009;**167**:516-522